D1155938

The QUANTITATIVE DESCRIPTION of the MICROSTRUCTURE of MATERIALS

CRC Series in
Materials Science and Technology

Series Editor
Brian Ralph
Brunel University
Uxbridge, U.K.

Published Titles

Control of Microstructures and Properties in Steel Arc Welds
Lars-Erik Svensson, The ESAB Group, Gothenburg, Sweden

The Extraction and Refining of Metals
Colin Bodsworth, Brunel University, Uxbridge, United Kingdom

Quantitative Description of the Microstructure of Materials
K. J. Kurzydlowski, Warsaw University of Technology, Warsaw, Poland
Brian Ralph, Brunel University, Uxbridge, United Kingdom

Forthcoming Titles

Grain Boundary Properties and the Evolution of Microstructure and Texture
G. Gottstein, Institute of Metallurgy and Metal Science, RWTH Aachen, Germany
L.S. Shvindlerman, Institute of Solid State Physics, Russian Academy of Science, Moscow, Russia

Grain Growth and Control of Microstructure and Texture in Polycrystalline Materials
Vladimir Novikov, Moscow Institute of Steel and Alloys, Moscow, Russia

The QUANTITATIVE DESCRIPTION of the MICROSTRUCTURE of MATERIALS

Krzysztof Jan Kurzydłowski
Brian Ralph

CRC Press
Boca Raton New York London Tokyo

Library of Congress Cataloging-in-Publication Data

The Quantitative description of the microstructure of materials / Krzysztof Jan Kurzydłowski, Brian Ralph.
p. cm.—(Materials science and technology)
Includes bibliographical references and index.
ISBN 0-8493-8921-6 (acid-free paper)
1. Materials. 2. Microstructure. I. Kurzydłowski, Krzysztof J.
II. Ralph, Brian. III. Series: Materials science and technology (Boca Raton, Fla.)
TA403.6.Q36 1995
620.1'1299—dc20 94-41304
CIP

International Standard Book Number 0-8493-8921-6
Library of Congress Card Number 94-41304
Printed in the United States of America 1 2 3 4 5 6 7 8 9 0
Printed on acid-free paper

Preface

This text came about from an interaction between the authors which was nucleated by our friend and colleague, Professor Maciej Grabski. If Maciej Grabski was the "nucleating agent" then the growth promoters for our collaboration on this text were TEMPUS and PECO initiatives from the European Union.

Initially, this text was to be produced by one of us, KJK, with the other, BR, acting as series editor. However, it soon became clear that we had complementary ideas and so a deeper collaboration was born.

The format we have chosen for this book is perhaps unusual and requires justification. Essentially the text is divided in two: the first part (Chapters 1 to 4) gives a comprehensive account of the entire quantification process. In this we have used an extensive set of examples in order to illustrate the methodology we propose. The second part of the text (Chapters 5 to 8) treats a series of significant examples in much more detail. Accordingly, we have chosen to repeat much material in the two parts of this book.

We feel that the first part of this book may be used as the basis for advanced courses on quantification, and both of us have taught such courses and tested the applicability of this material in the classroom. The second part of the text might then be seen as a guide to active researchers in the quantification field.

While the blame for any errors and omissions in this book lie clearly with us the authors, the input from our colleagues has greatly enhanced our activities. Amongst those who have been generous with their help and advice are: J.J. Bucki, D. Łazęcki, K. Rożniatowski, G. Żebrowski.

The following is a list of authors and publishers who have kindly let us cite from their works: J. Glijer, W. Kolczyński, W. Zieliński, L. Wojnar, J. Cwajna, and we are also grateful to the publishers of Scripta Metallurgica for agreeing to let us use modified versions of many illustrations previously published in Scripta Metallurgica.

Finally, we see this book as an update of the texts produced by Rhines and DeHoff and Underwood. In preparing our text we have tried to strike a balance between the traditional approach (so ably given by these authors) and the newly emerging techniques. We welcome views from our readers as to how well we have achieved this objective so that in future editions we may evolve to a higher level of presentation.

Table of Contents

Dedication

We respectfully dedicate this book to our long-suffering families:

Anna, Dominik and Michał;
Anne and Zoanna

K.J.K. and B.R.

chapter one

Introduction

1.1 Elements of the microstructures of materials

Materials science is a field of scientific pursuit that concentrates attention on the relationships between microstructure and properties of materials. The term microstructure in this context is used to describe a set of characteristic elements that can be identified in the arrangement of atoms and/or molecules forming a given material. In the modern approach, developed over recent years, materials are defined by their chemical composition and ideal defect-free arrangements of the atoms and molecules that are characteristic of a given phase. Based on such an approach, the elements of the microstructure of a material may be defined as distinctive defects of the ideal arrangement. This is a quite general definition and in particular applications one can define an almost unlimited number of different microstructural elements. Depending on the type of the material studied these elements may consist of single atoms, groups of atoms, and large aggregates of atoms or molecules.

As examples of microstructural elements one can recall impurities in crystalline materials, fibers in composites, pores in concrete, and cracks in a glass. Another example is given in Figure 1.1 which illustrates the main elements of the microstructure of a polycrystalline material.

Polycrystals are aggregates of large numbers of crystals (grains). They constitute a significant fraction of the materials used for a variety of applications. This group includes metals, ceramics, intermetallics as well as some types of polymer. The grains and the grain boundaries are one of the elements of the microstructure of polycrystals. The dimensions of the grains (and the grain boundaries) may vary from millimetres down to nanometres. Polycrystals contain fine scale defects such as dislocations and large scale defects such as surface coatings, cracks, and second-phase particles. The collection of microstructural elements encountered in polycrystals make them a possible case study for the problem of microstructural characterization and their example is widely used throughout this text. However, it should be stressed that the results obtained for polycrystals can be easily generalized by substituting specific defects in polycrystals by geometrically equivalent elements of other materials. For example, grains in polycrystals can be replaced by any particles, dislocations by fibers, etc.

There are a number of different classifications of microstructural elements used in different texts. However, the most general and most suitable for the

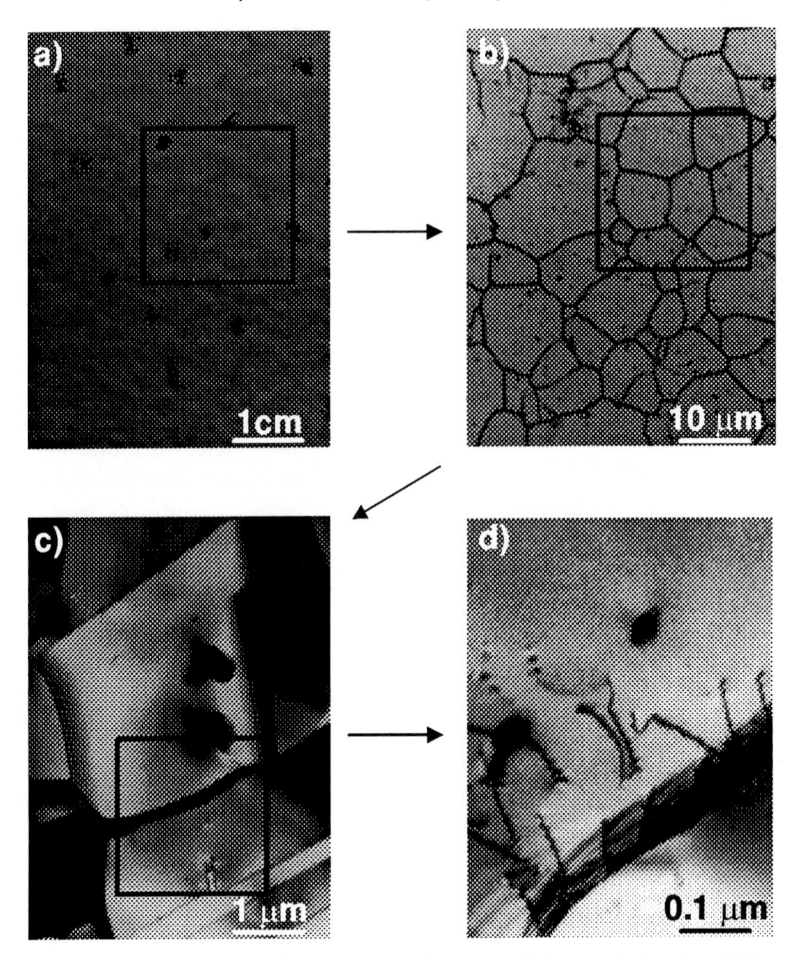

Figure 1.1 Microstructure of a polycrystalline material revealed using different experimental techniques: (a) light microscopy, low magnification image of a polished surface—dark areas are intersections of large inclusions; (b) light microscopy, higher magnification image of a polished and electrolytically etched specimen—the lines indicate the traces of the grain boundaries on the surface of observation; (c) transmission electron microscopy, low magnification projected image of a thin foil showing the presence of small precipitates and grain boundaries; (d) transmission electron microscopy, higher magnification projected image of a thin foil showing the presence of dislocations near a grain boundary.

present text seems to be the one based on the analysis of their dimensions. Microstructural elements can be broadly divided into:

a) point —0-dimensional
b) linear —1-dimensional
c) surface —2-dimensional
d) volume —3-dimensional

Table 1.1 Elements of Microstructure for Selected Types of Materials

Type of material	0-D elements	1-D elements	2-D elements	3-D elements
Polycrystals	Vacancies, impurities, alloying atoms	Dislocations	Stacking faults, grain boundaries, free surfaces, interfaces, cracks	Domains, grains, aggregates, second-phase particles, pores
Composites	Small voids, particles	Fibers	Interfaces, lamellar particles	Particulates, voids

Examples of the defects falling in these respective categories a) to d) are given in Table 1.1.

The proposed classification is in harmony with the senses developed in the process of everyday observations. Examples can easily be found of point, linear, surface, and volumetric artifacts. However, new findings from mathematics show that this categorization is incomplete and some additional features may be defined that have their dimensionality expressed in fractions. These features are called fractals and although they have been invented by mathematicians they have proved to be pertinent to materials science.

Fractals are curves, surfaces, or volumes generated by some repeated process involving successive subdivisions. Examples of fractals are snowflake outlines and the Sierpinski gasket shown in Figure 1.2. In order to explain the concept of fractal dimensionality it should be noted that the dimensionality parameter, d, appears in the relationship between the size of the object, s, (s is the volume for 3-D elements, area for 2-D, etc.), and its linear dimension, L:

$$s = L^d \tag{1.1}$$

The parameter d also describes the way a given size parameter scales. Points scale as L^0, linear dimensions scale as L, surfaces scale as L^2, and volumes as L^3. Fractals are the objects of the scaling factor that can be expressed as a fractional number, say 1.58. In other words, fractals are the geometrical features whose dimension is intermediate between 0 and 3. The Sierpinski gasket is an example of such an object. Consider the triangle ABC with the distance AB = a. The area of the gasket inside this triangle is three times the area of the subgaskets of each size a/2. This leads to a simple equation:

$$s_{ABC} = ka^d = 3k\left(\frac{a}{2}\right)^d \tag{1.2}$$

with d standing for the dimensionality of the figure. The solution of the equation is $d = \ln 3/\ln 2 \approx 1.58$.

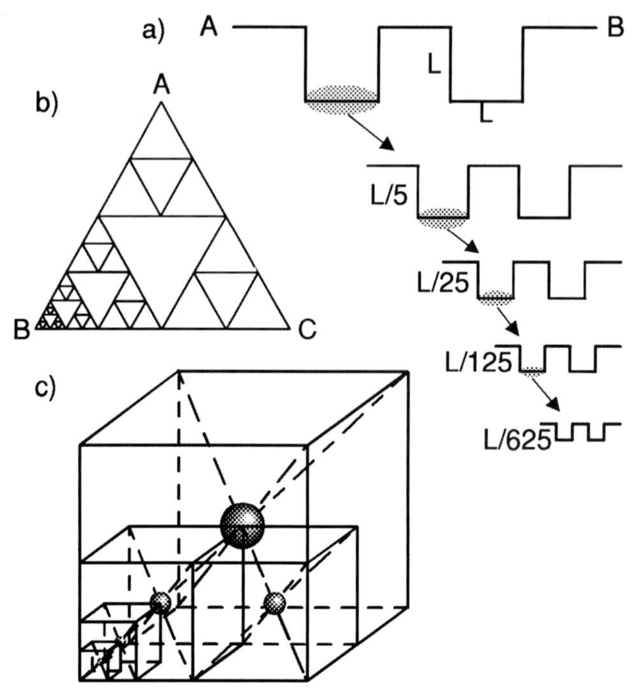

Figure 1.2 Illustration of the concept of fractal objects: (a) a line which, when examined under increasing magnification, is found to contain fine undulations; (b) "Sierpinski gasket" obtained by infinite number of triangles of diminishing size; (c) porous structure which is produced by incorporating an infinite number of pores.

Fractal objects have been found in the studies of fracture surfaces and some aggregates of particles. They are discussed in more detail in Chapters 4 and 5.

Before examples of different types of microstructural elements are given, it should be noted that, in general, all the defects extend into three dimensions and their representation by reduced-dimensionality is based on the comparison between the dimensions in different directions of the defects themselves or dimensions of the defects and of the specimen. As a result, classification of a given defect may change depending on the type of microstructure analyzed. For example interfaces are the microstructural elements typically considered to be 2-dimensional. This approach is based on observations indicating that the thickness of the interface region is much smaller than the other dimensions. However, in the case of nano-crystals of extremely small dimensions the interface regions may constitute a significant fraction of the material's volume and no longer should such interfaces be considered 2-D entities. This means that the definition of a given element's dimensionality depends on the context in which it is considered.

In principle, 0-D elements as such could not be revealed in the image of the microstructure. The decision whether or not to consider an object to be 0-dimensional usually depends on how far into its internal structure we intend

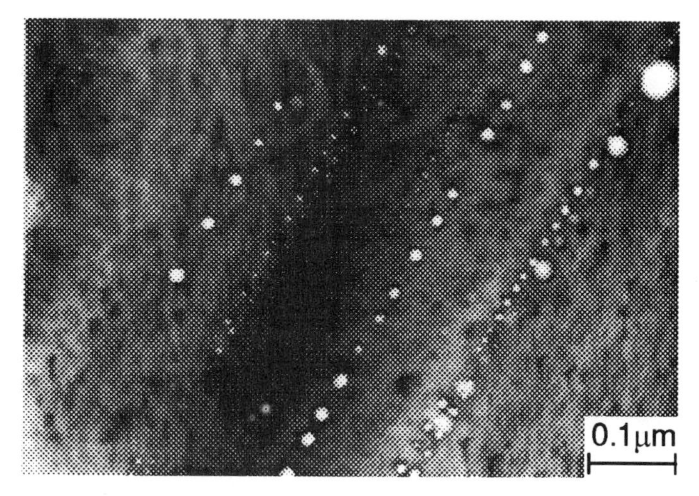

Figure 1.3 Transmission electron micrograph of pores in sintered tungsten.

to look. By assuming 0-dimensionality, we declare a lack of interest in the defect properties (size, shape, etc.) and emphasize the defects' concentration and positions. Typically foreign atoms, vacancies, molecules, particles, and voids are considered as 0-D elements depending on the size of other elements in the microstructure. As an example of 0-D elements a row of volume defects of nanometre scale in tungsten is shown in Figure 1.3.

Examples of linear elements in the microstructure of crystals (in poly- and mono- form) are dislocations, disclinations, and edges of surface defects (these defects are discussed in Chapter 5). For other materials one can recall elements such as molecular chains in polymers and fibers in composites. Segments of dislocations are shown in Figure 1.4 In the case of surface observations, line defects are imaged as points on their emergence from the surface (Figure 1.4a).

Two-dimensional defects are present in any material. This is related to the fact that the external surface, also called the free surface, by itself is such a defect. Other examples include grain boundaries separating crystals of the same type, stacking faults, and interphase boundaries. Transmission electron microscopy images of grain boundaries in polycrystalline materials are shown in Figure 1.1. In the case of observations carried out on cross-sections of materials, surface defects are seen as lines which are the intersections of the defects with the surface of observation. A 2-D network of the traces of grain boundaries revealed on the cross-section of a polycrystal is shown in Figure 1.5a.

A list of commonly observed 3-D defects include pores, particles of different phases, and elements of the same phase differing in crystal orientation. Grains forming a polycrystalline aggregate can be considered as 3-D elements. A grain randomly selected from the population of grains in a single-phase metal is shown in Figure 1.6. The grain geometry has been reconstructed from a stack of micrographs obtained from serial sections. This example illustrates that on sections through the material, 3-D features are revealed as 2-D elements.

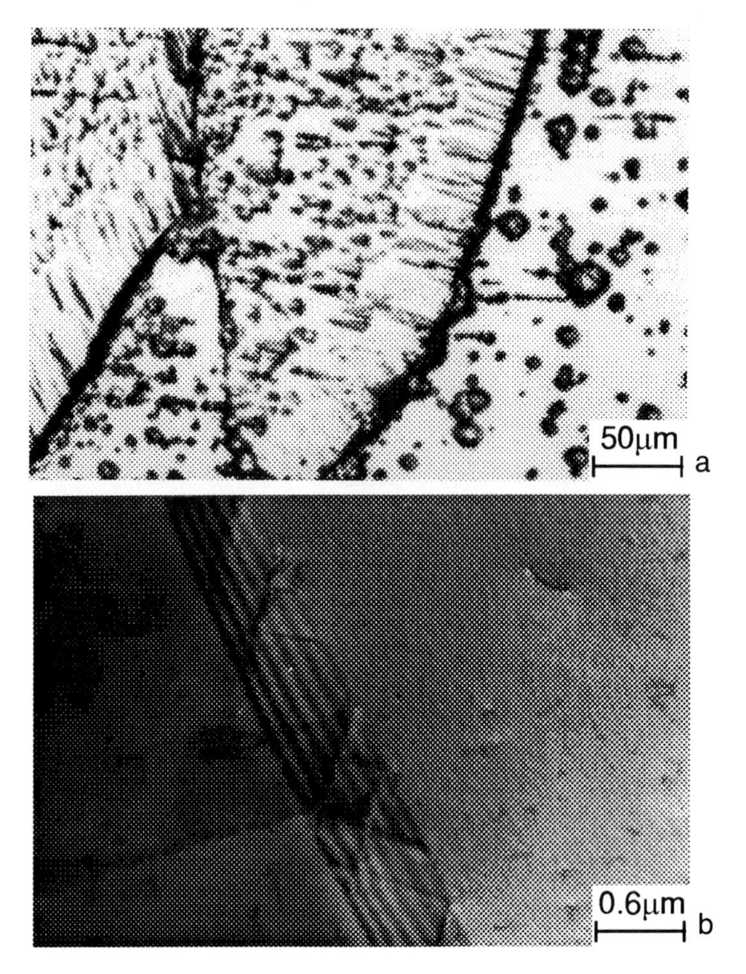

Figure 1.4 Dislocations in crystals: (a) light microscopy image of the points of intersections of dislocation lines with the surface of observation; (b) transmission electron microscopy projected image of dislocations in a thin foil.

Classifications of microstructural elements are made for the sake of generalization. Whenever possible, it seems to be much more sensible to derive appropriate formulae for the case of the general type of defect or microstructural element, than to consider separately the defects of the same geometrical gender found in different materials. For example, both dislocations in crystals and carbon fibers in composites can be considered to be linear elements although their dimensions differ by several orders of magnitude. The concept of generalization adopted in microstructural characterization techniques is based on the conjecture that the scale of observations (magnification)

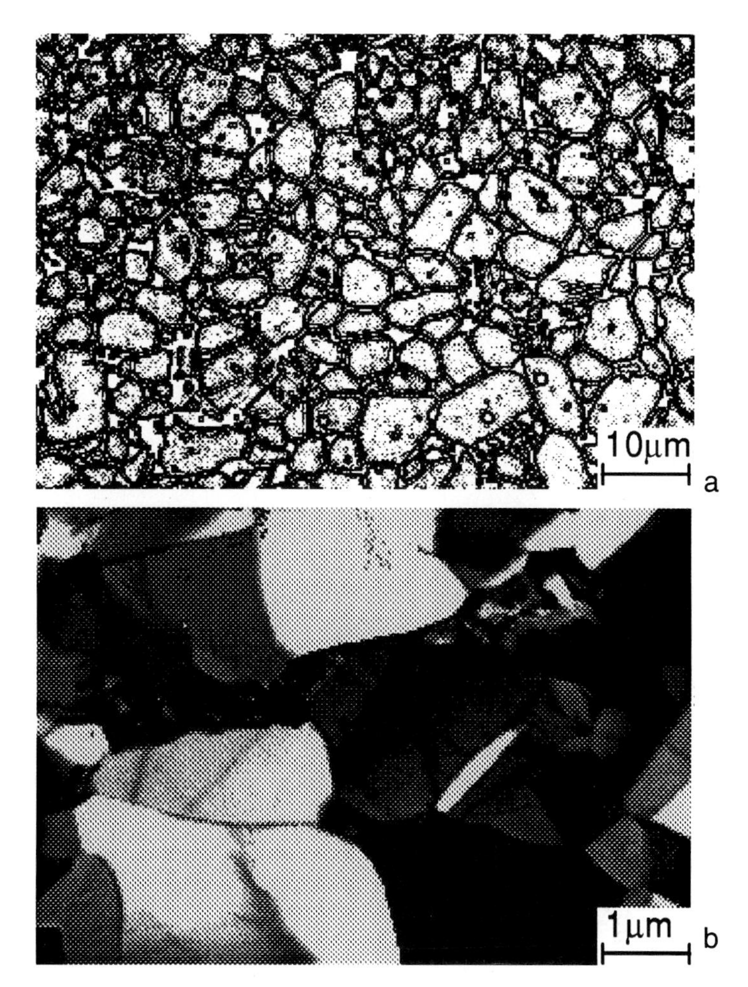

Figure 1.5 Grain boundaries in a polycrystal: (a) revealed on a polished and thermally etched section; (b) revealed in transmission electron microscopy observation of a thin foil.

and physical properties of the elements studied are less important than their geometrical characteristics, especially dimensionality.

1.2 Size and shape

Characterization of the microstructure of materials involves identification of the main microstructural elements present and a quantitative description of their sizes, shapes, numbers, and positions within the specimen of the studied material. In other words, this is a process that answers the following questions:

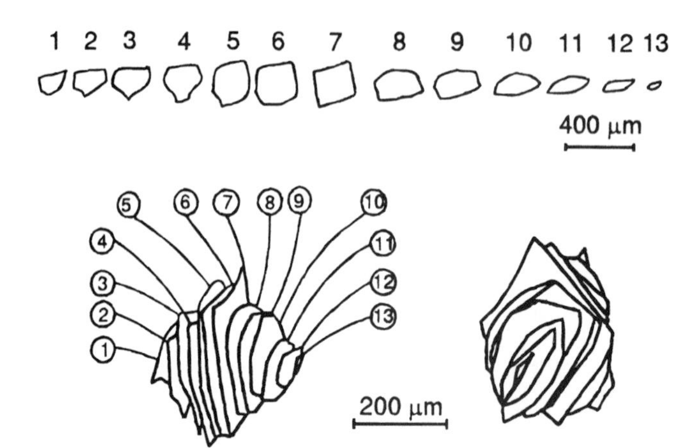

Figure 1.6 Geometry of a single grain observed on serial parallel sections 10 μm apart. There have been 13 sections studied. The reconstructed 3-D geometry of the grain is shown.

1. What are the elements in the internal structure of a given material that distinguish it from other materials of that kind (of say similar chemical composition)?
2. Where are these elements located and in what quantity?
3. What is their size and shape?

Answers to all these questions provide a comprehensive description of the material microstructure that can be used to explain its properties and to gain better control over its technological usage. However, the first of these questions is a domain of material physics. In fact, microstructural elements such as dislocations and grain boundaries by themselves have become subjects of extensive theoretical studies. In the present text the focus is placed on the two last questions.

Size and shape are the attributes of 1-,2-, and 3-D elements of the microstructure. The size of 1-D elements can be quantitatively described by their length, l, of 2-D elements by their area, A, and of 3-D elements by their volume, V. However, direct measurements of the length, area, or volume of microstructural elements are rarely possible. As a result, the measurements are usually done on sections or projections of the microstructural elements. The results of such measurements characterize the size of the elements studied but do not define their volume, area, or length.

A description of the shape of elements is less obvious and a quantitative description requires a larger number of parameters. One possible approach to shape description is based on shape classifications. A number of shapes have been catalogued and described by mathematicians and some of them are good approximate morphologies for microstructural elements observed in materials. As a result one can define, for example, spherical, rod-like, or disk-like particles, etc. However, this is not possible in the case of more complicated and less regular shapes observed, for example, in high-porosity materials. Special techniques need to be used then as described in Chapter 8.

1.3 The influence of microstructure on the properties of materials

One of the foundations of modern materials science is the recognition of the fact that the properties of materials are related to their microstructure. The existence of such relationships is illustrated in the rest of this section, and elsewhere throughout the text by a number of examples.

Materials containing two or more phases are widely used in technological applications. This is due to the fact that it is increasingly possible to "design" microstructures that deliver the combination of properties desired. Perhaps the best examples of this come from the field of composite materials where some of the simpler properties (such as stiffness/elastic moduli) are determined by taking account of the elastic moduli of the matrix and reinforcing phase, and the volume fractions of these two components. This leads to a very elementary law of mixtures, which gives the modulus of the composite. Whilst composite materials illustrate the designed microstructure approach most simply, many other, albeit more complicated, examples abound in other classes of engineering materials such as steels and nickel-based superalloys.

Example 1.1

An alloy of iron and chromium of a composition given in Table 1.2 has been studied. All specimens were initially annealed for 1 hour at 1300°C and subsequently for 1 hour at temperature, T_a, in the range from 400 to 1200°C. Figure 1.7a shows a microstructure representative of one of these annealing temperatures. These are two-phase microstructures containing particles of austenite (dark) γ-phase embedded in a ferritic, α-phase matrix. Depending on the annealing temperature, the amount of γ-phase varies from slightly above 5% of the volume (annealing above 1100°C) to more than 40% (annealing at 900°C). The variation of the γ-phase volume fraction, f_γ, as a function of T_a is depicted in Figure 1.7b.

The two phases observed differ significantly in the arrangements of their atoms and accordingly in their properties. The austenitic γ-phase has a face centered cubic (FCC) structure while the ferritic, α-phase has a body centered cubic (BCC) one. As a result, the γ-phase shows lower resistance to plastic deformation (a larger ductility) and lower values of the ultimate tensile strength, UTS.

Table 1.2 Chemical Composition (Main Constituents in Weight Percentage) of the Alloy Studied

C	Cr	Ni	Mn	Mo	Fe
0.04	25.3	4.7	1.34	1.48	64.0

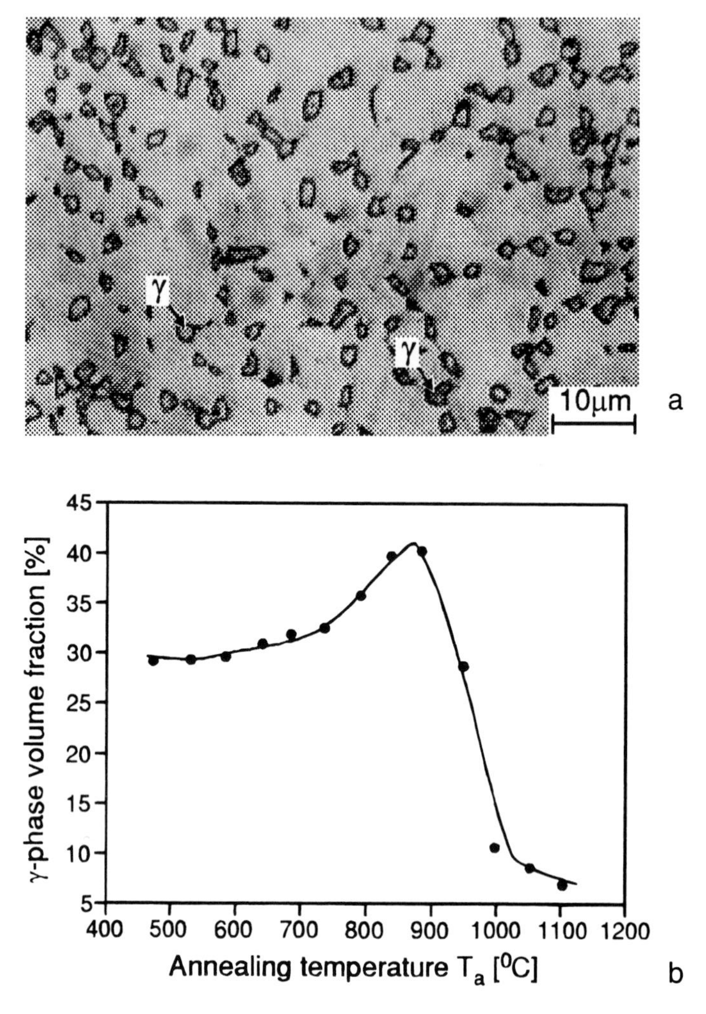

Figure 1.7 Microstructure typical of the dual phase steel studied (a) and the dependence of the volume fraction of austenite on the temperature of the annealing (b).

Figure 1.8a shows the effect of the annealing temperature on the value of the UTS of this material. The values of the UTS reach a minimum for the annealing temperature, T_a, close to 900°C. As the annealing treatments employed do not change the chemical composition of the material, the variations in its mechanical properties can only be explained in terms of the changes in the microstructure. As said before, one of the major changes in the microstructure in this case is the variation in the content of γ-phase.

Figure 1.8b shows the values of UTS plotted against f_γ (the volume fraction of austenite). This plot indicates that, as expected, the ultimate strength of the material decreases with increasing volume fraction of the γ-phase. However, the plot further shows that there are two relationships between UTS and f_γ: designated as I and II. The first holds for those specimens annealed at higher temperatures which result in a weaker effect of the austenite

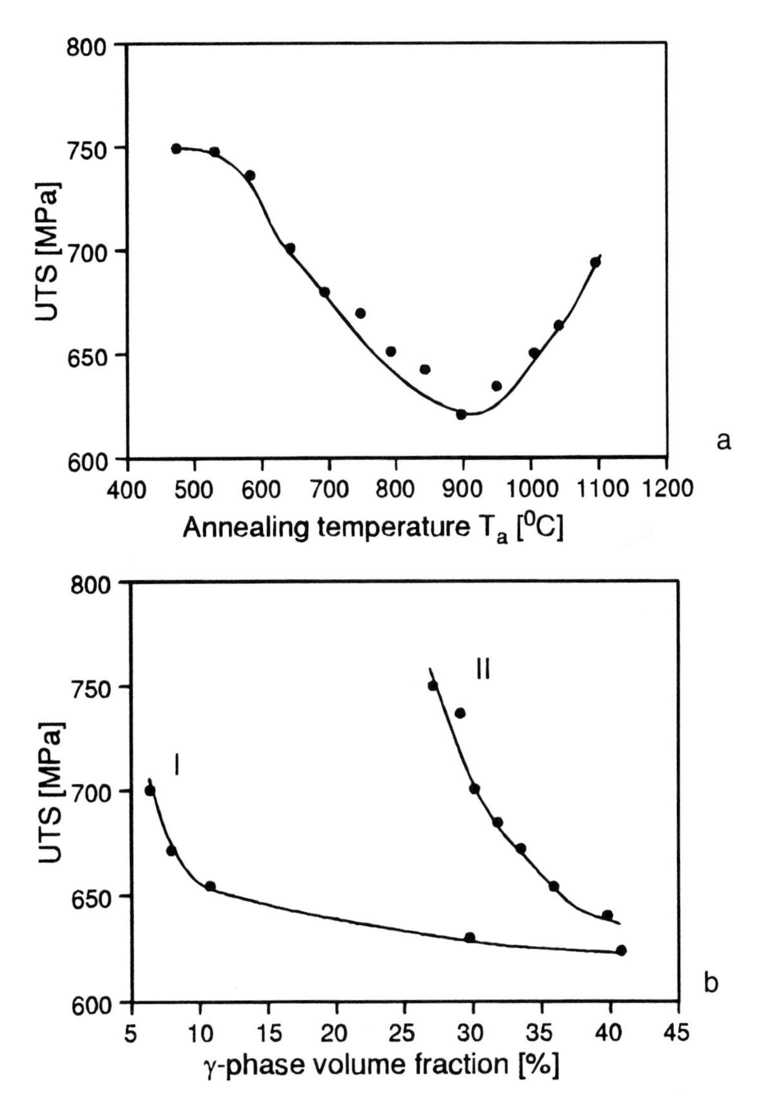

Figure 1.8 The correlation between the ultimate tensile strength (UTS) of the material and temperature of the annealing (a). By relating the values of UTS to the volume fraction of austenite, two characteristic relationships can be found between these two parameters (b).

volume fraction on the ultimate strength. By contrast, specimens annealed at lower temperatures exhibit a stronger sensitivity of the UTS on f_γ. For values of f_γ larger than 40%, the two relationships seem to merge.

The existence of these two relationships can be explained in terms of a redistribution of the alloying elements. Both phases differ not only in mechanical properties but they can also absorb different amounts of alloying elements. This leads to synergetic changes in their properties that are controlled by the annealing temperature.

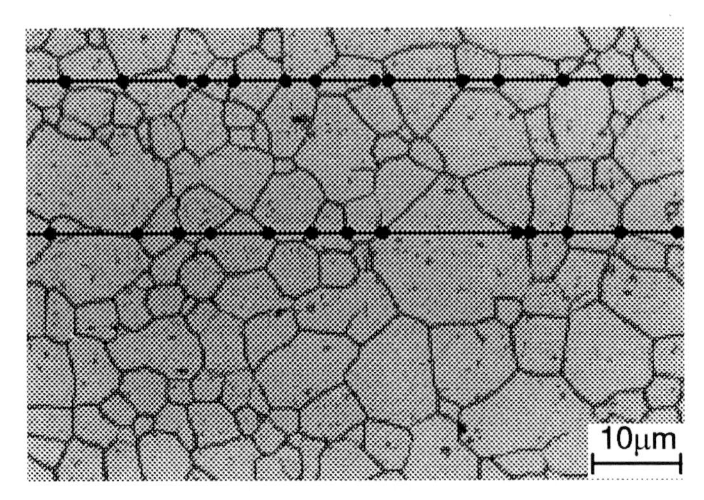

Figure 1.9 Schematic explanation of the technique for characterizing the grain size in terms of the mean intercept length.

This example illustrates a general rule that the properties of multiphase materials need to be correlated with the volume fractions of the phases. In the case studied the volume fraction of austenite is an important microstructural parameter that needs to be used to explain the variations in mechanical properties of the material subjected to different annealing conditions.

The properties of polycrystalline materials depend on the properties of grain boundaries and grain interiors, and on the amount of grain boundaries in a unit volume and, in turn, on the size of grains making up the polycrystalline aggregate. The grain size can be described and measured in different ways which are discussed thoroughly in Chapter 6. The simplest procedure, frequently employed in physical metallurgy, is based on the concept of intercept measurements, schematically explained in Figure 1.9. A system of test lines is superimposed on the image of the grain boundary network revealed on a cross-section of the polycrystal. The mean intercept length, l, is then calculated from a simple formula:

$$l = \frac{L_T}{N_i} \tag{1.3}$$

where L_T is the total length of the test lines and N_i the total number of their intersections with the traces of grain boundaries on the section. The value of l decreases with decreasing grain size and the parameter is assumed to be a desirable measure of the grain size and as such is correlated with the grain boundary-sensitive properties of polycrystals.

Resistance to plastic deformation is one of the grain size-sensitive properties of polycrystals. One of the simple measures of this resistance is the hardness test. Some hardness tests, such as Vickers and Brinell, yield a number that defines the resistance to plastic deformation in MPa. Figure 1.10 presents the results of Vickers tests carried out on a series of specimens of a

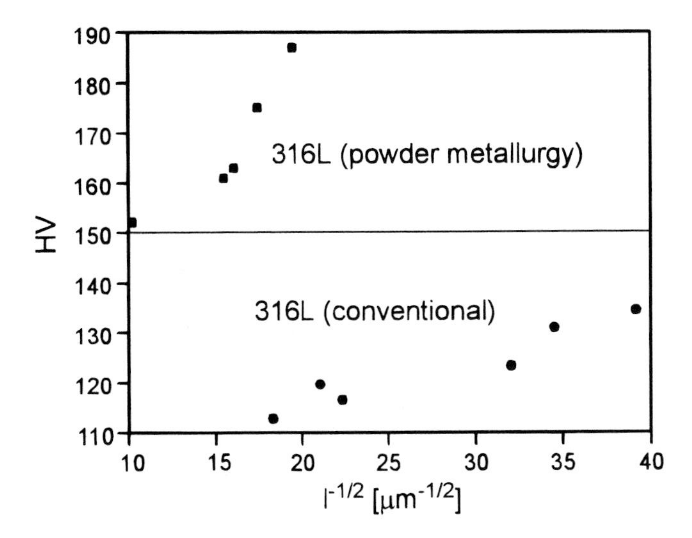

Figure 1.10 The dependence of material hardness, Vickers, on the inverse square root of the mean intercept length, $l^{-1/2}$, for an austenitic stainless steel produced by powder metallurgy and by the process of recrystallization.

single-phase FCC alloy with different grain sizes. The hardness is plotted against $l^{-1/2}$ in accordance with a linear relationship proposed:[1]

$$H = H_0 + K_H\, l^{-\frac{1}{2}} \tag{1.4}$$

This formula, known as the Hall-Petch relationship, is one of the most useful in the field of physical metallurgy. It can be used to predict the properties of a wide range of materials including metals, ceramics, intermetallics, and polymers and will be discussed in greater detail in Chapter 6.

Plastic deformation of crystals at temperatures lower than 0.2 of the melting point of a given material takes place via movement of dislocations that in the course of straining accumulate in the crystal lattice. The accumulation of the dislocations, which are line defects of the lattice, leads to an increase in the material's strength and this effect is termed *work hardening*. One of the fundamental relationships of physical metallurgy predicts the following equation:

$$\sigma = A\sqrt{\rho} + \sigma_0 \tag{1.5}$$

where σ is the flow stress, σ_0 is the friction stress for dislocation motion, ρ is dislocation density, and A is a numerical constant.

In this formula ρ depends on the amount of plastic strain and the grain size in the case of polycrystals. The friction stress, σ_0, on the other hand is affected by the processes of dislocation interactions with point defects and other elements of the microstructure. The value of the constant A depends on the value of the Burgers vector and the shear modulus.

Example 1.2

Studies of work hardening have been carried out on a 316L austenitic stainless steel of commercial purity. Specimens were strained in tension at room temperature to a total plastic strain, ε, in the range of 0.15 at a constant applied strain rate. The gauge sections of the deformed tensile specimens were cut into several pieces using a spark-cutting machine.

Thin foils for transmission electron microscopy, TEM, were prepared from the as-annealed and deformed specimens by chemical thinning. The foils thus prepared were examined with a scanning transmission electron microscope, STEM, operating at an accelerating potential of 200 kV.

Dislocation densities were calculated by counting the number of intersections of dislocations with randomly drawn lines on TEM photomicrographs taken in two-beam conditions. The corresponding foil thicknesses were determined by counting grain boundary fringes.

Photomicrographs typical of the as-annealed and deformed specimens are given in Figure 1.11a and Figure 1.11b. The density of dislocations has been measured using a technique based on counting the density of intersections with the surface of thin foils. Estimations of the dislocation densities within the grains for each of the specimens are plotted in Figure 1.12a against the amount of plastic strain. Figure 1.12b presents the results of Vickers hardness measured for the series of specimens studied. It can be noted that the results well agree with Equation 1.5 with the flow stress, σ, being replaced by the hardness, H.

Example 1.3

Figure 1.13 shows microstructures of the same material, a low-alloy steel, after two different heat treatments. Microscopic observations proved that the microstructures contained the same volume fraction of Fe_3C carbide (cementite), which appears, however, in two different geometrical forms: (a) pearlitic after treatment A, and (b) spheroidal after treatment B. This conclusion is rationalized by measurements carried out on the sections of the carbides revealed on cross-sections of the material. In these measurements the following two parameters of individual sections of carbide particles were studied:

a) area, A_i
b) perimeter, p_i

where i = 1,2, . . ., N and where N is the number of analyzed sections.

Particle area values, A_i, were used to calculate the equivalent circle diameters, $(d_2)_i$, and for each section studied its shape has been described in terms of the ratio κ_i, defined as:

$$\kappa_i = \frac{p_i}{(d_2)_i} \tag{1.6}$$

This ratio is equal to π for a circle and increases with an increase in the elongation of the figure.

A series of micrographs has been taken of the carbides under a constant magnification. Histograms of the measured values are plotted in Figure 1.14. A visible difference exists between the values measured for carbides after heat treatments A and B. This reflects differences in shape of the particles, which appear in the form of plates after heat treatment A and assume a spherical

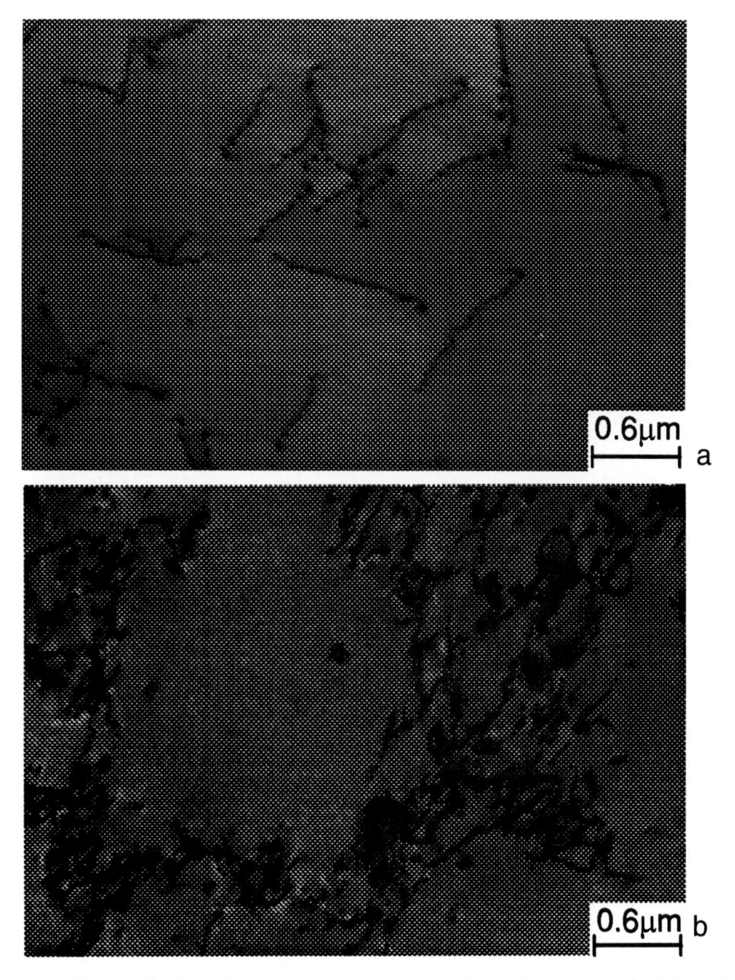

Figure 1.11 Transmission electron microscope projected images of dislocations in thin foils of 316L stainless steel: (a) as-annealed; (b) deformed to 3% total elongation.

shape after treatment B. Careful studies of other elements of the microstructure proved that this difference in the shape is responsible for significant differences in the properties of the two microstructures as documented by the data in Table 1.3.

Table 1.3 Typical Values of the Parameters Defining Mechanical Properties of a Low-Alloy Steel with Lamellar (A) and Spherical (B) Cementite Particles

	Low-alloy steel	
	Lamellar carbides	Spherical carbides
Yield point [MPa]	448	355
Tensile strength [MPa]	580	600
Impact strength [kJ/m^2]	215	294
Elongation [%]	12	16

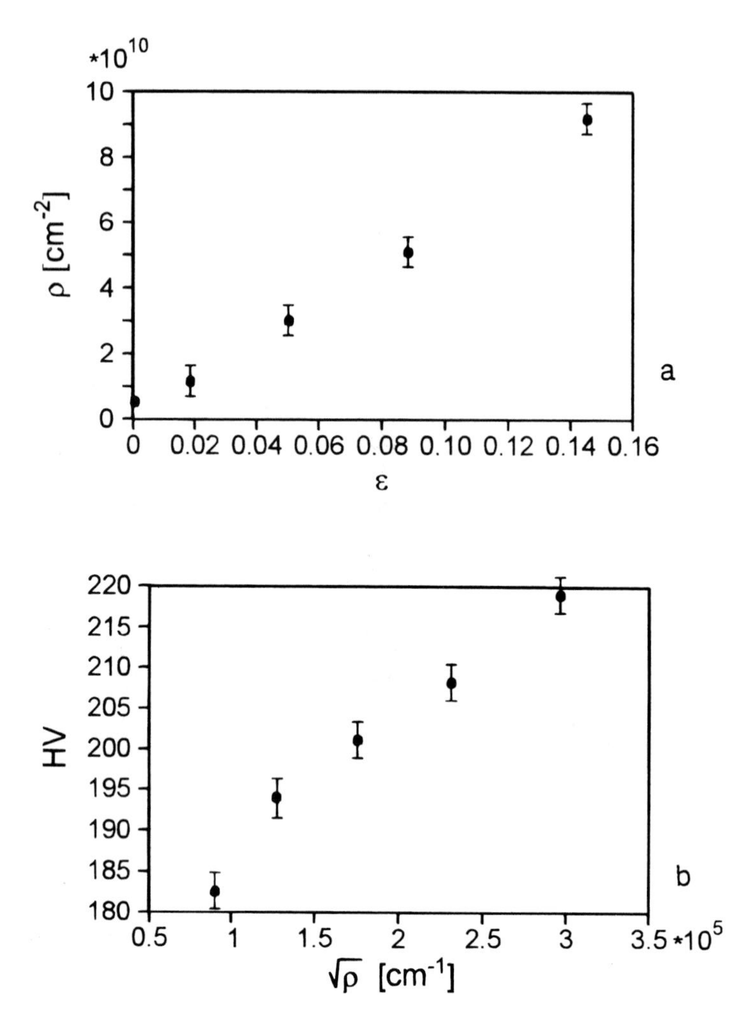

Figure 1.12 The relationship between the density of the dislocations and the strain; (a) the density has been found to be a linear function of the amount of plastic deformation. The relationship between the hardness and dislocation density (b).

The kappa ratio used in the present example is based on measurements of perimeter length. Mandelbrot,[2] has pointed out that such measurements in practice yield a series of estimations that may systematically differ as a function of the resolution or the length of the linear standard used to measure distance. If the line to be measured reveals an increasing number of finer details when examined under higher magnification the perimeter obtained would increase with an increase in magnification. In this case it is necessary to indicate the magnification as well as the perimeter measured. If the same measurements are made under different magnifications, the results can be used not only to describe the length of the line but also the character of the "oscillations" in its curvature. A concise form of such a description would provide

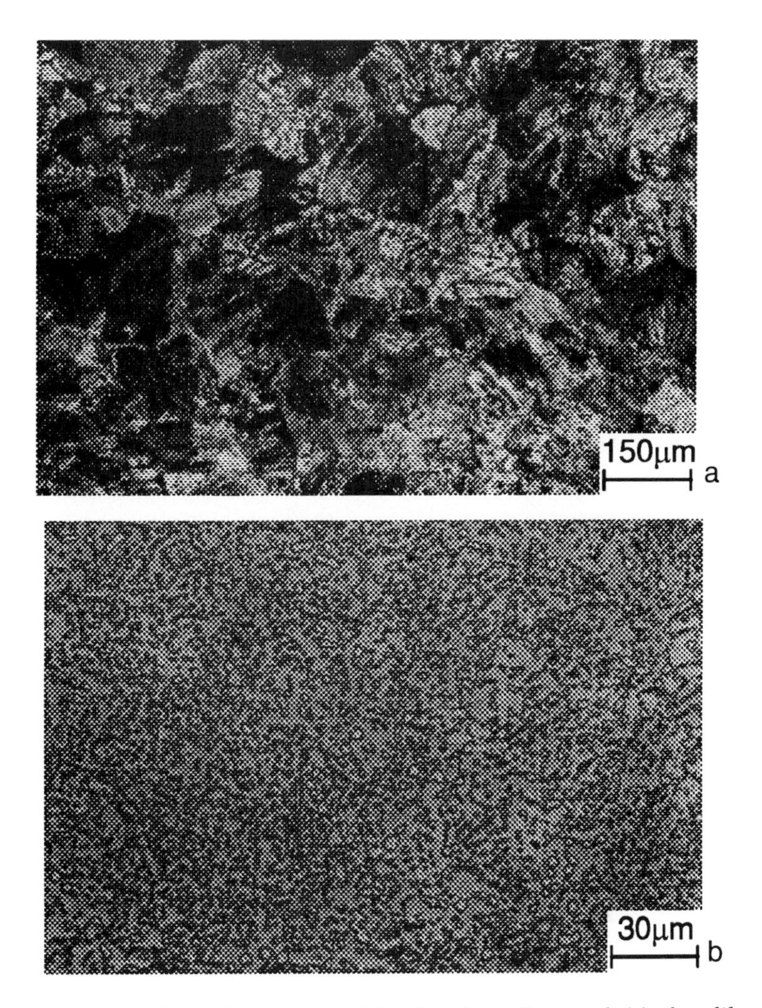

Figure 1.13 Morphology of carbide particles in a low-alloy steel: (a) plate-like carbides of pearlite; (b) spherical carbides produced in the same steel by a different heat treatment.

its line fractal dimension, which is explained in Figure 1.2. More about fractals can be found in Reference 3.

1.4 *Stochastic character of the geometry of microstructures*

Figure 1.1, which shows elements of the microstructure of a polycrystalline material, illustrates one of the important properties of the microstructures of materials—the stochastic character of their geometry (the term *stochastic* is used here to underline the fact that the geometry of the microstructural elements shows a degree of randomness, stochastic = random). Particles, grains,

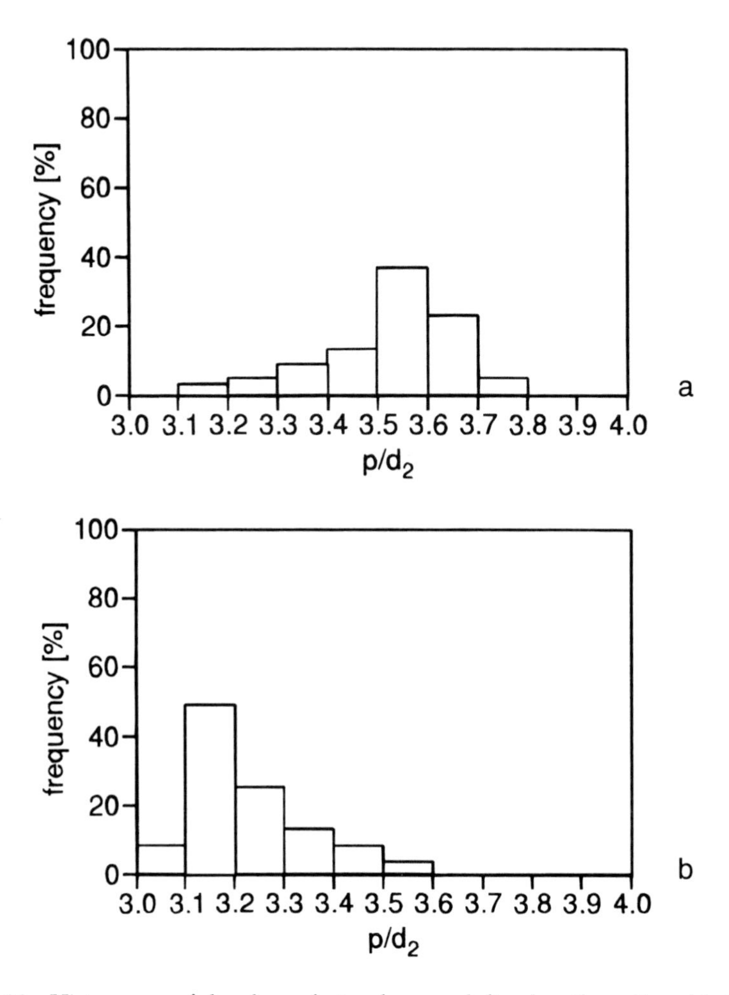

Figure 1.14 Histograms of the shape factor, kappa, defined as the ratio p/d_2 (p is perimeter and d_2 equivalent circle diameter), for the particles of carbide in a low alloy steel after two different heat treatments: (a) histogram for the microstructure exemplified by the microstructure shown in Figure 1.13a; (b) histogram for the microstructure shown in Figure 1.13b

grain boundaries, and dislocations differ in their size and shape. These microstructural elements form populations characterized by distributions of parameters defining them. This applies to all types of microstructural elements, with the exception of 0-D elements, such as vacancies and foreign atoms.

The stochastic character of the geometry of microstructural elements means that their description requires the application of mathematical statistical methods, which brings in with it the concepts of distribution functions, estimators, and intervals of confidence. Looking at some consequences of such a situation, it should be noted that as a result of their stochastic character the elements cannot be characterized by single numbers of an unconditional character. In order to describe the microstructure we are forced to use figures derived from a set of data, for example, the mean, median, standard deviation, etc. In addition to this, in most cases the populations of the microstructural elements are too big to be studied at the 100% level. Instead, we conduct our studies on samples, i.e., finite subsets of populations. This means that we usually have to deal with restricted information about the objects studied and are forced to design thoughtfully the procedures for sampling elements to be selected for the measurements from the whole population.

Despite the fact that we usually study a sample of the population, we always attempt to generalize the results by using them for estimating the parameters of the population. This can be done using a single value obtained from the measurements on a sample drawn from the population (a point estimate) or in the form of a confidence interval that contains all the values of the parameter that should be accepted under the conditions of committing an error with some probability α.

1.5 Observations of the microstructure of materials

The elements of a microstructure extend into three dimensions and are distributed over the volume of the specimen. This means that characterization of the microstructural elements should be based on some 3-D model for the material studied. On the other hand in an experimental approach they are commonly studied on 2-D cross-sections or via examination of thin slices. For example, metallographic observations are made on a cross-section by reflecting the light illuminating the specimen; scanning electron microscopy involves analysis of the signals coming from the surface activated by an electron beam; transmission electron microscopy provides information on the microstructure of slices with a thickness of a fraction of a micrometre. On the other hand a large number of properties are related to microstructural elements which are distributed over the volume of the material. In this situation, the required 3-D description of the microstructure is inferred from the 2-D images by means of the methods of quantitative stereology.

According to Webster's Dictionary *stereology* is ". . a branch of science concerned with the development and testing of inferences about the three-dimensional properties. . . . from a two-dimensional point-of-view . . ." In this context the term quantitative is used to indicate that the inferences mentioned in the definition are to be of a quantitative nature.

The term *stereology* was introduced in the early sixties; however, the foundations of the discipline existed earlier. In fact, the first accounts of a systematic stereological approach go back to the end of the first part of the 19th century when Delesse solved the problem of volume fraction measurements through measurements of aerial density of the profiles of features. In the 1920s, Wicksell developed a formula for estimating the tures. In the 1920s, Wicksell developed a formula for estimating the number of particles in the volume of a material from measurements of their section densities and in the 1940s, Soltykov derived a relationship that could be used to determine the specific surface area of particles in a given volume. These theoretical results in the field were not disseminated across the various branches of the science and characteristically the methods were discovered and rediscovered independently. The year 1961 is noted for the foundation of the International Society for Stereology. In subsequent years the Society has organized a large number of conferences, congresses, and seminars that have helped to accelerate progress in the theoretical foundations of stereology and to spread their knowledge among scientists working in different fields of biology, medicine, geography, geology, and materials science.

A quantitative description of the properties of 2-D images of microstructures is a prerequisite for the quantitative inference of the properties of the 3-D microstructure. In the past, a number of methods have been developed that allow an appropriate quantitative characterization of 2-D images by means of simple counting methods. Some details of such methods are given in Chapters 2 and 3. However, it should be pointed out that in recent years significant progress has been made in developing automatic computer-aided procedures as described in Chapter 4.

In an analytical approach the result of measurements are used to infer information on distribution functions characterizing the populations of microstructural elements. In this case, the measurements need to be designed according to the information sought. Different types of material can be studied and no restriction is placed on the procedures used as long as they are consistent with the basic requirements of stereology.

Example 1.4

Polycrystalline specimens A, B, and C contain grain boundaries revealed as traces on cross-sections shown in Figure 1.15. The population of grain boundaries can be characterized by the surface-to-volume ratio, S_V, which defines the total surface area of the grain boundaries in a unit volume. According to the basic stereological relationship:

$$S_V = 2P_L \tag{1.7}$$

where P_L is the density of the intersection points of the grain boundaries with lines randomly piercing the surface of the grain boundaries in the volume of a polycrystal.

The microstructures of specimen A and B do not show anisotropy in orientation of grain boundary segments and the P_L values, and in turn S_V, can be obtained from intercept length measurement carried out on lines of random orientations drawn on randomly made sections. In metallographic practice one section is frequently used and although it is not recommended to

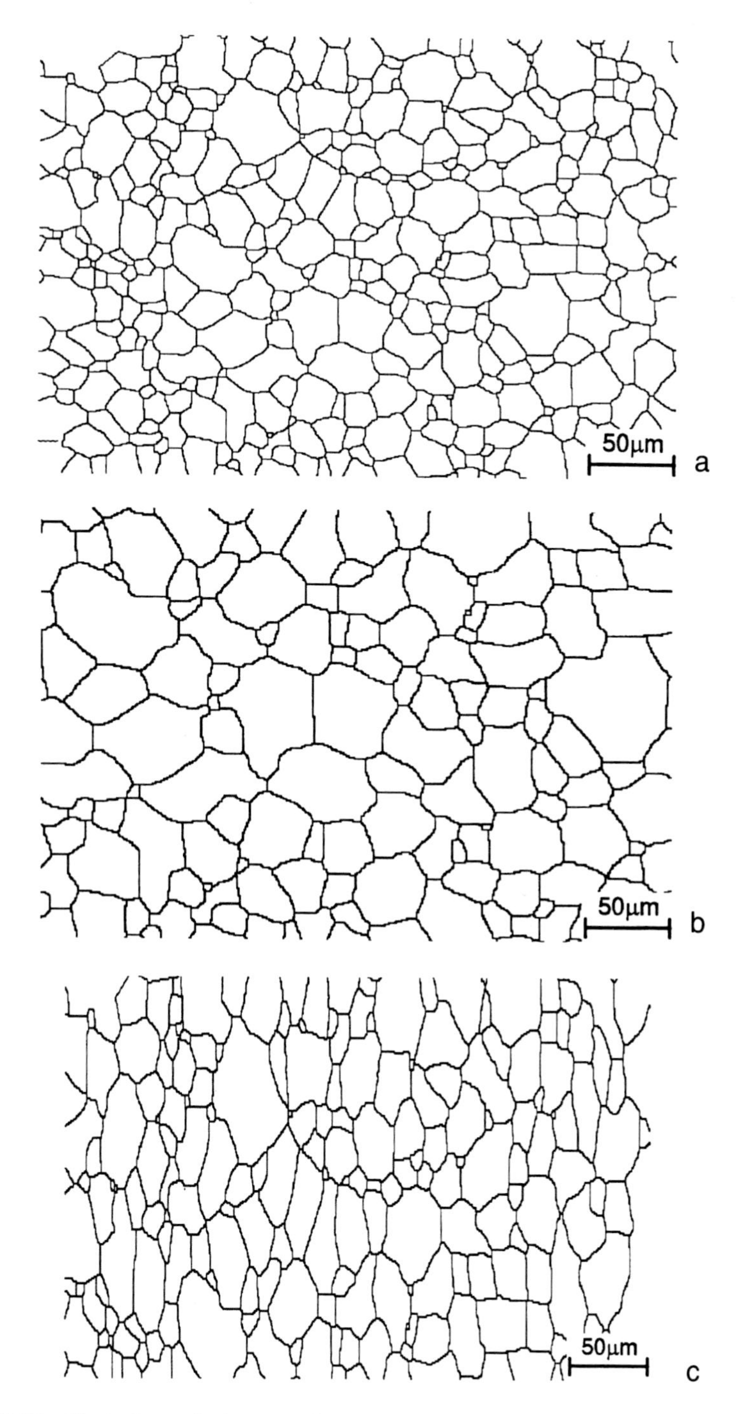

Figure 1.15 Grain boundaries revealed in as-annealed polycrystals (a) and (b) recrystallized at different temperatures. The microstructure in (c) shows the boundaries revealed in material (b) after tensile straining at a low temperature.

from measurements of the density of point of intersections, P_L, of the grain boundaries with the intersecting lines of known total length. For the microstructures shown in Figure 1.15 it can be found that:

$$(S_V)^A = 0.3\ \mu m^{-1} \quad \text{and} \quad (S_V)^B = 0.2\ \mu m^{-1}.$$

The microstructure of specimen C shows anisotropy of orientation of grain boundary segments. In this case the S_V value can be calculated using a more complicated procedure described in Chapters 2 and 3. This procedure requires a systematic study of sections parallel to some specified direction and intersecting the traces of the grain boundaries revealed by a system of cycloids. In the present case this procedure yields:

$$(S_V)^C = 0.35\ \mu m^{-1}.$$

The values of $(S_V)^A$, $(S_V)^B$ and $(S_V)^C$ can then be used to compare the surface density of grain boundaries in the respective specimens. In the analytical approach, an attempt is made to quantify the differences in the microstructures studied in terms of stereological para s that provide information about their true 3-D character.

The above example shows that quantitative data about the microstructures studied are obtained through measurements that are carried out on different types of sections (random sections, vertical sections) using testing lines of different natures (straight lines, cycloids). The positions of these sections and testing lines are randomly placed with respect to the microstructures studied but their geometry is a result of a precise analysis that takes into account the probability of intersecting the chosen or the other elements of the microstructure. The problem of proper selection of sections and test lines or other probing elements (points, quadrants, etc.) is one of the major issues of stereology.

Two general strategies can be adopted in attempts to quantify information on the microstructure of materials. In a descriptive approach the focus is on the data that are directly obtained from measurements performed on different variants of one type of microstructure. In this case usually no attempt is made to define the properties of the populations of the microstructural elements.

Example 1.5

Descriptive studies of the grain size in polycrystals can be made by intercept length measurements. Figure 1.16 presents histograms of intercept length measured for a series of specimens of the same material, a single-phase polycrystal, after different heat treatments. The rod-shaped specimens used had the same geometry and the measurements were carried out systematically on longitudinal cross-sections. Specimens A and B differed in their grain size as indicated by differences in the mean intercept length, E(l). On the other hand the histograms of normalized intercepts l/E(l) for these two specimens are similar. This is an indication of similarity in grain shape in both microstructures.

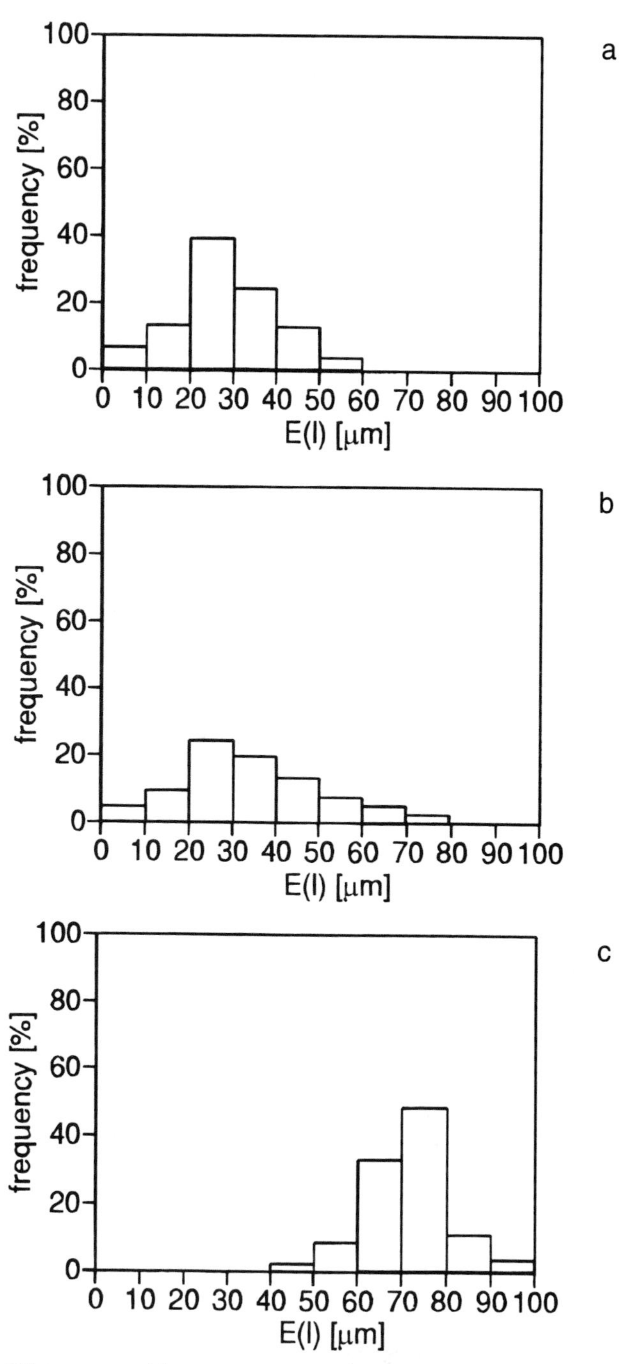

Figure 1.16 Histograms of the intercept length and normalized intercept length, l/E(l), for the microstructures shown in Figure 1.15a-c. (a) and (d) refer to the microstructure shown in Figure 1.15a, etc. For Figures d, e and f, please see following page.

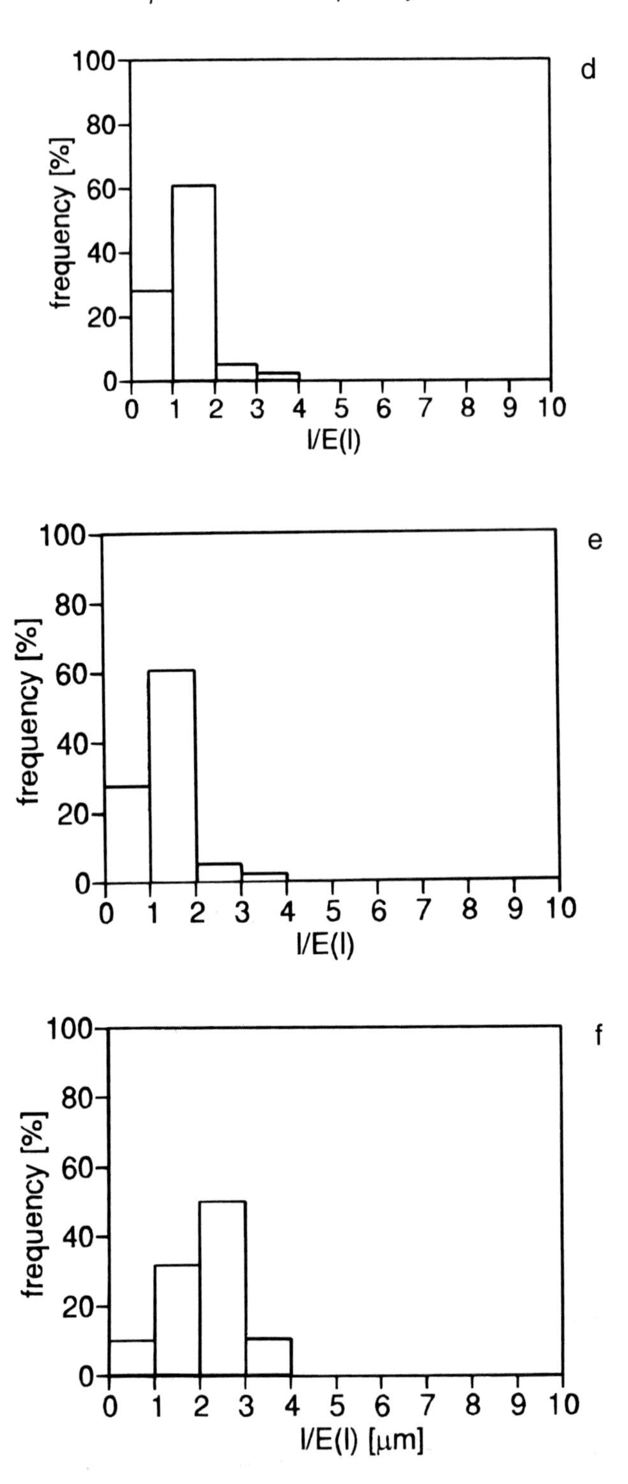

Figure 1.16 *(Continued)*

lar. This is an indication of similarity in grain shape in both microstructures.
Specimen C has been strained in a tensile test prior to measurement. The intercept length histograms are distinctively different from those for specimens A and B. This is related to a difference in grain shape. In a descriptive approach it is usually sufficient to state similarity or dissimilarity of the material microstructures studied.

In attempting to derive a quantitative description of a 3-D microstructure from measurements on its 2-D images, care should be taken that these images are statistically representative of the microstructure studied. This can be achieved by employing the proper stereological sampling procedures designed for specific situations; otherwise, the data obtained from the images studied will remain just the data for some images. It is still better to have these data than only images but it should be also remembered that with slightly more effort these data can be used to make inferences about the properties of the 3-D microstructure studied.

It was the aim of the present chapter to show that the properties of materials strongly depend on their microstructure. The properties are quantified through specific measurements of them and it is natural to do the same with respect to the properties of the microstructures. Microstructures are formed by specific elements arranged in 3-D space; however, these elements are studied in the form of 2-D sections. Stereology is a discipline which provides solutions to the question of proper sampling of 2-D sections and elements of microstructure revealed from the point of view of building a description of the 3-D microstructure studied. Although it does not answer questions about the role of different elements of the microstructure, quantitative stereology systematically becomes one of the important tools of modern materials science.

1.6 *About the book*

This book is prepared with the intention of covering all major areas related to a quantitative description of the microstructure of materials. The process of microstructural characterization starts with its imaging. Many techniques have been developed that can be used to produce images of various elements of the microstructure. They include: light, ultrasonic and electron microscopy, X-ray, electron and neutron diffraction, etc. Each of them deserves to be presented in the form of a book itself and in fact an excellent collection of such books has been published in recent years, e.g., References 4, 5, and 6.

Experimental techniques used to obtain images of microstructural elements work on the principle that these elements are distinguished by their physical properties. Different phases and grain boundaries can be observed in a light microscope due to their different optical properties or due to their different resistance to the applied etchant. As a result of the increased energy of atoms, dislocations in crystals are seen in transmission electron microscope images due to their strain fields, whilst segregation of chemical elements can be visualized by imaging information about X-ray emission of the atoms on a section of the material excited by temperature, light, or electrons.

This book is primarily focused on the question of a proper strategy for sampling the microstructure studied, or in other words, the question of a strategy for selecting representative images, and quantification of the information these images contain into a description suitable for applications in materials science. These are general questions and the answers presented are valid to any experimental technique.

Depending on its purpose, quantification of the information from images of microstructures is possible at different levels of generalization. First, some measurements can be carried out on one particular image yielding its quantitative description, for example, in the form of the area occupied by "black" dots (this can be done manually but with recent development in microcomputers it is likely to be done automatically on a digitized form of the initial image). Second, based on the description of the experimental technique employed for microstructural imaging, an attempt can be made to turn the information from the image into a description of the 3-D microstructure in the volume of material studied, for example, into a volume fraction of the phase depicted in the form of black patches. A third level of the analysis is achieved if a number of such images is used to produce a statistically correct description of the microstructure of the material: in the example used, into a phase content of a given material. The second and third levels involve stereology and are based on the principles of the theory of probability and mathematical statistics. This explains why quantitative stereology is the main subject in this text.

The subjects of quantitative stereology and image analysis have been covered in some excellent books such as References 7 to 10 and to these books those readers are referred who find themselves inspired to further reading. Working on this book, the authors have assumed that it should provide all the information required for those who want to practice the quantitative description of a materials microstructure (the authors view themselves as practitioners). The book starts with this introduction, which gives some examples explaining the needs for microstructural quantification. The next three chapters, Chapters 2, 3, and 4 present the foundations and advanced tools of quantitative stereology, including image analysis and some of the computer-aided methods. These three chapters, together with the introduction comprise the first part of the book. The second part offers applications of this philosophy in materials science. It is divided into four further chapters. Chapter 5 deals with the problem of line and surface microstructural elements, such as dislocations, fibers on the one hand and interphase boundaries, external surfaces, etc. on the other. Chapter 6 is devoted to the characterization of grain size in polycrystalline materials. Chapter 7 gives examples of the situation where microstructures undergo changes. It deals with the changes in grain geometry during plastic deformation and with grain growth. Chapter 8 is totally concentrated on the description of particulate systems, i.e., characterization of multiphase materials that contain isolated particles embedded in a matrix.

References

1. Armstrong, R. W., *The yield and flow stress dependence on polycrystal grain size, Yield, Flow and Fracture of Polycrystals*, Baker, T. N., Applied Science Publishers, London, 1983, 1.
2. Mandelbrot, B. B., *The Fractal Geometry of Nature*, W. H. Freeman, San Francisco, 1982.
3. Schwarz, H. and Exner, H. E., The implementation of the concept of fractal dimension on a semi-automatic image analyser, *Powder Technology*, 27, 207, 1980.
4. Loretto, M. H., *Electron Beam Analysis of Materials*, Chapman and Hall, London, 1984.
5. Briggs, A., *An Introduction to Scanning Acoustic Microscopy*, Oxford Univ. Press/Royal Microscopical Society, 1985.
6. Weinberg, F. (editor), *Tools and Techniques in Physical Metallurgy*, Marcel Dekker, New York, 1970.
7. De Hoff, R. T. and Rhines, F. N., *Quantitative Microscopy*, McGraw-Hill, New York, 1968.
8. Underwood, E. E., *Quantitative Stereology*, Addison Wesley, Massachusetts, 1970.
9. Weibel, E. R., *Stereological Methods*, Academic Press, London, 1980 and 1989.
10. Coster, M. and Chermant, J. L., *Precis d'analyse d'images*, Presses du CNRS, 1989.

chapter two

Basic concepts, definitions, techniques and relationships

2.1 *Populations, individual microstructural elements and distribution functions*

Microstructural elements such as particles, grains, dislocations, etc., of a given type differ in size and shape, and form populations, i.e., collections of elements about which information is sought in studies aimed at the characterization of the microstructure of materials. A commonly dealt with example of such a population is a set of particles in a two-phase material schematically shown in Figure 2.1.

Individual microstructural elements, e.g., particles, occupy some regions of the space with 3, 2, or 1 linearly independent directions. The elements can therefore be analyzed as sets of geometrical points. For example:

a) two particles, α and β, are two sets of points, Z_α and Z_β, in space;
b) two sections of grains, A and B, are two sets, Y_A and Y_B, of points on a plane;
c) two intercepts, I_A and I_B, are sets of points in 1-dimensional space (on a line).

Using this approach, the microstructure of a material can be defined as a system of sets of points Z_i, Y_j, I_k and in their analysis some elements of set algebra are helpful. In particular, it is desirable to consider operations of adding, subtracting, and finding common elements for two or more sets. These operations for any two sets, S_1 and S_2, assign other sets designated:

a) $UNI(S_1, S_2)$;
b) $DIF(S_1, S_2)$;
c) $COM(S_1, S_2)$;

respectively. The sets listed above have the following properties:

a) $UNI(S_1, S_2)$ contains all points that belong to either of the sets (S_1, S_2);
b) $DIF(S_1, S_2)$ contains the points which belong to S_1 and do not belong to S_2;
c) $COM(S_1, S_2)$ contains the points which belong to both sets (S_1, S_2).

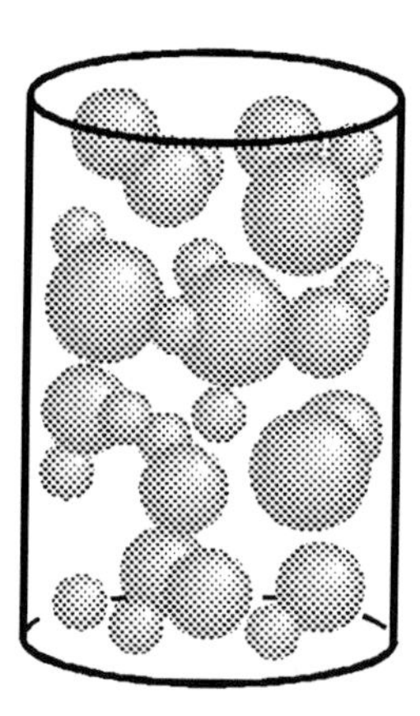

Figure 2.1 A hypothetical two-phase material with spherical particles of the second phase. The volume of the individual particles is designated V_i.

These definitions can be extended easily to a large number of sets with obvious properties as follows:

a) UNI(S_1, S_2 . . .);
b) DIF(S_1, S_2 . . .);
c) COM(S_1, S_2 . . .).

These operations are illustrated in Figure 2.2.

In the notation of set algebra the microstructure of a material is the sum of sets representing the microstructural elements:

$$Microstructure = UNI(Z_1, Z_2, ..Y_1..I_1...) \tag{2.1}$$

This set of sets is studied using systems of testing sets, TS, such as:

a) sets of points, O_k;
b) sets of lines, L_k;
c) sets of sections, S_k

and their combinations. The measurements of microstructural features are usually carried out on the intersections of the microstructure with the testing sets:

$$\text{Measurements} \Rightarrow \text{COM}(\text{Microstructure, TS}).$$

In a quantitative description of the microstructure, the properties of the sets representing the microstructural elements are studied. An important feature of these elements is their size. The size of the sets Z_k, Y_l, and I_m are quantified with respect to the size of some reference space. However, instead of counting the number of points in a given set representing an element of the microstructure (as is actually carried out in image analysis systems), the size is estimated using the principles of mathematical probability.

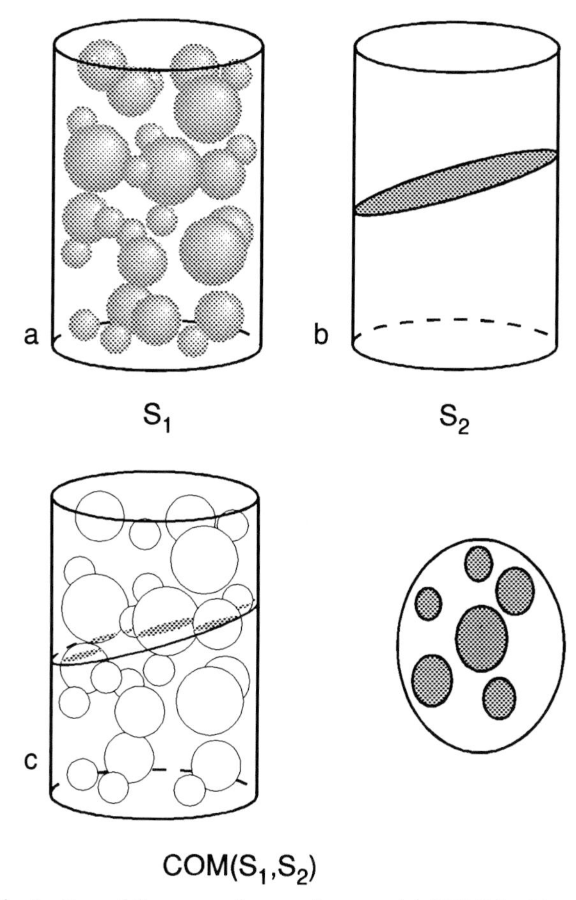

Figure 2.2 Illustration of the operation on the sets: (c) COM(S_1, S_2). The set S_1 is the set of particles shown in Figure 2.1; the S_2 is a plane sectioning the set of particles.

Consider two sets of points, Z_0 and Z_1, shown in Figure 2.3. The set Z_1 is formed by points that, by some imaging technique used, can be categorized as representing a microstructural element of interest. The set Z_0 is the matrix in which the element Z_1 is embedded. The relative size of the two sets in principle can be estimated by counting the number of both elements, i.e., the number of "points" that belong to each of them. However, this approach leads to difficulties involved in the definition of the size of the points into which the Z_0 and Z_1 sets are considered to be divided. Also, it requires counting a large number of points. An alternative, probability-based approach is based on a scheme of randomly selecting points from both sets. If points are randomly selected from the UNI(Z_1, Z_0) then the ratio of the points that belongs to Z_1, N_1, and Z_0, N_0, asymptotically approaches the ratio of the sizes of the sets:

$$\frac{N_1}{N_0} \approx \frac{size(Z_1)}{size(Z_0)}. \tag{2.2}$$

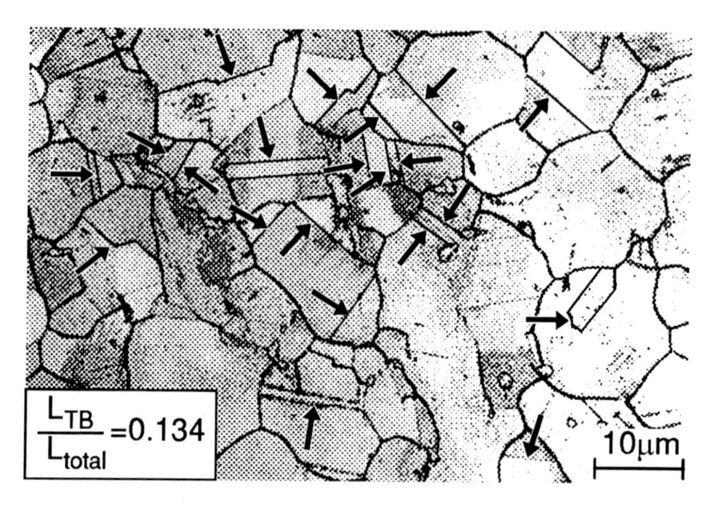

Figure 2.3 Two sets of microstructural elements: Z_0 is the set of grain boundaries revealed on section of a polycrystal; Z_1 is the set of twin grain boundaries which are distinguished by their flatness (arrowed).

If the size of the whole space studied, size UNI(Z_1, Z_0), is known, the size of any of the sets (Z_1, Z_0) can be estimated from the following:

$$size(Z_1) \approx \frac{N_1}{N_1 + N_0} size[UNI(Z_1, Z_0)] \tag{2.3}$$

This way of estimating the size of microstructural elements can be illustrated in the case of area estimation of particle sections revealed on a cross-section of a two-phase material (Figure 2.4). A set of N points is randomly positioned on the image of the phases and the points are counted separately that belong to:

a) phase α;
b) matrix.

The relative area of the particle sections, $(A_A)_\alpha$ is given as:

$$(A_A)_\alpha \approx \frac{N_\alpha}{N_{mat}} \tag{2.4}$$

The area A_α comes from the following:

$$A_\alpha \approx N_\alpha \, (M \, \rho_{TP})^{-2} \tag{2.5}$$

where M is the magnification, ρ_{TP} the density of the testing points (the number of points per unit area).

Microstructural elements usually extend into three dimensions while they are studied usually via observations on 2-D images. Such images contain sections or projections of the objects that do not directly describe the

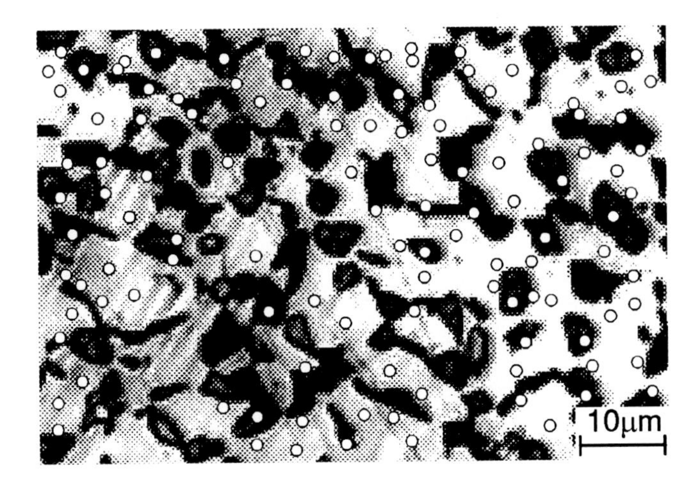

Figure 2.4 An example of a section of a two-phase material. The area of the particle sections is estimated by counting the number of points hitting the sections revealed in the field of observation. The field area is 0.1 × 0.1 mm; the density of the test points $10^4/\text{mm}^2$. There were a total of 32 points hitting the sections of the particles. This gives the following estimate of the particle section area: A = 0.0032 mm².

geometrical properties of the elements. There are two important probabilistic implications of this.

The first implication is that the objects seen in the images of the microstructure do not appear in numbers directly proportional to their real number. In fact, they appear in proportions that depend on both the size and the number of the elements of interest. Thus large particles are more likely to be seen on sections of a two-phase material than the smaller ones. This effect is illustrated for a 2-D case in Figure 2.5, which shows that large sections of grains are more frequently cut by intercepts than are small grains.

Second, the same size section or projection can represent on the image two elements of quite different properties (sizes). If an attempt is made to estimate the size of microstructural elements from the size of their sections, a concept of conditional probability has to be used. Looking at the image of an object we have to ask the question: what is the size (shape) of the object of interest that images in this particular way under the adopted conditions of observation?

The answer to this question is statistical. We cannot specify the size of any particular object imaged in the micrograph studied. However, it is possible to estimate some properties of the elements such as their total volume, surface, length, etc. These estimates are provided by the stereological relationships, which are derived from the principles of probability theory.

2.2 *Distribution functions*

In most cases the number of particles and other microstructural elements in the material studied is large and it is not possible to list the properties of each

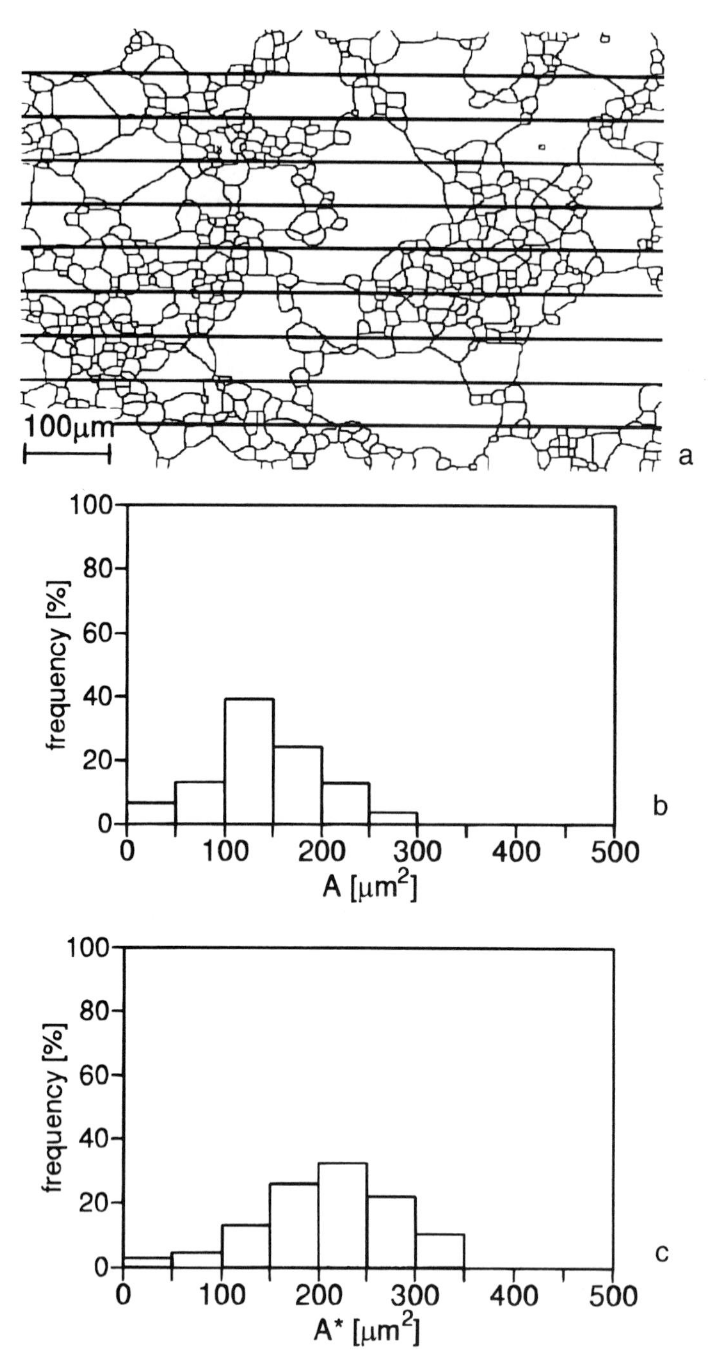

Figure 2.5 Schematic representation of the effect of the particle size on the probability of their intersection by a randomly drawn section: (a) system of grain sections revealed on a section of polycrystal with superimposed test lines; (b) histogram of grain sections area A; (c) histogram of areas of sections hit by test lines A*.

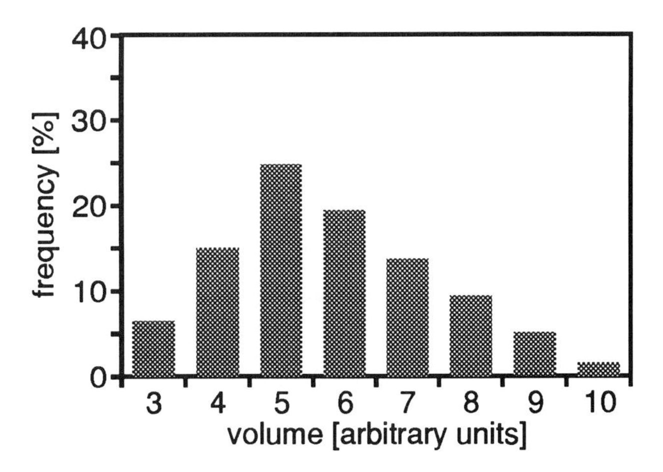

Figure 2.6 Bar plot of the distribution function of the volume of particles in the model two-phase material depicted in Figure 2.1.

individual particle. Instead, the particles are characterized by a distribution function, f(V), which defines the relative number of particles of size V. This distribution function can be of a discrete or continuous type. Discrete distribution functions are defined over discrete sets of variables; in the present example the volume of the particles. Their properties are visualized in the form of point or bar plots (Figure 2.6). On the other hand, continuous functions can be represented by line plots of f(V) vs. V. Experimental data usually is available in the form of discrete distribution functions. This is related to the fact that the very process of measurement is carried out with a certain precision that makes it impossible to detect very small differences in the size of the elements. Therefore, the measured quantities can usually be grouped into some classes quite naturally. This is illustrated by a graphical representation in the form of histograms, which consists of rectangles with areas being proportional to the number of measurements in a given interval.

Discrete distribution functions, for example, f(V), give directly the relative number distinguished by a specified value of the variable, in this case the number N(V) of particles of size V in a population containing N particles:

$$f(V) = \frac{N(V)}{N} \tag{2.6}$$

In the case of a continuous distribution function, f(V) is interpreted in the following way:

$$N(V_1, V_2) = \frac{1}{N}\int_{V_1}^{V_2} f(V)dV \tag{2.7}$$

where $N(V_1, V_2)$ is the relative number of particles in the size range from V_1 to V_2. The continuous distribution function has an important theoretical meaning. Such a distribution can be used to approximate the case of measurements made with increasingly high precision when the classes become infinitely narrow. Further, such distributions can be analyzed using the methods of calculus, which makes much simpler the study of their properties. As a result populations of microstructural features are frequently assumed to be of a continuous character.

The distribution functions are restricted by two formal conditions which require that: (a) the function be nonnegative

$$f(x) \geq 0 \tag{2.8}$$

and (b) be normalized, i.e.,

$$\int f(x)dx = 1 \tag{2.9}$$

In general, there are a number of such functions that have physical meaning and many such functions are discussed in various text books. In the field of materials characterization the most frequently used functions are the normal and log-normal distributions, which are discussed later in this chapter.

The distribution of a given variable over the elements of the population can also be described with the use of cumulative distribution functions. These functions for any number, S, define the number of elements in the population characterized by values of the parameter studied being smaller than S:

$$F(S) = \int_{-\infty}^{S} f(V)dV \tag{2.10}$$

The cumulative distribution function of particle volume for the material used in the present example is shown in Figure 2.7.

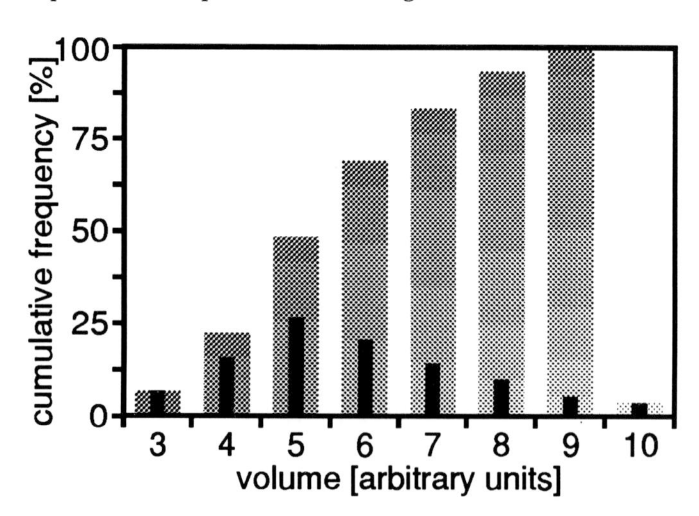

Figure 2.7 The cumulative distribution function of the particle volume for the material depicted in Figure 2.1.

2.3 *Parameters of the populations*

Distribution functions provide detailed information about the populations they describe. However, frequently such information is neither required nor available. As a result the populations are characterized alternatively by a set of parameters that measures tendencies in a given population. The list of commonly used parameters includes:

a) the mean value;
b) the variance;
c) the standard deviation;
d) the coefficient of variation.

Less frequently used are the skewness and the kurtosis. All these parameters are calculated by summation, over the whole population; giving expressions that contain the parameter studied and its mean value. Their definitions for a continuous distribution function f(x) are given in Table 2.1 and for discrete functions in Table 2.2.

The mean value of the variable x, E(x), is called a measure of central tendency in the distribution of x. The variance, VAR(x), and the standard deviation, SD(x), are measures of the dispersion or scatter of the values of the random variable around the mean, E(x). This dispersion is also measured by the dimensionless coefficient of variation, CV(x), defined as the ratio of the standard deviation to the mean. Low values of CV(x) indicate that the values of

Table 2.1 Definitions and Interpretation of the Parameters Used to Characterize Populations of Random Variables for Continuous Distribution Functions

Parameter	Function	Formula
Mean value E(x)	Measure of the central tendency	$\int_{-\infty}^{\infty} xf(x)dx$
Variance VAR(x)	Measures of dispersion	$\int_{-\infty}^{\infty} [x - E(x)]^2 f(x)dx$
Standard deviation SD(x)		$\sqrt{Var(x)}$
Coefficient of variation CV(x)		$\frac{SD(x)}{E(x)}$
Skewness SK(x)	Measure of symmetry	$\frac{E(X - E(x)^3)}{SD(X)^3}$
Kurtosis K(x)	Measure of peakiness	$\frac{E(x - E(x)^4)}{SD(x)^4}$

Table 2.2 Definitions of E(X) and VAR(X) for Discrete Distribution Functions

Mean value E(x)	$(x_1 f_1 + x_2 f_2 + ... + x_i f_i + \ldots)$
Variance VAR(x)	$\ldots + (x_i - E(x))^2 f_i + \ldots$

the variable are concentrated around the mean. The coefficient of skewness is another dimensionless parameter used to characterize random variables. It is equal to zero for symmetrical distributions and positive for the distribution skewed to the right (negative for the distribution skewed to the left). Another dimensionless parameter, the kurtosis, describes the "peakiness" of the distribution. The larger the value of the kurtosis, the less flat the distribution function. The normal distribution, which may be used as a reference distribution, is characterized by a kurtosis equal to 3.

In addition to the mean value the central tendency in a population is sometimes characterized by the median or the mode. The median, Med(x), is the value of the parameter for which the cumulative distribution function is equal 1/2:

$$\int_{-\infty}^{Med(x)} f(x)dx = \frac{1}{2} \tag{2.11}$$

The mode, Mod(x), defines the value of the parameter that occurs most often. In other words, it is the value that is measured for the maximum number of elements or maximum number of points in the distribution function:

$$\max\{f(x)\} = f[Mod(x)] \tag{2.12}$$

If there is more than one maximum in the distribution then it is of multimodal character. Basic definitions are schematically explained in Figure 2.8.

Example 2.1

A population of particles in a hypothetical microstructure consists of three types of spherical features A, B, and C. Their respective diameters are as follows:

$$d_A = 2, d_B = 4, d_C = 10.$$

The particles are distributed randomly in the volume of the material in the proportion $N_A{:}N_B{:}N_C = 3{:}6{:}1$. Simple computations show the following properties of this population:

$$E(V) = 26.6; SD(V) = 16.5; CV(V) = 1.61; Mod(V) = 33.5.$$

It should be noted that although the parameters such as the mean, standard deviation, skewness, and kurtosis in most situations are sufficient to

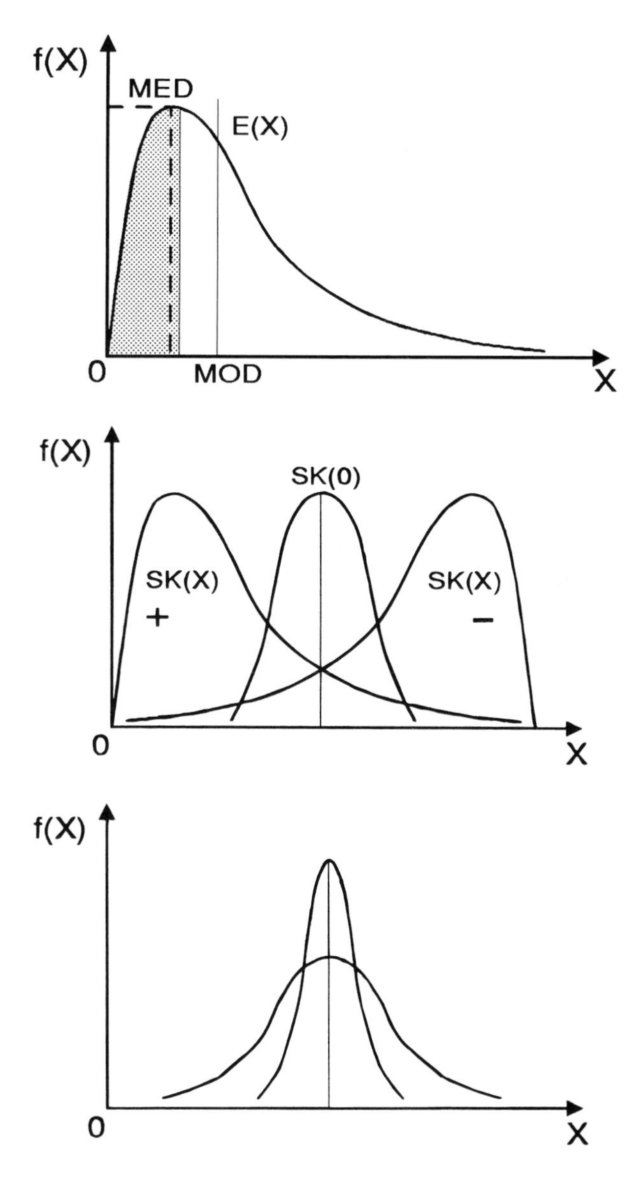

Figure 2.8 Graphical explanations of the basic parameters used to characterize a population. (Figure is continued on the next page.)

characterize the populations with the required precision, they do not define them uniquely as do the respective distribution functions.

2.4 Normal and log-normal distributions

The normal distribution has a bell-shaped distribution function given by the following formula:

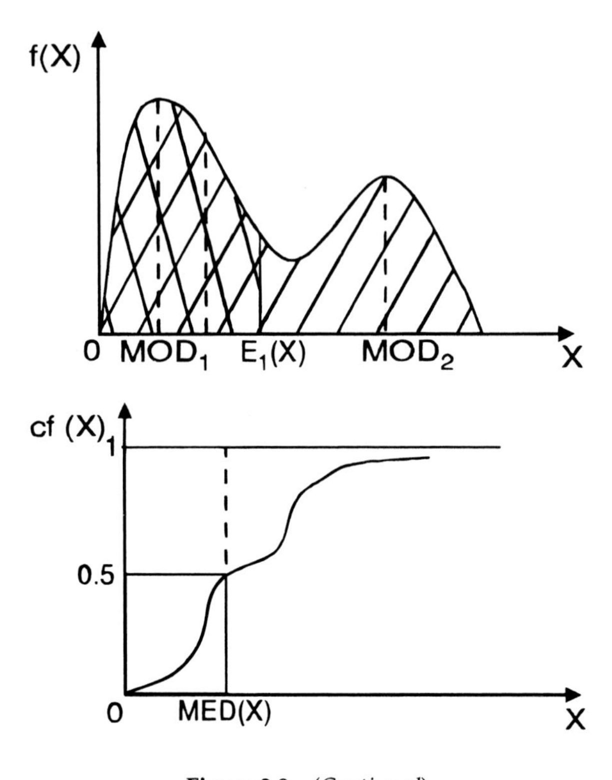

Figure 2.8 (*Continued*)

$$f_{nor}(x)=\frac{1}{SD(x)\sqrt{2\pi}}\exp\left[\frac{-(x-E(x))^2}{2SD(x)^2}\right] \tag{2.13}$$

where E(x) and SD(x) are constants equal, respectively, to the mean and the standard deviation, and x varies from minus to plus infinity. For the normal distribution:

a) more than 68% of the values of x are in the range of values from E(x) – SD(x) to E(x) + SD(x);
b) more than 95% of the values of x are in the range E(x) – 2SD(x) to E(x) + 2 SD(x).

Characteristically, the skewness of a normal distribution is equal to 0 and its kurtosis to 3.

The normal distribution is sometimes called the Gaussian distribution. Its importance is based on both its role in theoretical analysis and the fact that it is frequently observed in experimental situations. It has been proved, under quite general assumptions, that the normal distribution approximates the distribution functions that are the sum of a large number of constant distribution

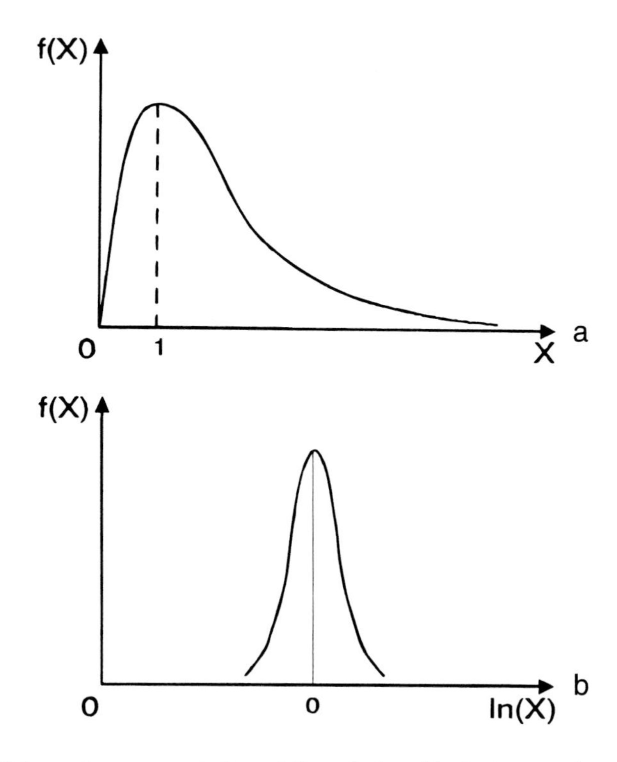

Figure 2.9 Schematic representation of the relationship between a log-normal and normal distributions: (a) a lognormal distribution of a variable x; (b) the same distribution plotted against $\ln(x)$ yields a normal distribution.

functions. In particular, this distribution typically describes the experimental error produced in well-designed experiments. This distribution also approximates well some other important distributions such as the binomial and the Poisson.

The log-normal distribution is closely related to the normal one. A random variable, x, is of log-normal character if a variable y, defined as the logarithm of x:

$$y = \ln x \tag{2.14}$$

is of normal character. The distribution function of a log-normal variable is thus given for $x > 0$ by the following equation:

$$f_{\ln}(x) = \frac{1}{SD(x)\sqrt{2\pi}} \exp\left[-\frac{(\ln x - E(x))^2}{2SD(x)^2}\right] \frac{1}{x} \tag{2.15}$$

The relationship between the normal and log-normal distributions is shown in Figure 2.9. Since a normal distribution can be associated with the conditions where the outcome of an experiment is a large number of independent, additive events and due to the fact that logarithms obey the rule

$$\ln(a \times b) = \ln a + \ln b \tag{2.16}$$

the log-normal distribution function can be expected to occur in those cases where the outcome of an experiment is a result of independent, multiplicative events. This distribution function by its nature is also more likely to occur for variables that are restricted to be bigger than a given critical value z_0:

$$\begin{aligned} f(z) &\geq 0 \quad \textit{for} \quad z \geq z_0 \\ f(z) &= 0 \quad \textit{otherwise} \end{aligned} \tag{2.17}$$

This is the case for geometrical measurements that yield positive values of the variable x. For a parameter x of a geometrical nature there is a zero chance to find values of x smaller than 0:

$$f(x) = 0 \quad \textit{for} \quad x < 0 \tag{2.18}$$

A distribution function of geometrical properties is likely to be of log-normal character.

The formal use of a log-normal distribution requires that the variable studied be positive and dimensionless. On the other hand, in experiments we usually measure parameters that are expressed in specific physical units. In order to overcome this discrepancy, the results of the measurements need to be restructured to make them dimensionless. This can be done by: (a) normalization and (b) standardization. Normalized variables are calculated according to the following equation:

$$z = \frac{x - E(x)}{SD(x)} \tag{2.19}$$

Standardized variables are obtained by dividing the values of the variable by the units of measurement:

$$x = \frac{x}{[x]} \tag{2.20}$$

where [x] gives the units of x.

In the case of log-normal distributions it is sometimes more sensible to use the mean values, standard deviations, and higher-order parameters such as skewness for the logarithm of the variable x, lnx. One can then concentrate attention on the values of E(lnx), SD(lnx), etc. instead of E(x), SD(x). These parameters for a log-normal distribution remain in simple relationships:

$$E(x) = \exp\left[E(\ln x) + \frac{SD(\ln x)^2}{2}\right] \tag{2.21}$$

$$SD(x) = \exp\left[2E(\ln x) + 2SD(\ln x)^2\right] \tag{2.22}$$

Log-normal distribution functions have been reported in a number of studies of the microstructure of materials. These functions have been found to approximate well the distribution of the volume of particles and grains,[1–3] the grain area,[4] and the number of grain edges.[5] Examples and more detailed analyses of this type of data are given in Chapter 6.

Some more properties of the log-normal distribution are illustrated by the following example.

Example 2.2

A population of grains in a recrystallized material is described by a log-normal distribution function of grain volume, V. It has been found, for polycrystals of aluminum, that E(lnV) = 3 and SD(lnV) = 1. As said before, 68% of grain are expected to be characterized by lnV in the range E(lnV) – SD(lnV), E(lnV) + SD(lnV). Simple calculation show that approximately 16% of grains are characterized by lnV ≥ 4 or a volume 2.7 times larger than the mean volume.

2.5 *Sampling*

In most cases of practical importance while characterizing the microstructures of materials it is not possible to deal with all the features of interest since their number is too large. For example, the number of particles in a cubic millimetre of a dispersion strengthened alloy or the number of grains in a polycrystalline specimen can easily exceed one million, which makes it virtually impossible to examine all of them one by one. Moreover, in the process of materials characterization an attempt is made to generalize the findings by suggesting that the object of study was not a set of specimens but a certain material of unlimited volume, defined by its chemical composition and the treatment. As a result we are usually forced to perform or consider the measurements carried out on *samples of the populations studied*. The term *sample* is used here in a mathematical sense and means a finite subset of elements that represents the properties characteristic of the population. Examples of samples are particles extracted from an alloy or grains revealed on a cross-section of a polycrystal.

Samples of microstructural features are taken out of their population in the material studied through the process called sampling. The previous remark means that sampling is an important step of almost any investigation of the microstructure of a material and relevant to understanding its properties.

The process of sampling has become an area of study for mathematical statistics (see, for example, Reference 6). Ideally the process should proceed along some general lines defined by the following steps:

Step 1
setting the objective of the study
Step 2
defining the population of the relevant microstructural features

Step 3
defining the parameters to be measured and the precision required
Step 4
division into sampling units and defining the frames
Step 5
selection of the samples: random choice against scanning
Step 6
measurements of the selected parameters via geometrical probing
Step 7
data analysis, graphical representation, correlation with the properties

In step 1 the question has to be answered: What is the purpose of the investigations? Is it to produce a general characteristic of the material or an artifact? Is it aimed at explaining some of the material properties? Are the conclusions to be generalized for other materials/technologies? Is the specimen an object by itself or a representative piece of the material that can be obtained in larger quantity according to a specified technology? What is the required precision of the expected conclusions? The subsequent steps should be correlated to these objectives.

In step 2 the material is analyzed from the point of view of relevant microstructural elements. These might be pores, particles, grain boundaries, etc. Frequently, a combination of the elements might be used, for example, particles and grain boundaries. The analysis should address such questions as:

a) What is the dimensionality of the microstructural features?
b) In what way can they be revealed or mapped?
c) Are they expected to be isotropic or anisotropic?
d) Do they differ in size, shape, and physical characteristics?
e) How uniformly are they distributed in the volume?

In giving answers to these questions it is usually helpful to develop a model of the microstructure studied. This model can later be verified or refined in the process of the measurements. While giving an answer to question b) it has to be remembered that there is a fundamental requirement that the objects of interest may be separated from the rest of the microstructural elements. This point cannot be stressed too highly. A requirement here is that the specimen preparation technique produces systematic contrast from all the features under examination. Since we are also going to stress plane section sampling this also means that techniques such as deep etching have to be avoided. Our system of data collection requires that all the features (particles, grain boundaries) be imaged and in their correct sizes. Essentially, even if a manual data collection technique is adopted, we need to get to a binary image (the basis of all automatic image analysis techniques) where points delineating features or grain boundaries are seen as 1s and all other points as 0s (or vice versa).

The next step (3) is the selection of the parameters to be measured. These parameters, on the one hand, are determined by the dimensionality of the features (particles—volume, dislocations—length) and on the other, by the purpose of the investigations. They may range from the parameters that describe size to those that define the shape of the objects, again depending on the purpose of the studies. Their selection should be preceded by reexamination of the theoretical models for the pertinent material properties. It is important to avoid employing an unnecessarily large number of parameters, as this may lead to diversion of attention from the key elements. As for the degree of precision needed in the measurements, this relates to the precision of estimation of a given parameter for the population. At this stage we have to decide exactly how we want to measure individual features (usually what magnification is to be used) and how many features are to be included in the analysis (how many images are to be studied). In giving answers to these questions it should be remembered that in most cases the overall precision of estimation is not so much limited by the precision of measurements on individual images, but by the number of images measured.

In the next step (4) the identified population of microstructural elements is divided into units to be sampled. These units are samples of the material that need not overlap and entirely cover the population of interest. In the case of microstructural characterization the population is made up either of a given set of specimens or is defined indirectly by specification of the composition and technological history of the material. In the first case, the sampling units are different pieces of one specimen, while in the second, these are different specimens of the material that may be produced by, in general, various manufacturers.

The sampling units are defined according to a specified scheme of frames. The samples can be of volumetric character (bulk materials) as well of aerial form (thin layers, etc). In the case of materials characterization the frames should specify not only the dimension of the sampling units but also give a rule for accepting or rejecting any element intersected by the boundaries of the units. This latter condition is related to the requirement that each element of the microstructure can be attributed to one and only one sampling unit.

The measurements are subsequently carried out on a set of sampling units selected from the whole set covering the population. Various selection strategies, step 5, can be adopted that can be divided broadly into two categories:

a) systematic scanning over a specified volume, area, etc.;
b) random sampling.

In both cases the object of interest, an artifact or specimens of a material of volume V is divided in a real or thoughtful experiment into N nonoverlapping blocks of volume V_b. The blocks are then numbered from 1 to N in a way that is not restricted by any further conditions. The only requirement is that the numbers be kept unchanged during the study. In the case of systematic scanning, measurements are carried out on all of the blocks, n, so that n = N. Typically the results of measurements are analyzed then as a function of the block position with respect to the studied artifact geometry.

For the case where measurements are carried out on a smaller number of blocks (n < N) the blocks subjected to examination should be selected randomly. This means that any of the blocks should have the same chance of being examined.

There are a whole variety of methods assuring random selection of the samples. In one of the simplest the blocks are numbered and the numbers of the blocks, between 1 and N, are selected either with the use of a special computer program or of random number tables, which can be found in most text books on mathematical statistics. Another option for random selection of the blocks is to ascribe to them random numbers, for example, by tossing them on the table, and then picking up every second, third, or fourth, depending on the fraction of the blocks to be examined.

An efficient way of sampling has proved to be the method that combines random selection with systematic positioning, the so-called random systematic sampling technique. In this method, the first sample is positioned randomly, for example, by random selection of its coordinates. The subsequent samples are selected by an adopted pattern. This might be a square grid of points, fields, etc., (see, for example, Reference 7 and references therein).

The requirement of random sampling is an important, frequently underestimated element in the process of the characterization of specimens for microstructural observations. The fact that its importance is not properly addressed in metallographic practice is based on the commonly made assumption that the microstructure of a random distribution of microstructural elements may be sampled less carefully and in the extreme case of a totally random microstructure, any sampling is acceptable. However, the microstructure of materials usually shows a considerable degree of order and specimens from it should always be selected randomly. Otherwise the results of statistical inference are likely to be biased or the error of estimation will be difficult to evaluate. Examination of a material should never be limited to just one area. Also, in order to illustrate average features of the microstructure, one should not attempt to show so-called typical or characteristic micrographs because these are usually chosen based on the individual preferences of the investigator. Instead, it is better to randomly select the area for examination or photorecording, unless the intention is to show some specific features observed in the microstructure.

There is no simple rule for the choice between systematic scanning and random sampling. Systematic scanning provides information on the spatial gradients of the microstructural features. However, this is usually a much more expensive approach and is justified probably only in the case where the material is expected to be highly inhomogeneous and/or its properties dependent on local events such as the growth of cracks.

The actual measurements are carried out on the selected samples. These measurements include geometrical probing, step 6, which is discussed extensively below. The last step (7) is the analysis of the data collected and its interpretation. Some procedures and illustrative examples are given below and in Chapters 7 and 8.

2.6 Size of the samples

Analyzed on an atomistic level, the microstructure of materials is likely to be of inhomogeneous character. On the other hand, some materials if studied over sufficiently large blocks of volume, V_b^*, yield similar values of their microstructural parameters such as grain size, particle contents, etc. The results of these measurements are thus invariant under translations. The materials showing this property could be called homogenous at a scale of V_b^*. This concept has a practical meaning if the scale of homogeneity is significantly lower than the size of the artifact studied: $V_b^* << V$.

In the case of homogeneous materials the differences among the values of a given parameter, p, measured for different blocks are due to chance variations, and for a large number of blocks studied the mean values for the blocks converge to the mean value for the population.

In contrast to the previous situation, inhomogeneous materials show a systematic dependence of their microstructure on the position of the block taken. Examples of homogeneous and inhomogeneous microstructures are given in Figure 2.10 and Figure 2.11.

Example 2.3

Figure 2.11 shows a microstructure on a cross-section of a weld of two plates of austenitic stainless steel. The welded plates were exposed to a corrosive environment. This resulted in corrosion of some of the grain boundaries, distinguished by a more pronounced contrast. The number of such boundaries

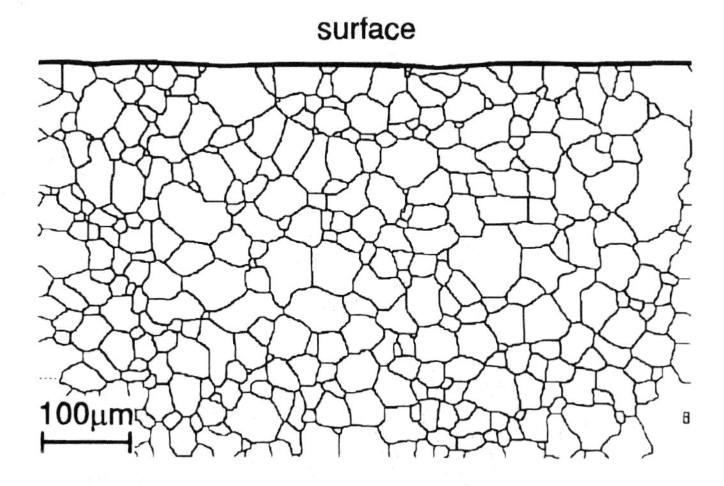

Figure 2.10 An example of a macroscopically homogenous microstructure of grain boundaries in a polycrystalline material.

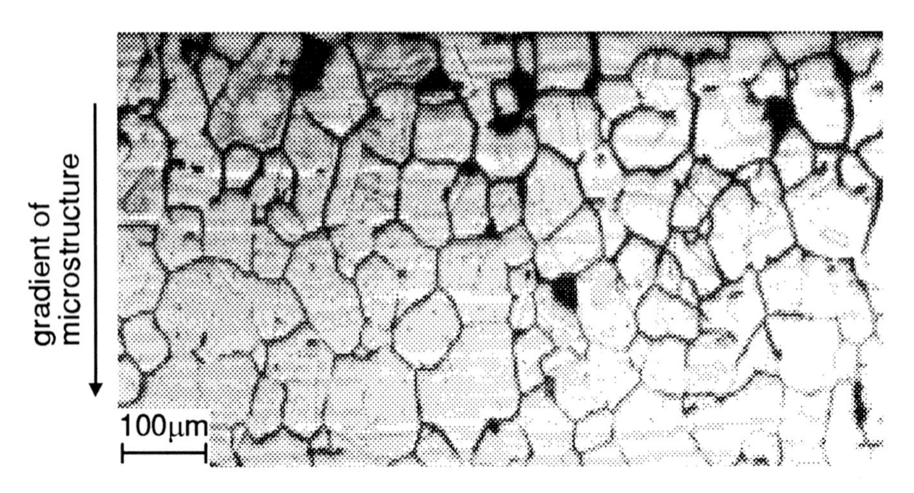

Figure 2.11 An example of a macroscopically inhomogeneous microstructure of the corroded grain boundaries revealed in a weld of an austenitic stainless steel specimen that has been subjected to an aggressive environment of an acid solution.

in the volume of the material is a function of (x, y), i.e., the distance from the weld and the distance from the aggressive environment.

This example illustrates that an inhomogeneous situation may be encountered in an otherwise homogenous material.

The concept of material homogeneity is related to the scale of the blocks and in turn to the scale of observations or to the magnification. A material that has been declared to be inhomogeneous in one type of experiment can be found to be of satisfactory homogeneity if examined at a lower magnification. As a result, in a general case the characterization of a material microstructure should be carried out at different levels of magnification or the level should be determined based on the scale which is adequate to explain a given property. Elastic constants, for example, are determined by the details revealed at an atomic scale. In the case of flow stress, the pertinent microstructural elements extend to micrometres. On the other hand the process of fracture is related to elements that usually achieve dimensions from hundreds of micrometres to millimetres.

In studies of a new artifact (material), one should follow a step-in approach. This means that we start with an examination of the whole artifact, disregarding whether it is an ingot, a sintered disc, or a rolled plate. In the first stage, the measurements can be carried out on a subset of K blocks selected randomly from the whole set of blocks and the mean values of the parameter, p_i, for each block measured. In the case of an homogeneous material and N > 30, these values are expected to show a normal distribution. In addition, the standard deviation of the mean values for the blocks, $SD(p_i)$, decreases inversely with the square root of N.

If the set of blocks initially taken does not yield conclusive results, more blocks need to be investigated and the distribution of p_i examined again. At

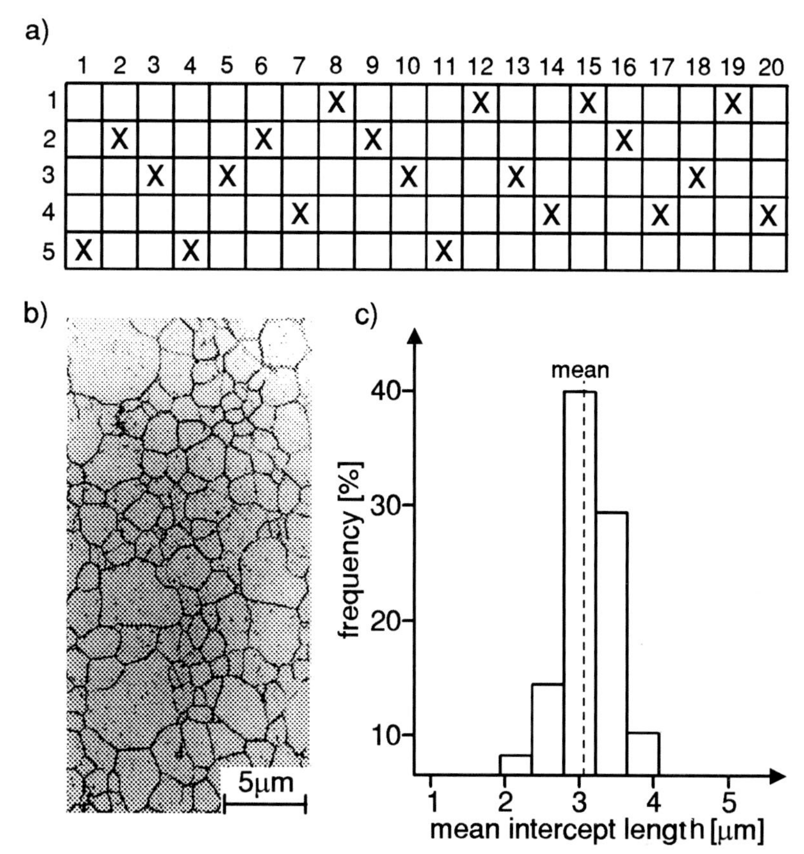

Figure 2.12 An example of a random selection of the fields of observations (a) on the surface of a strip of an austenitic stainless steel specimen. The microstructure of the material is exemplified by the micrograph shown in (b). The distribution function of the mean values of the intercept length for the fields studied is given in (c).

the end of the measurements the artifact microstructure can be described adequately by the mean value of p, E(p).

Example 2.4

The procedure described can be illustrated by the results obtained on a polycrystalline material in the form of a strip cold rolled and subsequently recrystallized, which is schematically shown in Figure 2.12.

The whole artifact was divided by a grid of lines into blocks of equal size. Measurements of the mean grain size by means of random secants were carried out on the surface of the blocks starting initially with 15 blocks randomly selected from the set. After computations of the mean size for these blocks, 35 more were taken and measurements repeated. It was observed that for over 30 blocks, the mean intercept length for a block, $E(l)_b$, does not show significant variation. The histogram of the $E(l)_b$ values is shown in Figure 2.12c. This is a Gaussian type distribution with the coefficient of variation, CV, close to 0.1.

2.7 Microstructural homogeneity

Consider an artifact of volume V divided into N nonoverlapping blocks of volume V/N each numbered systematically with numbers from 1 to N. The measurements of a given microstructural element, say particle volume, are carried out for these blocks and the results are presented in the form of a series of vectors, V_i, giving the volume of particles in the i-th block:

$$V_i = \left[V_{i1}, V_{i2}, \ldots, V_{i,N(i)}\right] \tag{2.23}$$

For each vector the mean value, $E_i(V_i)$, can be defined over the N(i) particles found in the block. Also the grand mean, E(V), for all the particles sampled can be computed. If the blocks of the material are small, as happens in most cases of practical importance, the mean values of the particle volume for blocks are different. The question arises now as to whether or not this scatter is due simply to chance or a proof of the nonuniform distribution of particles. In other words, one has to decide if it is possible to describe the material with one parameter, defining the volume of particles, or should one use a set of N mean volumes characteristic of each block specifically positioned within the artifact. These questions may be answered with the help of the method called analysis of variance. The idea behind it is to compare variations in the volume of individual particles within a block with the variation between blocks. To this end the following variations are defined:

a) total variation defined as:

$$VAR_T(V) = \sum_{i=1}^{i=N} \sum_{j=1}^{j=N(i)} \left[V_{ij} - E(V)\right]^2 \tag{2.24}$$

b) variation within blocks defined as:

$$VAR_w(V) = \sum_{i=1}^{i=N} \sum_{j=1}^{j=N(i)} \left[V_{ij} - E_i(V_i)\right]^2 \tag{2.25}$$

c) variation between blocks defined as:

$$VAR_b(V) = \sum_{i=1}^{i=N} N(i)\left[E_i(V_i) - E(V)\right]^2 \tag{2.26}$$

These three variances are related by a simple rule:

$$VAR_T = VAR_w + VAR_b \tag{2.27}$$

Also, if the differences in the volume of particles in the blocks are a result of chance, the variances VAR_w and VAR_b are chi-square distributed with the degrees of freedom determined by the number of measured particles and the number of blocks. In the end, the homogeneity of the material is tested by calculating the ratio:

$$F = \frac{\left(\frac{VAR_b}{N-1}\right)}{\left(\frac{VAR_w}{N^* - N}\right)} \tag{2.28}$$

where N^* is the total number of particles measured. The ratio F should be compared with the value of the so-called F-distribution function, known also as the Fisher distribution function. The F-distribution function describes the distribution of the ratios of variances for samples drawn from two normal populations. Its value depends, for a given level of significance, on two additional parameters defining the number of independently drawn elements in both samples. In this case their parameters, termed degrees of freedom, are $N-1$ and N^*-N, respectively. In this way the variation in the results of measurements for elements in a given block of the material is compared with the variation for different blocks. If the latter is significantly larger in terms of the F-distribution function the material cannot be considered to be homogeneous.

It is a frequently encountered situation that the microstructures of two materials, or a material after different treatments, are to be compared. In such a case, two sets of blocks are randomly selected from the two materials, say A and B, and in each block a certain number of elements is measured. The results of measurements can be presented in the form of two sets of vectors, V_i^A and V_i^B, giving the volume of particles in the i-th block of A and B material, respectively:

$$V_i^{A/B} = \left[V_{i1}^{A/B}, V_{i2}^{A/B}, \ldots, V_{i,N(i)}^{A/B}\right] \tag{2.29}$$

For each vector the mean values, $E_i^A(V_i)$ and $E_i^B(V_i)$, can be defined over the N(i) particles found in the block. Also, the mean $E^A(V)$ for all the particles sampled in material A and $E^B(V)$ for B can be computed as well as the grand mean for both materials, taken as being two samples of the same microstructure. This analysis is an extension of the previous one and the variations are then computed for the values measured:

a) between the blocks in material A and B, VAR_b;
b) between the material A and B, VAR_{AB};
c) total variation defined, VAR_T.

The appropriate formulae and the test for evaluating the statistical significance of the differences in these variations can be found in textbooks on statistics (for example, Reference 8). The basic idea again is to compare the ratios of the different variations (VAR_b/VAR_E, and VAR_{AB}/VAR_E, where $VAR_E = VAR_T - VAR_{AB} - VAR_b$) with the F-distribution values.

The mathematical model rationalizing this approach is based on the assumption that the results of measurements in a block, $E_{jk}(V)$, are given as:

$$E_{jk}(V) = E(V) + \alpha_j + \beta_k + \varepsilon_{jk} \tag{2.30}$$

where E(V) is the mean for the material, α_j and β_k account for the effect of treatment and block, respectively, and ε_{jk} is due to chance or error. This model is used further to test the hypothesis that $\alpha_j = 0$ and $\beta_k = 0$ and, depending on the outcome, the materials can be considered to differ significantly or not.

Example 2.5

Consider a two-dimensional situation with the particles forming patterns shown in Figure 2.13. The fields of observations were divided into 6×6 blocks of a constant area and the number of particles in each strip was measured. The following results were obtained:

Microstructure A:	Microstructure B:
31, 28, 27, 35, 29, 34	26, 8, 29, 6, 25, 11,
25, 27, 24, 35, 36, 32	29, 7, 31, 5, 26, 10,
28, 26, 28, 34, 32, 31	28, 8, 30, 4, 27, 8,
29, 34, 33, 33, 29, 28	30, 9, 28, 9, 32, 5,
25, 26, 30, 25, 26, 25	27, 5, 28, 10, 33, 4,
24, 29, 31, 30, 28, 27	25, 10, 7, 8, 30, 6.

These values of the number of particles yielded the following variances.

Microstructure A: $VAR_T(n) = 4398.3$, $VAR_w(n) = 5126.8$, $VAR_b(n) = 27.2$
Microstructure B: $VAR_T(n) = 2327.8$, $VAR_w(n) = 2583.1$, $VAR_b(n) = 795.6$

The problem of material homogeneity can also be discussed in terms of ergodicity of the random variable describing quantitatively a given element of the microstructure.[9] Two useful terms are used in this treatment: stationarity and ergodicity. A random variable is stationary if it yields the same values when measured on large samples of the material. Ergodicity means that measurements can be limited to one particular batch of the material.

The concept of ergodicity will be further illustrated for the specific case of a random variable, f(x), defined as the area fraction of particles A_A. This function can be replaced by any other microstructural variable.

Consider a microstructure containing second-phase particles of a mean area fraction $E(A_A)$ and variance $VAR(A_A)$, and a process of area fraction measurements that consists in evaluating the area fraction of second-phase particles over a region of area, A. The mean area fraction, $E(A_A)$, can be estimated from the following formula:

$$est(A_A) = \frac{1}{A}\int A_A dA \tag{2.31}$$

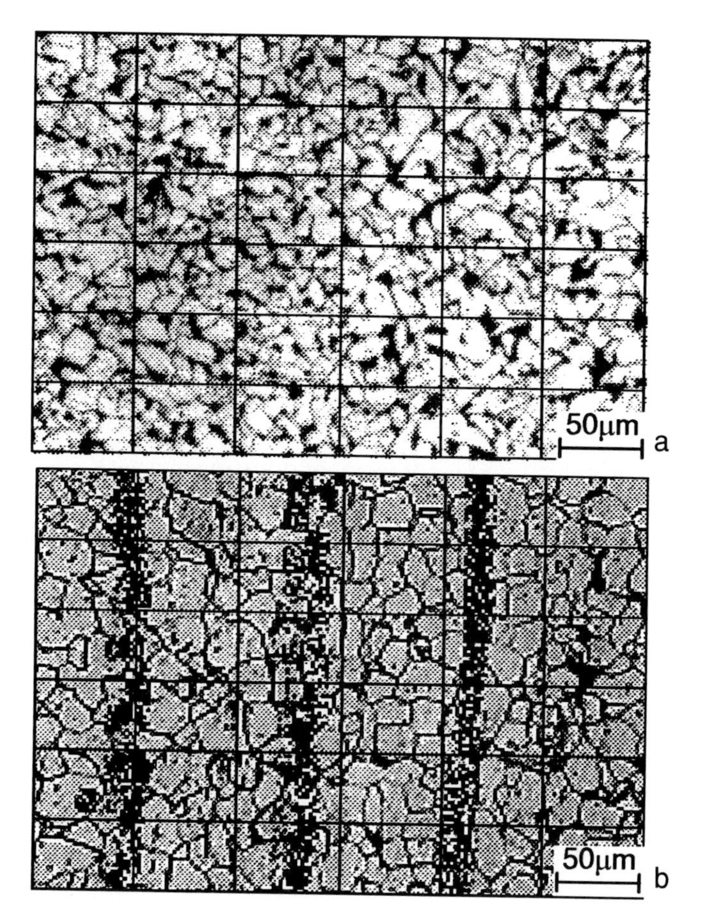

Figure 2.13 Microstructures representing uniform (a) and nonuniform (b) distributions of particle sections in a two-phase material.

where the integral is taken over the area A. This is a commonly adopted procedure and it can be shown to be an unbiased estimator of the mean value:

$$E[est(A_A)] = E(A_A) \tag{2.32}$$

On the other hand, the values measured for different regions of observation provide different estimations of the parameter and the variance VAR(est(A_A)) is given as:

$$VAR(est(A_A)) = \frac{VAR(A_A)\, A_o}{A} \tag{2.33}$$

where A_o is the integral range defined by the formula:

$$A_o = \lim A \to \infty \left[A \frac{VAR(est(A_A))}{VAR(A_A)} \right] \tag{2.34}$$

where the limit is for A approaching infinity. This quantity, if it has a finite value, can be used to characterize the scale of the phenomenon, in this case the distribution of the second-phase particles, while A gives the scale of the observations. Also, if an integer N is defined such that:

$$N \approx \frac{A}{A_o} \tag{2.35}$$

the variance of the estimation can be expressed as follows:

$$VAR\left[est\left(A_A\right)\right] \approx \frac{VAR\left(A_A\right)}{N} \tag{2.36}$$

This is equivalent to the formula which can be obtained if the area of observations is divided into N independent domains of size A_o.

The theory described above can be used in practical situations for characterizing the distributions of particles and other microstructural features. To this end, the dispersion variance of the estimated variable (x_A in the example used) should be plotted against the size (area, volume, etc.) of field for which the mean value is measured, the so-called support size (A in the example used). It can be shown that for ergodic functions the plot in a double logarithmic system is a straight line with a slope –1. Such a plot can also be used to estimate the value of the expression of $VAR(A_A)A_o$, or its equivalent (in the general case).

If the data points do not fall on such a line it is an indication that the size of the domains may not be sufficiently large compared to the integral range or that the microstructure exhibits inhomogeneous properties. In this case a poor precision of the estimation with just the mean value is to be expected. Here it might be more appropriate to characterize the material by a plot of VAR(est(x)) vs. support size rather than by any single value obtained with a particular size of the field of observations.

2.8 General characterization of material microstructures

The general examination of material microstructures can be described schematically by the following sequence:

1. material and treatment;
2. samples for macroscopic observations (cracks, surface defects);
3. samples for light microscopy (inclusions, particles, grains larger than a micrometre);
4. samples for electron microscopy (fine particles, grain boundaries, dislocations).

This is a multilevel or cascade sampling approach[10] which results in four sets of specimens $S_{1,i}$, $S_{2,j}$, $S_{3,k}$, $S_{4,l}$ for each component of the sequence. Each set of specimens at a higher magnification level is usually contained by the set of specimens at the lower magnification level. As a result, the object of interest at one level becomes a matrix or reference space at the next level.

The idea of cascade sampling is exemplified by studies of grain boundaries and dislocations in polycrystalline materials deformed in a tensile test. In this case the sequence of multilevel investigations would consist of:

1. annealing and subsequent tensile straining of the specimens;
2. observations of possible localization of the plastic deformation in the form of Lüders bands, etc.;
3. light microscopy observations—estimation of the total area of the grain boundaries;
4. transmission electron microscopy—estimation of the grain boundary dislocation density, etc.

In this example, the grains in step 3 are the object of interest, while in step 4 they become a reference space for calculation of grain boundary dislocation density.

In multilevel observations specimens are randomly selected at each level of investigation. In the example used the process starts from a random selection of the strained specimens, proceeds through random selection of fields for light microscopy observations, random selection of the slices for TEM observations, and ends with random selection of the observation area during electron microscopy studies. The number of specimens at each level should be decided based on the variation of a given parameter between the specimens. In the absence of any *a priori* information on this variation, the required number of specimens can be assessed from pilot experiments.

The question about the required number of samples is related to the problem of the proper strategy in measuring geometrical features of microstructural elements revealed in a field of observation. One possibility is to measure, with the highest feasible precision, all the elements in a field of observation employing, for example, an automatic image analysis system. An alternative approach is to randomly sample some elements or utilize appropriate manual counting methods. In answering this question it should be recognized that the precision of estimation of a given parameter depends mainly on the differences, or more precisely, coefficient of variation, between different specimens selected from the material studied. From this point of view it is not essential that all the objects included in one specimen are measured and analyzed. As a result, manual point counting might yield the same precision as computer-aided planimetric measurements.[11] Moreover, beyond some critical density of test points there is no gain in the precision of the measurements with increasing the number of measurements on a given field of observation. When dealing with a collection of samples it is much more efficient to use the available time to measure less precisely more pictures from a large number of sections than to achieve very high precision on a few fields of observation. However, high-precision measurements are essential in such studies as serial sectioning and in the case of some particular microstructures, for example in the case of 2-dimensional polycrystals.

The advantages of the use of automatic image analysis are mainly related to the speed of measurement. This is particularly evident in the case of measurements of more complex geometrical parameters such as particle perimeter. In considering the advantages of a computer-based system for planar measurements an important remark should be made with regard to the basic principle of quantitative analysis, which requires that **all objects of interest on the studied image should be identified**. In manual measurements this requirement is usually met, although using subjective criteria employed, by the person executing the measurements. In the case of the computer-aided approach the criteria of object recognition are more objective but failures to identify certain objects are more likely to occur.

2.9 *Geometrical probing*

After selecting from the volume of material a set of specimens for microstructural investigations, the second stage of study involves intersecting the specimen with a geometric probe. The need to introduce the concept of geometrical probing comes from the fact that measurements are made either on slices or sections of the specimens. This means that instead of analyzing directly the 3-D microstructure of materials, the investigations are carried out on its representation obtained according to specified geometrical rules that define position and orientation of the section or slices. Further, in dealing with images of the microstructure investigated, it is usually beneficial to reduce the number of objects revealed by selecting for measurement those which are hit by a superimposed system of lines or points or their combination. This is again geometrical sampling; however, in this case it is rationalized through the principles of stereology, which make it possible to estimate parameters of the 3-D microstructure studied from its 2-D images.

There are four basic probing procedures commonly used in materials science applications. These are

a) slicing;
b) sectioning;
c) line sampling;
d) point sampling.

These probes are illustrated in Figure 2.14. They can be used separately; however, combinations of a to d are also possible and in fact such combinations are extensively used in standard and modern techniques (see also Chapter 3).

Preparation of thin foils for transmission electron microscopy is an example of slicing. Another example is the procedures used for specimen preparation of polymers or ceramics studied by transmitted light microscopy. In all these cases the measurements are subsequently effected on projected images

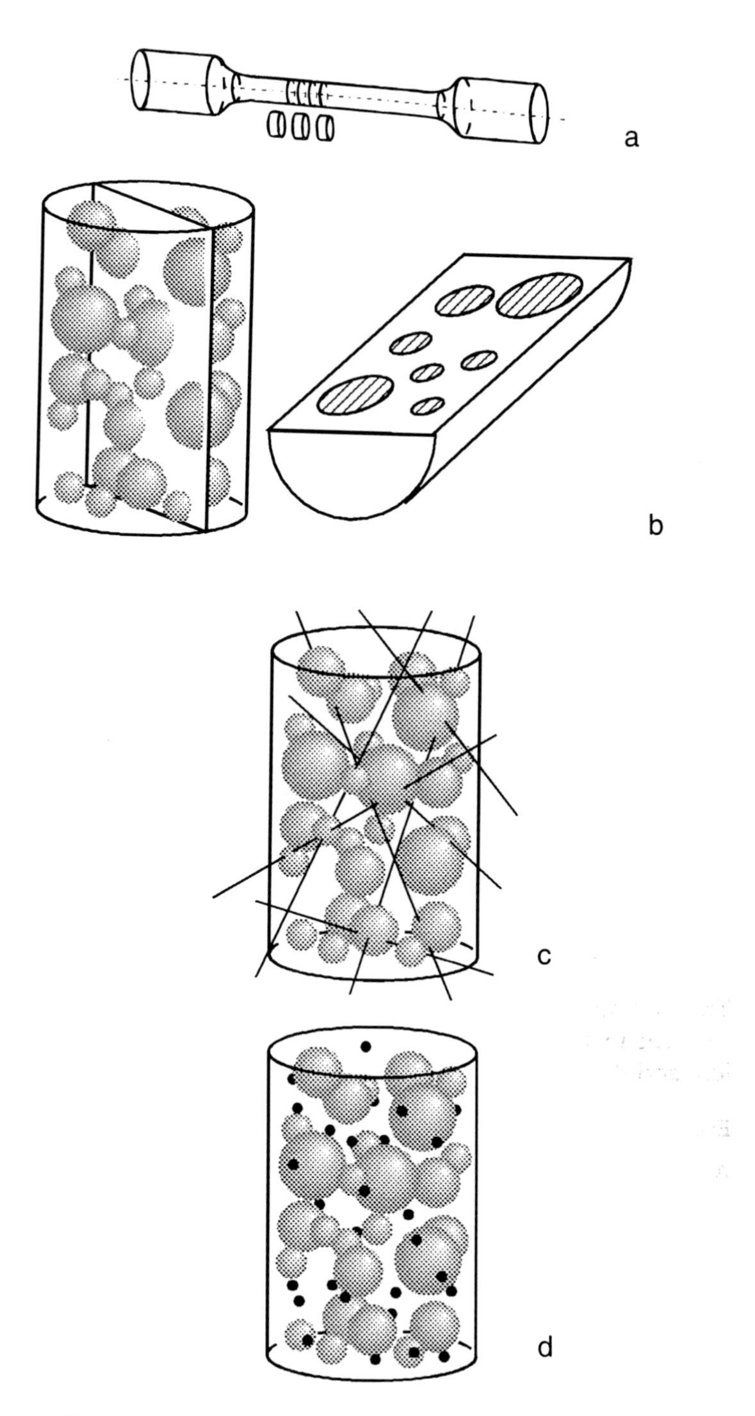

Figure 2.14 Illustration of the basic methods of probing a material microstructure: (a) slicing; (b) sectioning; (c) probing with lines; (d) probing with points.

of the elements contained in the volume of the slices. From a geometrical point of view this probing is characterized by:

a) orientation and position of the slice surfaces with respect to some reference system; if the surfaces are parallel and the material is homogeneous then two parameters of the normal, **n**, to the surface are involved;
b) thickness, t, which in general can be a function of some spatial coordinates.

Materials characterization mainly relies on sectioning specimens. A section is geometrically defined by its position and orientation with respect to some reference system, which usually, but not necessarily, is a system of Cartesian coordinates. The position of a section can be determined by its distance, d, from the specimen center of gravity, or some other characteristic point. The section orientation is given by its normal, **n**. The couple (d, **n**) uniquely defines the geometry of this section. There are three independent parameters that can be changed to produce a set of sections $(d, \mathbf{n})_i$ of a studied specimen:

a) d_i—the distance from a specified point;
b) θ_i—the angle with the Z axis;
c) ϕ_i—the angle between the projection of **n** along the Z, $\mathbf{n}_z$, and X axes.

The sectioning can be done in a systematic way throughout the studied object. In the so-called serial sectioning method the following assumptions are made:

a) $d_i = d_1, d_2, \ldots d_n, \ldots;$
b) $\theta_i = \text{const.};$
c) $\phi_i = \text{const.}$

The method of serial sectioning allows 3-D reconstruction of the features studied. The method of such reconstructions has been described by Rhines, DeHoff, and their co-workers.[12,13]

Example 2.6

An example of serial sectioning is given from a study performed on an Ni-2% Mn sample. The specimens were annealed at 1373 K ($0.8T_m$ where T_m is the melting point) for 1 h. This yielded a mean grain size of 200 μm (intercept length). Four holes, $\phi = 0.5$ mm, were drilled into the sample for alignment of areas on successive sections. Rhines et al.[13] suggest that the sections' thickness (t_i) should equal about 1% of the maximum size of the largest grains. In this study, the average grain size was 200 μm. The largest linear grain dimension in the observed regions proved to be 800 μm. Thus, it was decided that the thickness, t_i, should be in the range 2 to 8 μm. The specimens were polished and etched. Polishing and etching were repeated when a layer of material of a thickness smaller than 8 μm had been removed. The inspected areas were examined with a metallographic microscope. In these studies two areas were examined of a size 1.57 × 1.17 mm each. The microstructure was observed at a magnification of 100. The micrographs were recorded and later traced onto transparent foils. The data interpretation carried out on 100 sections was simplified by tracing the twins and grain networks separately into different sets of foils.

Auto CAD PL, a computer-aided design program, was used to reconstruct the 3-D shapes of the grains revealed on the sections. The 2-D images of the twins were digitized into Auto CAD as a set of points. The program connected the points into 2-D images and these images were stacked to generate the full 3-D geometry of grains. One of the reconstructed grains is given in Figure 2.15, measurements of grain sizes (volume, surface, etc.) were carried out using an image analyzer equipped with a macrostage. Specially developed programs were used to acquire and process the data in the form of grain cross-sectional areas, A, and grain boundary trace length, L. The volumes, V_G, of individual grains were determined by assuming that a linear change in area occurred between sections and by utilizing the thickness data from the equation:

$$V_G = t\sum_{i=1}^{N} A_i \tag{2.37}$$

where t is the thickness of the layer, A_i the area of the grain in i-th layer, and N the number of intersections of a given grain.

2.10 Random and oriented sections

Sectioning by its nature is a probing operation of a directional character. Section orientation, specified by angles Θ and ϕ, can be either random or in a close relationship with the geometry of an artifact studied. The term *random sections* in this case means the section normal vectors, $\mathbf{n}_i$, are isotropic. Such

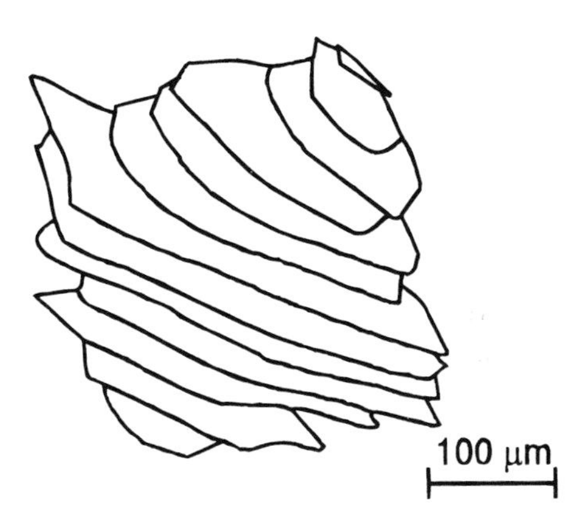

Figure 2.15 A view of a grain whose geometry has been reconstructed from the serial sectioning of a polycrystalline material.

vectors, if attached to a common origin, form a set of constant density points on a unit sphere. The practical ways of selecting these vectors are discussed in Chapter 3. Random sections are used if no *a priori* information is available as to the character of the material or microstructure studied. Such sections provide an unbiased description of the material without any further assumptions. In fact, so-called isotropic uniform random (IUR) sampling with sections is one of the principles of quantitative stereology. It requires that sections used for materials characterization are randomly positioned and their normals are assumed to be isotropic in orientation. A section chosen according to the IUR principle cannot be of any predetermined orientation with respect to the geometry of the specimen.

In most cases the difficulty in producing IUR sections is related to their orientation and not their position, which can be changed in the process of specimen grinding and polishing. As a result, it is usual in materials characterization to use specially oriented sections instead of randomly oriented ones. These sections are usually parallel or perpendicular to some specific directions determined by the geometry of the specimen. They can be effective for different reasons in two cases:

a) the microstructure is of isotropic character;
b) the microstructure exhibits a known type of symmetry in the arrangement of the defects or features resulting from directional treatment of the material such as rolling, drawing or uniaxial straining (such probes provide information on selected aspects of the microstructure, for example, the grain shape in the longitudinal section of the strained specimen).

As a simple rule specimens can be sectioned with oriented sections without violating the need for IUR sectioning if their microstructure is isotropic. Isotropy in this context means that the microstructure shows the same properties in any direction. In the case of isotropic microstructures observations can be made on a single cross-section provided the area of observation is sufficiently large.

On the other hand anisotropic microstructures can be probed with sections directed with respect to the anisotropy axes, assuming that the directions of anisotropy are known beforehand, and appropriate corrections can be made for systematic differences in the images representative of sections of different orientations. Observations are commonly restricted to three perpendicular surfaces and the hypothesis about isotropy of the microstructure is tested based on these results as illustrated in Figure 2.16. This gives an acceptable level of statistical confidence (see Chapter 3) and the precision demanded by materials modelling at present.

2.10.1 *Counting number of objects in observation field*

In order to ensure objectivity in the studies, it is essential that the elements are included or excluded from the measurements consistently, according to

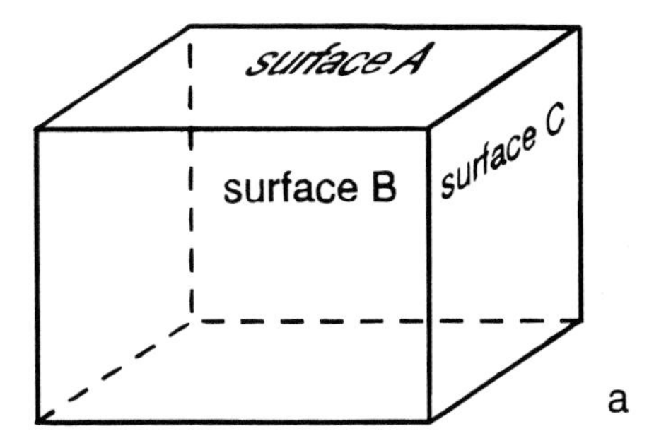

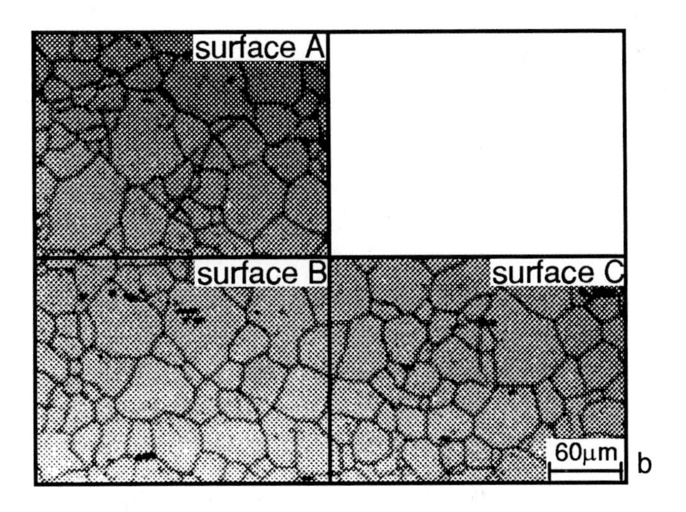

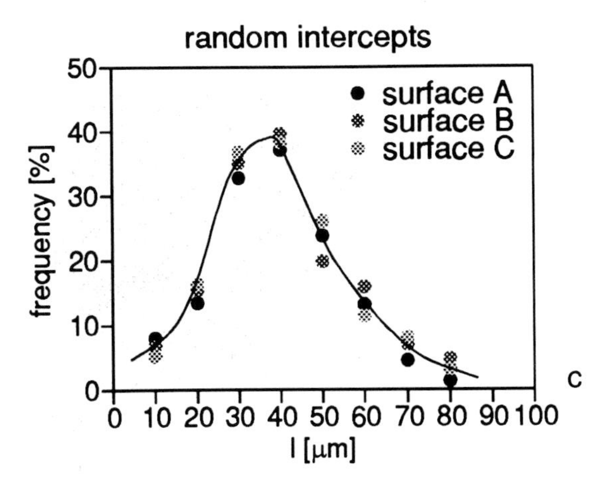

Figure 2.16 A set of micrographs representing the microstructure of a material examined on three perpendicular sections: (a) a schematic of the specimen and the orientation of the sections; (b) the micrographs characteristic of the images observed on the sections; (c) the intercept length distributions for the sections indicated in (a).

some specified rule. This rule should be set beforehand and no exceptions should be made in the course of the studies. An integral test system that can be used is described in Reference 14. The system shown in Figure 2.17 defines the area of observation as the interior of a frame that has two exclusion edges. The unbiased counting rule requires one to count all elements in the micrograph that have any part within the frame and do not intersect any exclusion-edge or extended exclusion line. The integral system may also contain sets of points of varying density, and sets of lines. All these can be used for probing the microstructural elements, and the unbiased counting of the number of elements and intersections.

An alternative approach assumes that the objects of interest, X_i, are assigned to specific points, Q_{ij}, (the so-called assigned specific points, ASP). Such points might be, for example, the centers of gravity in the case of particles or the tangent points to the upper horizontal line. In both cases there is one specific point, Q_{i1}, per object. However, this need not be a rule and more points may to used. Consequently, the objects are analyzed within the frame if the frame contains their specific points (see Figure 2.18).

In general, the best way to avoid frame effects is to increase the area of observations by taking micrographs systematically of larger areas. As the ratio of the elements intersected by the final frame (drawn around the total area examined) to the elements inside decreases, the problem of the frame effect becomes less important. One advantage of using many of the automatic image analyzers is that handling the frame effect is built into the architecture used. Frequently, more than one frame is used so that a "guard frame" surrounds a "live frame", the latter being the one in which measurements are made. The size of the guard frame is then adjusted so that the largest feature in the image has a width (i.e., linear size) that is smaller than the width of the guard frame.

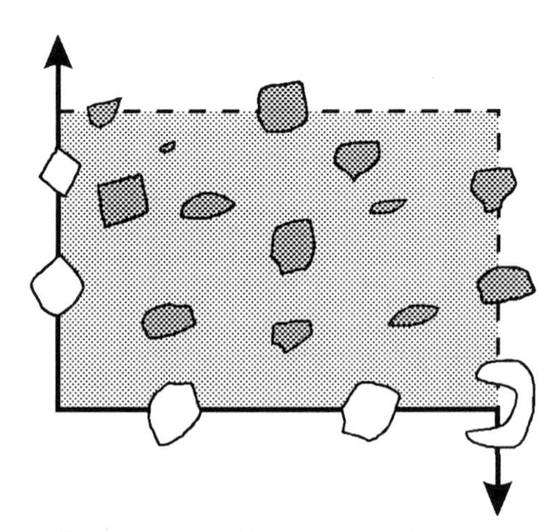

Figure 2.17 A model frame proposed by Jensen and Gundersen.[14] An object is accepted as belonging to the field of observation if it is entirely inside the frame or if the parts outside the frame are not cut by the exclusion edges.

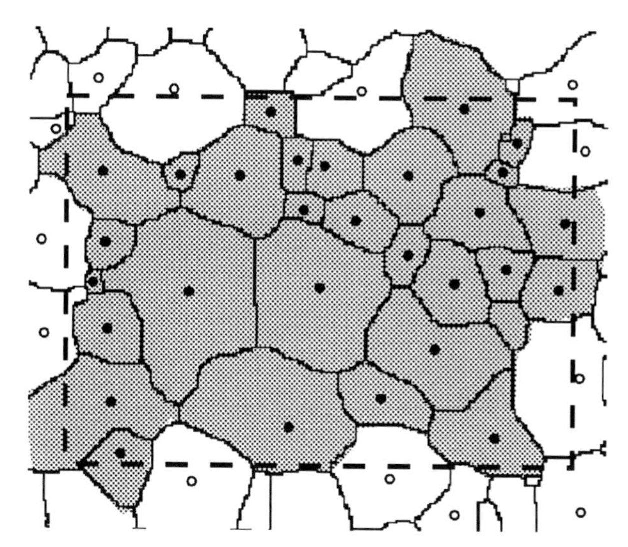

Figure 2.18 Schematic explanation of the assigned specific point method.

By this means, features that only just appear at the edge of the live frame may be sampled properly.

2.11 Random oriented lines

The lines used for geometrical probing are defined by:

a) d—their distance from a known point;
b) **t**—a tangent vector.

Similarly to the case of section planes, the test lines can be formed into a random set, or sets of parallel lines inclined at a given angle to the reference direction. Randomly oriented and positioned lines are used for estimating important parameters of 3- and 2-D microstructures as discussed later. In the case of 2-D microstructures a pattern of randomly oriented and positioned lines can be computed easily. Examples of such patterns are shown in Figure 2.19. As for 3-D microstructures, the test lines are drawn on sections of the microstructure and a random orientation of the lines is assumed as a result of placing them on randomly oriented sections. In this case the need for random orientation of **t** is achieved via random orientation of sections normals, **n**.

It should be noted that, in the general case, the test lines do not need to be straight nor the test sections planes planar. In fact, modern methods, described in Chapter 3, heavily rely on the use of curved test elements such as cycloids.

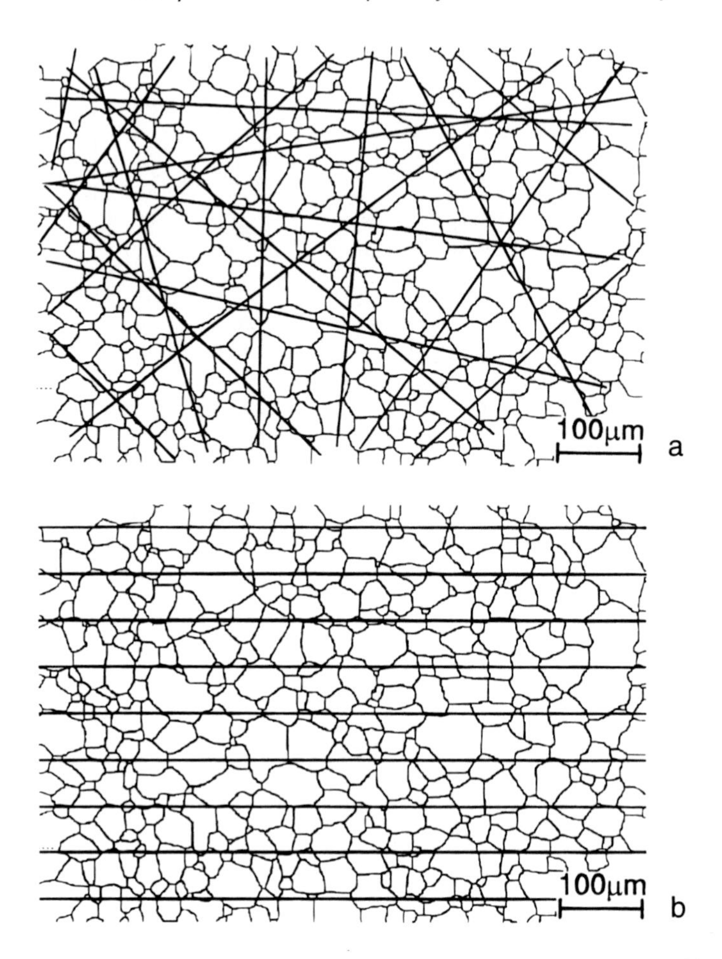

Figure 2.19 Pattern of randomly oriented lines on a plane (a) and positioned lines (b).

2.12 Point testing

Test points are used to quantify features of 3-, 2-, and 1-D microstructures. The system of test points is defined by their density P/A. If the points form a regular array, they can also be described by their pattern (e.g., square grid) and a characteristic distance d_o (see Figure 2.20). Test points can either be randomly generated (Figure 2.20a) or positioned randomly but as a regular grid of points (Figure 2.20b). The latter method usually proves to be more efficient. If the points are imposed on sections of random orientation they form a set randomly oriented in space.

The process of microstructural observation consists of the examination of selected fields under a microscope or (micrographs of the microstructure). The fields of observation are usually much smaller than the size of the artifact studied and some elements are likely to be partly included while others are well inside the field. One case where this is not true is the case where we are

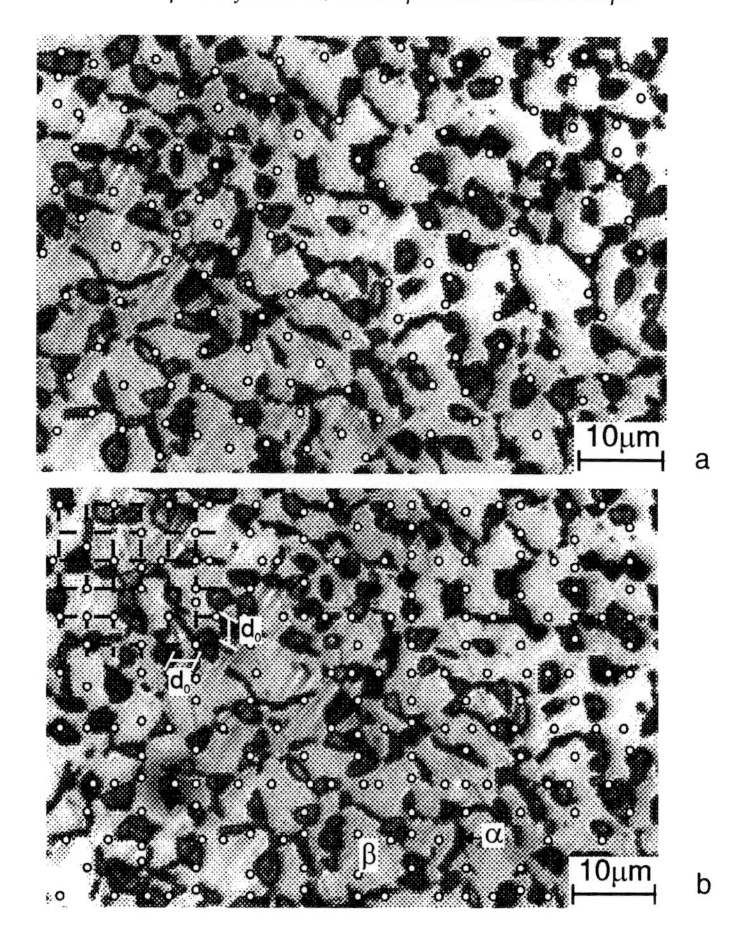

Figure 2.20 Example of randomly generated testing point pattern on a two-phase material (a) and randomly positioned pattern but with a regular grid of points with a characteristic distance d_0 (b).

concerned with the overall distribution of some impurity. Thus, for instance in the case of the distribution of sulfur in steels, it is common to section a complete ingot and then take a sulfur print, which reveals the macroscopic distribution of this element and can be indicative of the mechanisms occurring during casting.

2.13 Number and weighted distributions

Sampling with the use of geometrical probes generally differs significantly from the scheme employed in simple situations such as taking balls from a box containing constant size balls, distinguished by their colors. The analogy between geometrical sampling and taking balls from a box can be exemplified with the case of a set of unequally sized balls, where small balls are less likely to be chosen. Geometrical probing of a microstructure results in a

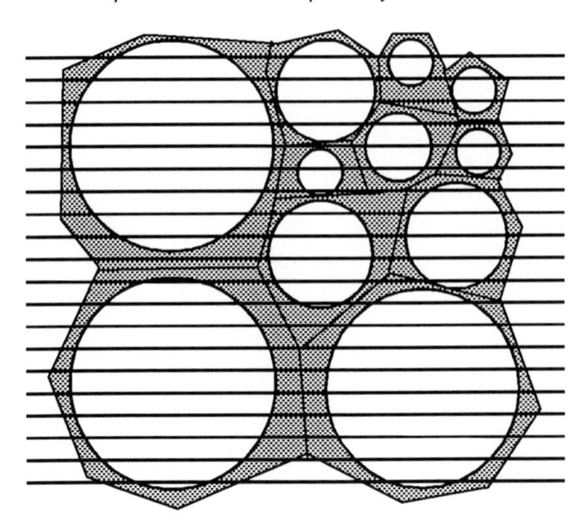

Figure 2.21 Example of sampling of small and large features where larger circles are "hit" more frequently with test lines.

tendency for larger grains or particles to be sampled more frequently than the smaller ones. This effect is illustrated by the example shown in Figure 2.21.

The fact that geometrical probing of the microstructure is characterized by a tendency for larger objects to be sampled more frequently makes it necessary to distinguish between **number** and **weighted distribution functions**. The number distribution function provides information on the numbers of objects of a given size/property. The weighted distribution on the other hand is influenced by both the number and size of elements of a given size. It should be noted that the two distribution functions differ and accordingly **the population revealed on a cross-section differs from the population of elements in the material's microstructure**.

An analysis of the situation of probing with planar intersections leads to the conclusion in the case of 3-D objects, as for instance particles or grains, that the weighted, f_W, and number f_N distribution functions remain in the following relationship:

$$f_w(V) = \frac{H\, f_n(V)}{\int H f_n(V) dV} = \frac{V^{\frac{1}{3}} f_n(V)}{\int V^{\frac{1}{3}} f_n(V) dV} \tag{2.38}$$

where H is particle height, which is assumed to be proportional to $V^{1/3}$.

Example 2.7

For the population of spherical particles defined in Example 2.1 the number distribution function is given as: $N_A{:}N_B{:}N_C = 3{:}6{:}1$. If the particles are sampled by sectioning it, one would obtain the following distribution function of the particles:

$$(N_A)^*{:}(N_B)^*{:}(N_C)^* = 3{:}12{:}5.$$

It should be noted that 25% of the particles sectioned are the largest particles, C, which make up 10% of the population.

2.14 Notation

Stereology is a branch of science that has developed over the years a commonly used specific system of symbols. This set contains first five letters, namely A, L, P, S, and V that are used as abbreviations for the words *area, line, point, surface,* and *volume*:

A planar surface
L line
P point
S surface or interface (potentially curved)
V volume element

These symbols are used at the same time to identify the type of object studied and its size as follows:

P number of points
L length of a line
A area of a planar object
S area of a curved surface
V volume

These symbols, together with the symbol N which is reserved for *number*, are then used in combinations:

$$X_Y \tag{2.39}$$

where X indicates the type of object and Y the way it is probed or the space in which it is measured. As a result one can define the following symbols:

P_P number of points per test points (point fraction)
P_L number of points per test line length
P_A number of points per test area
P_V number of points per test volume

N_L number of features (not necessarily points) per test line length
N_A number of features per test area
N_V number of features per test volume

L_L length of objects per unit test line (linear fraction)
L_A length of objects per unit test area
L_V length of objects per unit test volume

A_A area of objects per test area (area fraction)
S_V surface per unit test volume

V_V volume of objects per unit test volume

There is some ambiguity in the above list of symbols related to the fact that two symbols are used in the case of 2-D elements with a distinction being made for planar elements (usually employed in probing) and curved surfaces (which are microstructural elements). It should be noted also that the list does not contain symbols such as A_L, which as yet have no geometrical interpretation.

The single symbols are used at the same time to indicate the respective variables for a given microstructural element. For example, V may designate volume of particles, volume of grains, volume of pores, etc. In this case the mean values of these parameters are frequently specified by symbols with bars such as $\bar{P}, \bar{L}, \bar{A}, \bar{S}, \bar{V}$.

However, this system fails if one is forced to introduce higher-order distribution parameters such as standard deviation, skewness, and kurtosis. In order to avoid complications involved in a double system of symbols in the present text the mean values and other parameters are designated in the following way:

E(X)	mean value of X
VAR(X)	variation of X
SD(X)	standard deviation of X
CV(X)	coefficient of variation (CV(X) = SD(X)/E(X))
SK(X)	skewness of X
K(X)	kurtosis of X

The same convention is used for any variable, X, including logarithmic variables so that X may be replaced by lnV, lnA, etc.

2.15 Volume fraction determination

The parameters defined in the preceding section have a variety of applications in the characterization of the microstructure of materials. The volume fraction, V_V, is one of the fundamental quantities used for characterizing 3-D elements of the microstructure of materials. It can be applied to multiphase materials as well as single-phase materials containing grains of different orientations, pores or voids, etc. This is a dimensionless parameter that determines the **fraction of the total volume** filled with a given constituent α—$(V_V)_\alpha$ (for the material with the microstructure depicted in Figure 2.20 $(V_V)_\alpha$ is equal to 0.361). As such, it provides global information on phase content. However, it provides information neither about the size nor the shape of the particles.

The volume fraction can be obtained from measurements on planar sections of the material. In fact, one of the basic relationships of stereology states the following:

$$(V_V)_\alpha = (A_A)_\alpha = (L_L)_\alpha = (P_P)_\alpha \tag{2.40}$$

This set of equations can be derived using the methods of geometrical probability for the general case of microstructures containing particles of different shapes and sizes, isotropic or not, providing that in the latter case the microstructure is sampled randomly with either a set of points, lines, or sections in space. This relationship between the fractions gives us freedom in selection of the geometrical probing method. In practice it means that we can determine the volume fraction by a simple point counting technique. An application of these equations to the characterization of a two-phase material is given in Example 2.8. Applications of the equations to 2-dimensional and 1-dimensional microstructures are obvious.

Example 2.8

The microstructure shown in Figure 2.20 was measured using a grid of parallel lines and a grid of points. The image was also digitalized and planimetric measurements of the particle section areas were performed. The following results were obtained:

$$(A_A)_\alpha = 0.354$$
$$(L_L)_\alpha = 0.369$$
$$(P_P)_\alpha = 0.375$$

It may be noted that the same microstructure may yield different estimates depending on the method used. However, the precision of estimation is mainly influenced by the number of measurements.

2.16 Density of two-dimensional features in space

The surface properties of 3-D objects and 2-D features in 3-D microstructures are described in terms of the parameter S_V, which defines the surface-to-volume ratio. In single-phase materials there is one type of boundary and one value of S_V. In two-phase granular materials one can expect three types of interfaces: $\alpha\alpha$, $\alpha\beta$, $\beta\beta$ (where α and β indicate the two phases observed). Accordingly, one can define three values of specific surface area: $(S_V)_{\alpha\alpha}$, $(S_V)_{\alpha\beta}$, $(S_V)_{\beta\beta}$. The parameter S_V has dimensions of m^{-1} inverse metres.

The value of S_V can be estimated from one of the basic stereological relationships. It can be shown that for a system of surfaces, S_i, with any configuration in space and a system of IUR test lines, L_i, of known length, L, the following relationship is valid:

$$S_V = 2P_L \tag{2.41}$$

where P_L is the number of intersections of S_i and L_i, divided by L.

Example 2.9

The microstructure shown in Figure 2.20 was tested with a randomly placed system of parallel lines. The total length of lines, L, was 10 mm. The number of intersections of the particle boundaries with the test line, N_i, was measured to be equal to 245. This yields the following estimate of the particle surface area in unit volume:

$$(S_V)_\alpha = 2\,N_i/L = 0.049\ \mu m^{-1}.$$

Example 2.10

The microstructure shown in Figure 2.19 is typical of a single-phase polycrystalline material. A set of parallel lines of the total length, L, equal to 10 mm, was randomly superimposed on the image of the grain boundaries. The total number of intersections, N_i, was found to be 1832. (See below for the additional rules of counting.) This yields the following estimate of grain boundary surface area in unit volume:

$$S_V = 2\,N_i/L = 0.367\ \mu m^{-1}.$$

In practical implementation, the test lines are imposed on sections of the microstructure and the intersections are counted in a natural way using some additional rules such as:

a) each tangent point is assigned a count of 0.5 and in the case of the surface of 3-D objects.
b) if the test line ends inside the object the intersection with its surface is counted as 1.5.

The examples discussed above illustrate that the relationship for S_V estimation can be used for single-phase as well as for multiphase materials. In the case of single-phase, space-filling cells (grains) the density of intersecting points, P_L, approaches the inverse of the mean intercept length $E(L)^{-1}$. For these materials the equation can be rewritten in the following form:

$$S_V = \frac{2}{E(L)} \tag{2.42}$$

In the case of multiphase microstructures a distinction should be made for the different types of interface present with the number of intersections with the probing lines counted separately. Then for a given type of interface one can use the following:

$$(S_V)_{\alpha\alpha} = 2(P_L)_{\alpha\alpha} \tag{2.43}$$

The equation given above can be used for a system of grains and particles and also for single particles. Consider the body shown in Figure 2.22 of volume, V, and the surface area, S. The body is pierced by a randomly oriented line in three-dimensions. If a system of such lines is used the mean length of the intercepts inside the body, E(L), can be computed. Since each intercept gives two intersection points $P_L = 2/E(L)$ and for a single body its surface and the volume are related to the mean intercept length in the following way:

$$S_V = \frac{4}{E(L)} \tag{2.44}$$

This equation can be used to show that the mean intercept length for a sphere of radius R is equal to 4/3R.

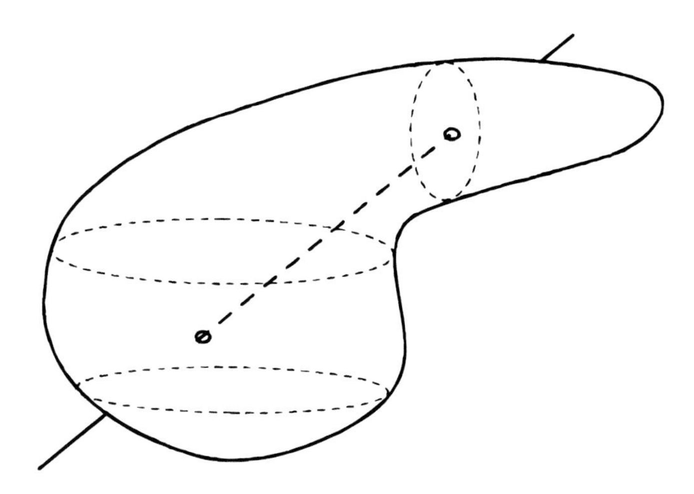

Figure 2.22 A 3-D object pierced with a randomly placed and oriented line.

The number of intersecting points is an additive quantity. As a result, the last equation can be extended for a system of particles by replacing the mean intercept for a single body with the mean intercept for a system of particles.

2.17 *Density of lines on a plane*

The same type of dependence as for the S_V ratio can be derived for the density of lines, L_A, in 2-D structures. The dimensions of L_A are m^{-1} inverse metres. Suppose that there are two systems of lines on the area A and that at least one of these is of a random nature. It can be shown that the average number of intersections between the lines in the different systems, N_i, is given by the following:

$$N_i = \frac{2L_1L_2}{\pi A} \tag{2.45}$$

where L_1 and L_2 are the total lengths of the lines in the two systems. If one takes one set of lines to be probing lines and the second the lines measured, their density is given as:

$$L_A = \frac{\pi}{2} P_L = \frac{\pi}{2} \frac{1}{E_2(L)} \tag{2.46}$$

where $E_2(L)$ indicates the mean intercept length measured in two dimensions.

Once again, as in the case of the surface-to-volume ratio, the equation above can be used for a system of features in a 2-D microstructure and for individual objects. For an isolated feature the equation can be used to determine the ratio of the perimeter, L_p, to the feature area, A:

$$E_2(L) = \frac{\pi A}{L_p} \tag{2.47}$$

This yields for a circle of radius R the following formula:

$$E_{circ}(L) = \frac{\pi}{2} R \tag{2.48}$$

For a system of features this equation is valid if the perimeter, area, and the mean intercept are taken as the means for the whole system.

2.18 Density of lines in space

The parameter L_V is used to define the density of lineal elements in a microstructure. A classical example of its application is a description of dislocation density in crystalline materials. Other examples are the density of grain boundary edges and the density of thin fibers in composites. In certain cases it might be necessary to distinguish between the densities of different linear elements (for instance screw vs. edge dislocations, different fibers). This would lead to the use of different L_V values: $(L_V)_1$, $(L_V)_2$, etc. The dimensions of this parameter are m^{-2} inverse square metres. Stereological considerations lead to a simple formula relating the density of line, L_V, with the number of intersection points with cross-sections, P_A. It can be shown that L_V and P_A remain in a simple proportion:

$$L_V = 2P_A \tag{2.49}$$

This relationship is widely used in measurements of dislocation densities by means of surface observations of etch pits (see Chapter 5).

The parameters discussed so far have clear interpretations for any microstructure extending into three dimensions. However, in the case of thin, layer-type materials the microstructures have 2-D character. The parameters A_A, L_A, N_A, and P_A can be used as they have the same meaning for 2-D structures as they do for 3-D structures. In this way A_A defines area fraction, L_A density of lines on the surface, N_A and P_A densities of the specified features. This concept can be extended for the case of 1-D microstructures (e.g., bamboo wire structure) with L_L, N_L, and P_L defining the linear fraction and the linear densities, respectively.

2.19 Density of elements and average properties

The parameters P_V and N_V describe the number of elements (N) or point-like elements (P) in a unit volume of the material. Their dimensions are m^{-3} inverse cubic metres. They can be estimated by means of a method discussed further, which requires estimation of the distribution function of the size of the individual objects. Modern methods for their estimation are discussed in Chapters 3 and 8.

If the parameters describing the fraction of volume, area, line, and surface-to-volume ratio are measured together with the number of the elements, N_V, one can calculate the mean volume, area, length, and surface of the microstructural elements. These parameters:

$$E(V) = \frac{V_V}{N_V}$$
$$E(S) = \frac{S_V}{N_V}$$
$$E(A) = \frac{A_A}{N_A} \tag{2.50}$$
$$E(L) = \frac{L_L}{N_L}$$

characterize the size of the elements studied. In the case of 3-D elements, mean volume and surface may be used directly. For 2-D elements, the mean area is used and the mean intercept for 1-D objects.

2.20 *Modification for anisotropic lines and surfaces*

As has been stressed already, the basic stereological formulae for V_V, S_V, and L_V require testing with uniform isotropic random sections, lines, etc. In the case of isotropic microstructures this condition is met by any set of test elements. However, in the case of anisotropic microstructures, which are frequently observed in a large number of materials, random isotropic sampling is not an obvious choice. The problem of such a sampling procedure has been solved recently with the advances in stereology discussed in Chapter 3. In classical stereology the problem is approached by making assumptions with regard to the anisotropy of the studied objects.

Microstructural features are formed during materials processing and subsequent treatments. There are three types of texture (preferred orientations) that are frequently found in the microstructures of materials:

a) fiber;
b) planar;
c) planar-linear textures (see Figure 2.23).

A fiber texture is characterized by the existence of a preferential alignment of the elements along a direction, **t**. A planar texture is distinguished by preferential orientation in the directions perpendicular to a specified normal direction, **n**. In the case of planar-linear texture, there are two characteristic directions: normal, **n**, and additional direction, **t**, normal to **n**. The different types of the texture are shown schematically in Figure 2.23.

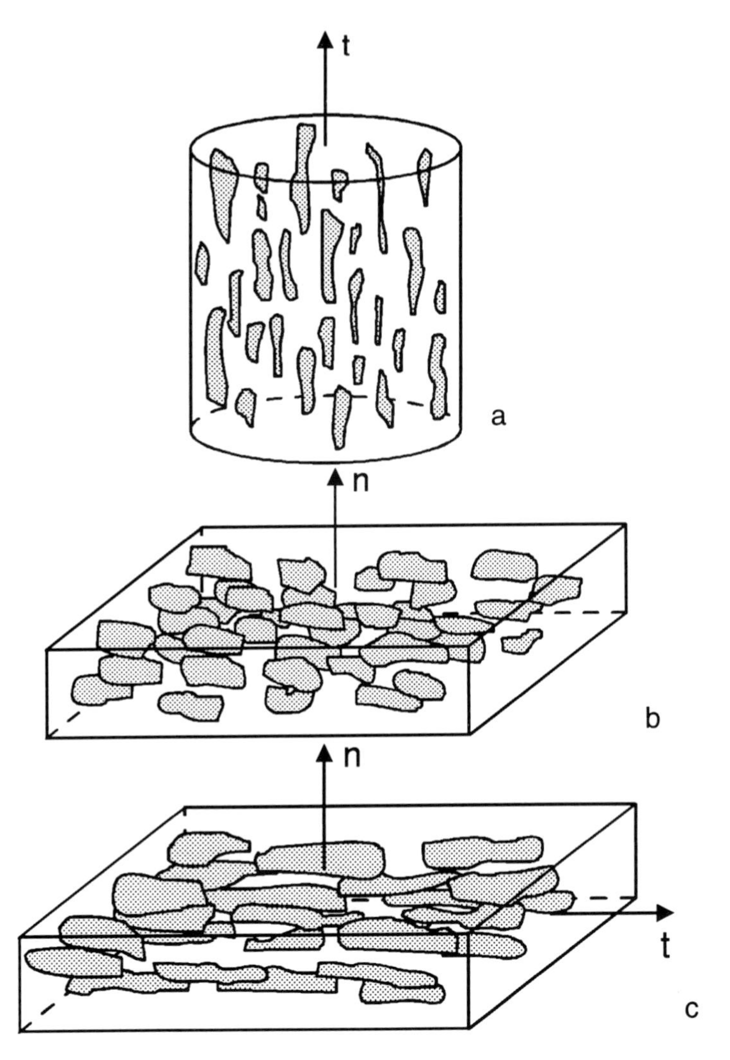

Figure 2.23 Schematic representation of different types of texture in the geometry of microstructural elements.

Example 2.11

Figure 2.24 presents two sets of lines representative of the traces of grain boundaries revealed in polycrystals after different treatments. The images were selected so that one set of grain boundaries does not show preferential orientation while the other is visibly of anisotropic character with the elongation axis indicated. Two systems of parallel lines were placed randomly on the image of the grain boundaries as shown in Figure 2.24 giving the density of intersections points, P_L. The following values were measured: A: $(P_L)_1$ = 548, $(P_L)_2 = 532$; B: $(P_L)_1 = 486$, $(P_L)_2 = 541$ (arbitrary units). The differences for the microstructure A are statistically insignificant. However, the difference for B indicates anisotropy in the geometry of the grain assemblies.

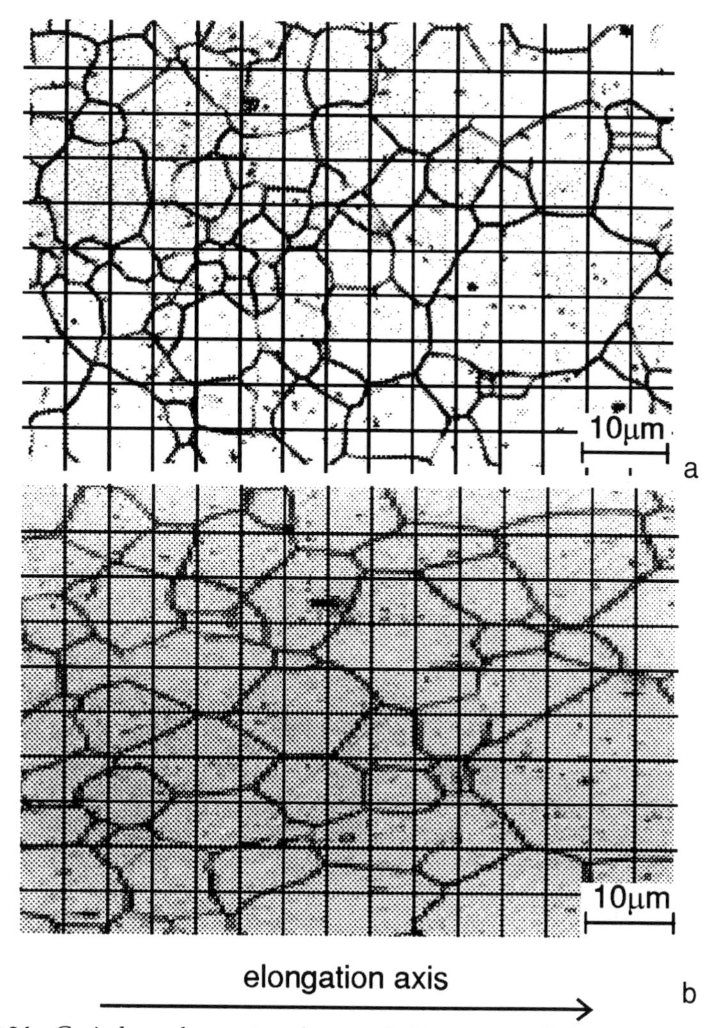

Figure 2.24 Grain boundary networks revealed in an annealed (a) and strained (b) polycrystals of α-Fe.

The anisotropy of the system of lines on a plane can be tested and described by probing with oriented line probes as shown schematically in Figure 2.25. For an isotropic structure the density of intersection points, P_L, does not show any systematic dependence on the orientation of the probing lines at an angle θ, to some reference direction. By contrast, the dependance of $P_L(\theta)$ has a well-defined maximum and minimum if a preferential orientation of the lines exists as in Figure 2.26. This finding can be visualized also by a plot of the mean intercept length, E(L), in linear coordinates or in the form of a rose diagram (Figure 2.27).

For the purpose of this analysis, the lines revealed in Figure 2.25 can be divided into two subsets: (a) isotropically, randomly oriented segments of total length L_r and (b) preferentially oriented along $\theta = 0$ of total length L_o.

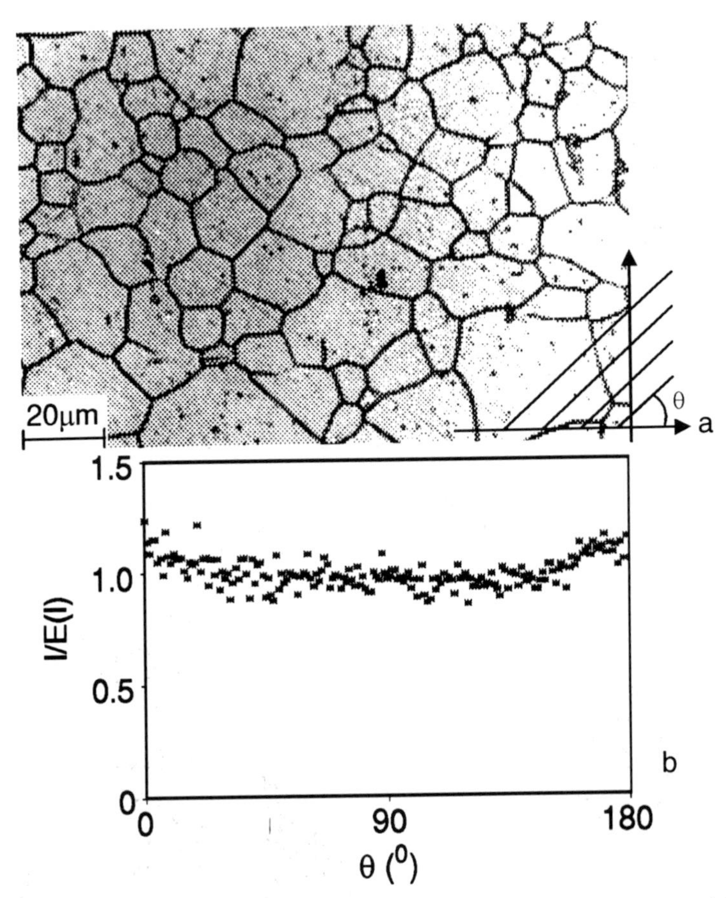

Figure 2.25 Probing with an oriented system of lines (a) and the density of the intersections points of the test lines with the grain boundaries as a function of the angle θ (b).

This division has purely theoretical meaning. It simply describes the fact that certain orientations of the segments appear more frequently than would be expected for a random distribution of line segments. The length of randomly oriented segments can be related to the density of the intersection points in the way described before. If one probes an area A with systems of lines making an angle of $\theta = 0$ (parallel probes) the number of intersections is given by the following:

$$(P_L)_{\theta=0} = \frac{2L_r}{\pi A} \tag{2.51}$$

The probe lines perpendicular to the elongation axis ($\theta = 90°$) intersect both oriented and random segments. The density of the intersection points

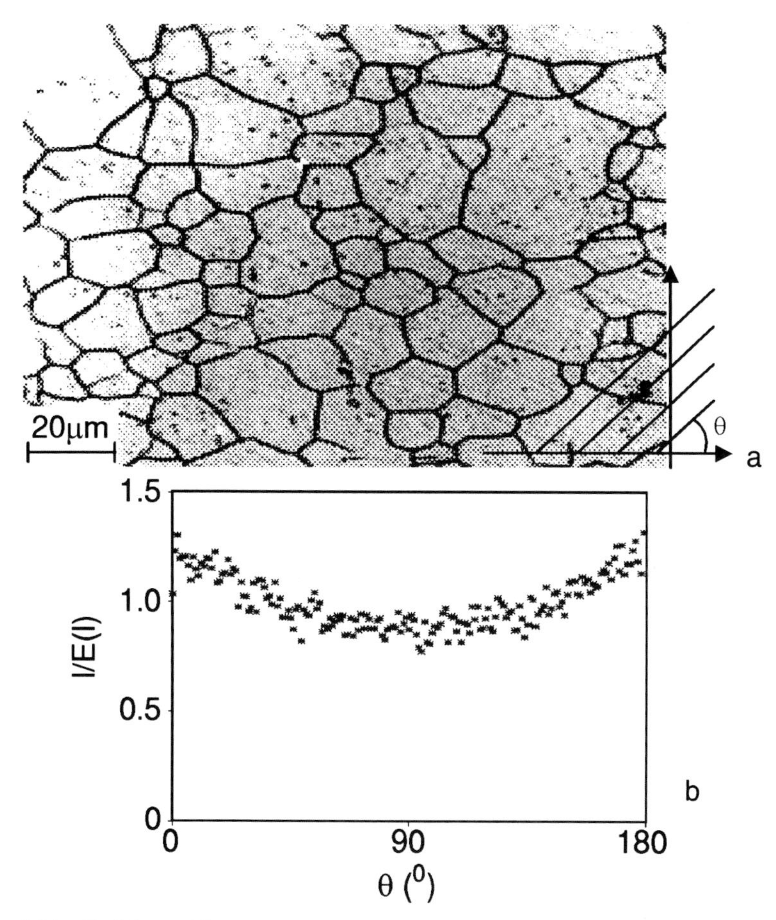

Figure 2.26 Probing with an oriented system of lines (a) and the density of the intersection points of the test lines with grain boundaries a function of the angle θ (b) for an anisotropic grain assembly.

with random segments does not depend on the orientation of the probe lines and is the same as for the parallel probes:

$$\left[(P_L)_{\theta=90^\circ}\right]_r = \frac{2(L_A)_r}{\pi} \tag{2.52}$$

On the other hand, the oriented segments are intersected with a higher probability by perpendicular probes and the density of their intersections is given by:

$$\left[(P_L)_{\theta=90^\circ}\right]_o = (L_A)_o \tag{2.53}$$

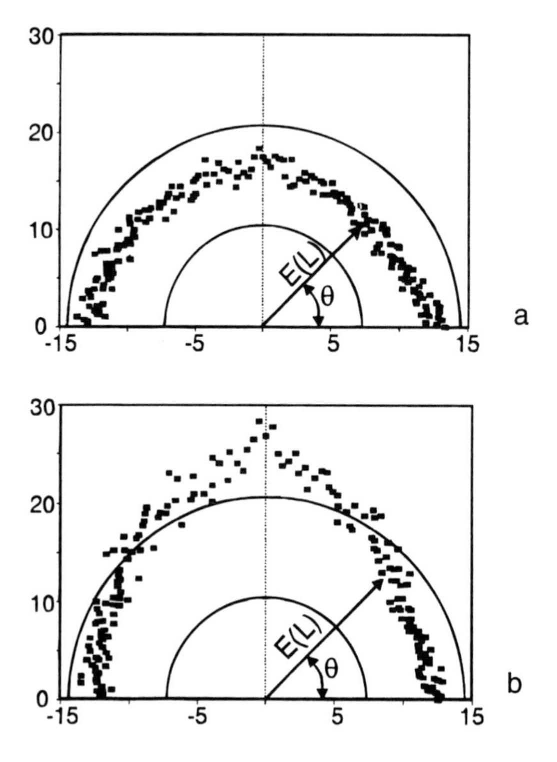

Figure 2.27 Rose diagram for the $P_L(\theta)$ function (a) for Figure 2.25a and (b) for Figure 2.26a.

By combining the equations one can obtain the following formulae:

$$(L_A)_o = (P_L)_{\theta=90^\circ} - (P_L)_{\theta=0} \tag{2.54}$$

and

$$(L_A)_r = (P_L)_{\theta=90^\circ} + \left(\frac{\pi}{2} - 1\right)(P_L)_{\theta=0} \tag{2.55}$$

In the light of this analysis the ratio of the total length of the oriented segments to the total length of the lines is a convenient measure of the degree of orientation. This ratio in the case of lines in a plane is designated $\Omega_{\text{lines}(2)}$ and is given by:

$$\Omega_{lines(2)} = \frac{(P_L)_{\theta=90^\circ} - (P_L)_{\theta=0}}{(P_L)_{\theta=90^\circ} + \left(\frac{\pi}{2} - 1\right)(P_L)_{\theta=0}} \tag{2.56}$$

The ratio is equal to zero for completely random structures and 1.0 for 100% oriented.

An alternative approach to the description of the preferential orientation utilizes the concept of a Fourier series. If the density of line segments making an angle θ with a reference direction is given by $L_A(\theta)d\theta$ then it can be shown that the density is related to the density of intersections with the test lines inclined at an angle ω and is given by the following equation:

$$P_L(\omega) = \int_{\omega+\frac{\pi}{2}}^{\omega-\frac{\pi}{2}} L_A(\theta)\cos(\theta-\omega)d\theta \quad (2.57)$$

For a given P_L function the equation can be solved using Fourier transforms. Some simple solutions are listed in Table 2.3 and described in Example 2.12.

Example 2.12

The grain boundary microstructures shown in Figure 2.24 were studied with systems of parallel lines inclined at θ to the indicated axis. For each value of θ in the range 0 to 90° the density of intersection points, $P_L(\theta)$, was determined. It was found that this function can be approximated by the following formula:

$$P_L(\theta) = A_0 + A_1 \sin(\theta).$$

Table 2.3 List of Some of the Solutions of Equation 2.57

$L_A(\theta)$	$P_L(\omega)$
A_0	$2A_0$
$A_1 \sin(\theta)$	$\frac{\pi}{2} A_1 \sin(\omega)$
$A_2 \sin(2\theta)$	$\frac{2}{3} A_2 \sin(2\omega)$
$A_3 \cos(\theta)$	$\frac{\pi}{2} A_3 \cos(\omega)$
$A_4 \cos(2\theta)$	$\frac{2}{3} A_4 \cos(2\omega)$
$A_5 \cos^2(\theta)$	$A_5\left(1 + \frac{1}{3}\cos(2\omega)\right)$
$A_6 \sin^2(\theta)$	$A_6\left(1 - \frac{1}{3}\cos(2\omega)\right)$

The $L_A(\theta)$ function defines the density of grain boundary segments of orientation θ with respect to some axis. The $P_L(\omega)$ functions describes the experimental density of the intersection points with a grid of lines inclined at the angle ω to the same axis.

The following values of the constants A_0 and A_1 have been determined:

Microstructure a: $A_0 = 0.08864$ $A_1 = 0.13810 \mu m^{-1}$
Microstructure b: $A_0 = 0.80639$ $A_1 = 0.01489 \mu m^{-1}$

Using the results given in Table 2.3, it can be expected that the orientation function of the grain boundary segments is described by the following function:

$$L_A(\theta) = B_0 + B_1 \cos(\theta).$$

The measured values of A_0 and A_1 can be used to estimate the following values of B_0 and B_1:

Microstructure a: $B_0 = 1.71482$ $B_1 = -0.07247 \mu m^{-1}$
Microstructure b: $B_0 = 1.27035$ $B_1 = -0.03331 \mu m^{-1}$

Similarly to the procedure for lines on a plane, lines in space elongated preferentially in a direction parallel to **t** can be divided into classes:

a) L_r segments of random orientation;
b) L_t segments parallel to **t** appearing in excess by comparison with the length of such oriented segments expected for a random distribution of total length L_r.

The density of randomly oriented segments can be measured by observations made on cross-sections parallel to the vector **t**. Using one of the formulae given earlier one can obtain an $(L_V)_r$ value in the following way:

$$(L_V)_r = 2(P_A)_p \tag{2.58}$$

In the case of an isotropic specimen probed with oriented lines, the same density of intersections would be measured for the cross-section perpendicular to **t**. Thus, a difference in the two respective densities:

$$(P_A)_t - (P_A)_p \neq 0 \tag{2.59}$$

is an indication of the texture in the orientation of line segments. The total length of the lines in such a system is a sum of the two classes of segments and this leads to the following:

$$(L_V) = (L_V)_p + (L_V)_t \tag{2.60}$$

Using the last equation one can introduce a dimensionless parameter $\Omega_{lines(3)}$:

$$\Omega_{lines(3)} = \frac{(P_A)_t - (P_A)_p}{(P_A)_t + (P_A)_p} \tag{2.61}$$

This parameter is equal to zero for texture-free, isotropic lines and increases its value with an increase in their alignment. It is equal to 1.0 for a completely ordered system of lines.

2.21 Anisotropic distribution of surfaces

Anisotropic distributions of surfaces in space are examined with the use of directed lines as the probing elements. It should be remembered that for isotropic, random orientations of surface segments the number of intersections with oriented lines per their length, P_L, is proportional to the surface-to-volume ratio, S_V; with the proportionality constant being equal to 2. The problem of oriented surfaces can be solved in a manner similar to that used previously for lines. It is based on the discrimination of surface segments in an oriented system into the randomly oriented and oriented segments of the surface-to-volume ratios $(S_V)_r$ and $(S_V)_o$, respectively. In the case of a fiber texture, all the required measurements should be conducted on cross-sections parallel to the fiber axis. All such cross-sections are equivalent, viewed in the direction perpendicular to the fiber orientation on the section perpendicular to **t**. By orienting the probing lines along the preferential orientation, **t**, it is possible to estimate $(S_V)_r$:

$$(S_V)_r = 2(P_L)_t \tag{2.62}$$

where $(P_L)_t$ is the number of intersections points per unit length of the lines parallel to **t**.

In the subsequent step, the system is probed with lines perpendicular to the fiber direction, **t**, and the number of the intersections per line length, $(P_L)_p$, determined. The difference between these two numbers of intersections is a measure of the linear anisotropy of the surface orientation. More detailed considerations show that the total surface-to-volume ratio, S_V, is given by the following formula:

$$S_V = \frac{\pi}{2}(P_L)_p + \left(2 - \frac{\pi}{2}\right)(P_L)_t \tag{2.63}$$

and consequently, the dimensionless ratio of the anisotropy by:

$$\Omega_{fiber} = \frac{(P_L)_p - (P_L)_t}{(P_L)_p + \left(\frac{4}{\pi} - 1\right)(P_L)_t} \tag{2.64}$$

Also, in the case of a planar texture (see Figure 2.28) it is sufficient to confine the studies to one of the cross-sections: a section parallel to the normal, **n**. Probing lines for planar texture are oriented in a direction parallel to and perpendicular to **n**. However, in this case the total surface-to-volume ratio is

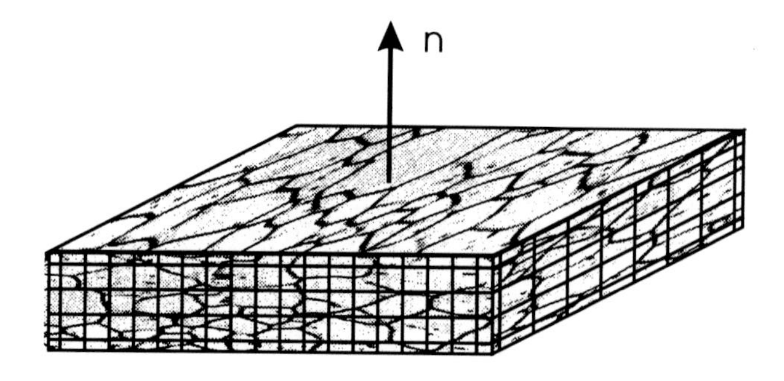

Figure 2.28 An example of a planar texture produced by a compression test on a two-phase material.

given by:

$$S_V = (P_L)_p + (P_L)_n \tag{2.65}$$

and the dimensionless ratio of the anisotropy by:

$$\Omega_{planar} = \frac{(P_L)_p - (P_L)_t}{(P_L)_p + (P_L)_t} \tag{2.66}$$

Planar-linear textures are characterized by two vectors: **n** and **t**. Studies of these textures require measurements to be carried out on two types of cross-sections: (a) longitudinal, **L**, which is normal to the vector product of **t** × **n**, and (b) transverse, **T**, perpendicular to **t**. Transverse sections are probed with lines perpendicular to **n**, i.e., parallel to **L** and longitudinal sections are probed with lines parallel to **t**, in order to obtain the values of $(P_L)_{Tn}$ and $(P_L)_{Lt}$, see Figure 2.29. The third probing is to be done on either the transverse or longitudinal sections in a direction parallel to **n**. This probing makes it possible to measure the value of $(P_L)_{Ln}$. With this one can obtain the following equations:

$$(S_V)_r = 2(P_L)_{Lt} \tag{2.67}$$

$$(S_V)_{planar} = (P_L)_{Lt} - (P_L)_{Tn} \tag{2.68}$$

$$(S_V)_{linear} = \frac{\pi}{2}\left[(P_L)_{Tn} - (P_L)_{Lt}\right] \tag{2.69}$$

The total surface-to-volume ratio is the sum of the respective segments showing random, linear and planar orientation:

$$S_V = (S_V)_r + (S_V)_{linear} + (S_V)_{planar} \tag{2.70}$$

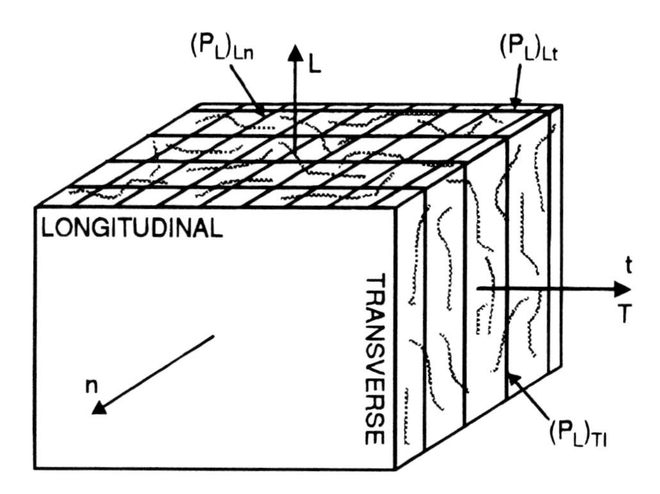

Figure 2.29 An example of a planar-linear texture produced by rolling a polycrystalline material.

By combining the last four equations, one can derive obvious formulae for anisotropy factors defining the ratio of the surface segments showing linear, planar and planar-linear orientations. For example, in the last case it requires calculation of the following:

$$\Omega_{pl-lin} = \frac{(S_V)_{linear} + (S_V)_{planar}}{S_V} \tag{2.71}$$

Example 2.13

The formulae derived can be used to characterize the anisotropy of the grain geometry in a polycrystalline material after compression of a material as shown in Figure 2.28.

2.22 *Estimation of the density of volume elements, N_V, by a method of size distribution reconstruction*

One of the methods for estimating the density of volume elements has been proposed by Saltykov (see, for example, Reference 15) and is known in the literature under his name. The method is based on two important assumptions:

a) all the elements (grains, particles) are of the same shape;
b) the function of the normalized grain/particle section area distribution function $f(A/A_{max})$ is known and this function reaches the highest frequency for values close to A_{max}.

The procedure is of an iterative character. The distribution of grain/particle volume is broken into a finite number of classes, $V_i(i = 1,2,3, \ldots, k)$, and the distribution of functions $f^i(A/A_{max})$ of the grain areas are established for each class. In the last stage the numbers of grains in each class of the volume are calculated in a process of "backwards discrimination" of the grain area distribution function divided into the same number of classes A_i.

Let us assume that the polycrystal analyzed contains in a specified volume V, N_i grains of a volume V_i. It may be assumed that the number of grain areas, N_k, in the class, V_k, is directly related to the number of grain areas n_k in the class A_k. However, this means that a certain specific number of grain areas in the class A_{k-1} are a result of cutting of N_k grains of volume V_k. This number can be calculated from the $f^k(A/A_{max}{}^k)$ function and subtracted from n_{k-1}. Then in the same way as before the number of grains of V_{k-1} volume can be calculated and the process is repeated to the point of calculation of the number of grains in the "1" class. The process is schematically explained in Figure 2.30. It should be stressed that the method is based on assumptions of the grain shape and relies on an ability to detect the biggest cross-sections of the biggest grains. The errors involved in this transformation are additive.

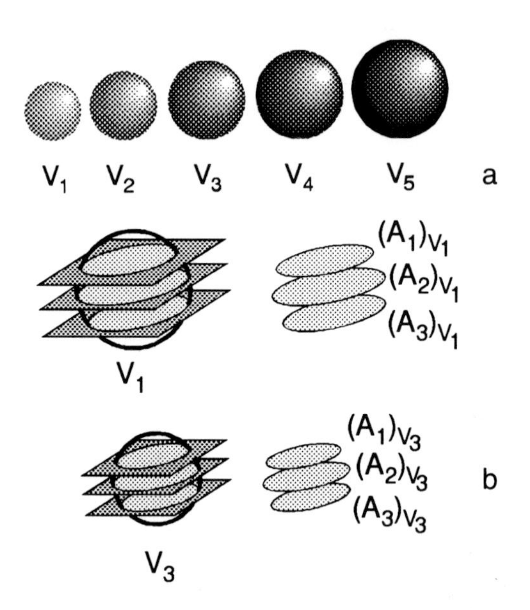

Figure 2.30 A schematic explanation of the Saltykov method. For a given shape of particles, in this case spheres (a) the distribution of particle section area is obtained as a function of the particles volume (b). The particles section areas are measured as a section of the material studied (c) and the experimental distribution function *f*(*A*) is obtained (d). In the last step (b) and (d) are used to obtain the volume distribution function *f*(*V*) (e). (Figure is continued on the facing page.)

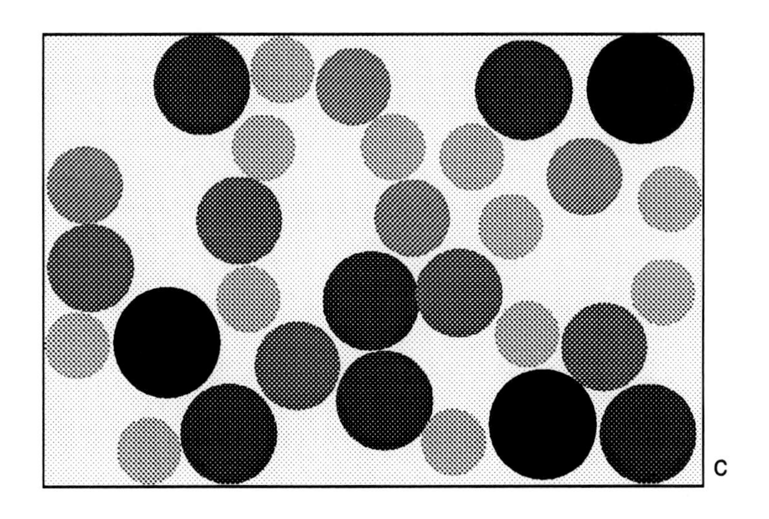

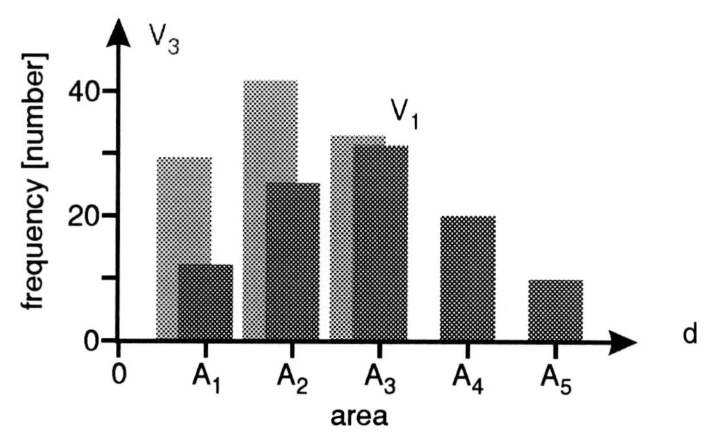

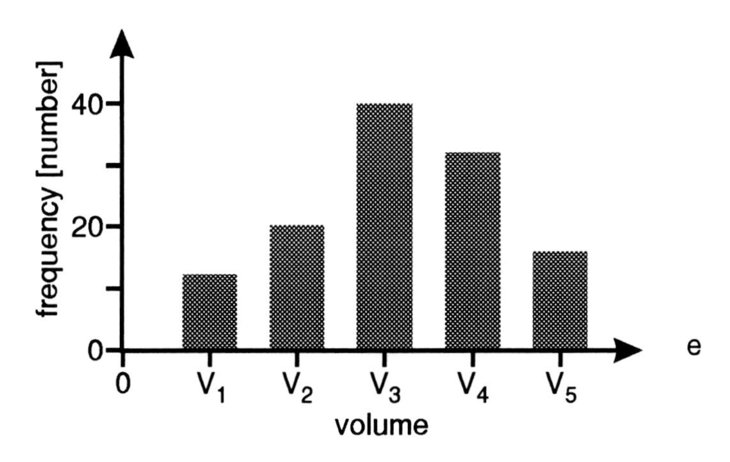

Figure 2.30 (*Continued*)

2.23 Description of particle agglomeration

Multiphase materials usually contain significantly different volume fractions of the constituents and one of the phases can be considered to be a matrix while others are particles distributed over its volume. In such a situation measurements required for the characterization of the system can be simplified by taking into account the relationship between the number of intercepts with the interface boundaries, $(P_L)_{\alpha\beta}$, and the number of α particles intersected by the testing lines, $(N_L)_\alpha$:

$$(P_L)_{\alpha\beta} = 2(N_L)_\alpha \tag{2.72}$$

In dilute systems where particles do not touch one another the parameter $(S_V)_{\alpha\alpha}$ can be determined from the following:

$$(S_V)_{\alpha\alpha} = 4(N_L)_\alpha \tag{2.73}$$

In certain materials on the other hand, a tendency is observed for particles to be in contact and to form aggregates of a low number of elements. This tendency can be quantitatively described by a parameter C called contiguity. The parameter C defines the fraction of total particle surface area that is shared by the particles of the same phase. In the case of α particles distributed in a β matrix, C is given as:

$$C = \frac{2(S_V)_{\alpha\alpha}}{2(S_V)_{\alpha\alpha} + (S_V)_{\alpha\beta}} \tag{2.74}$$

The constant 2 in this formula underlines the fact that if particles are viewed as separate entities the interface shared by two particles should be counted twice.

Values of C can be obtained from line testing using the equation, which is a simple consequence of one given earlier:

$$C = \frac{4(P_L)_{\alpha\alpha}}{4(P_L)_{\alpha\alpha} + 2(P_L)_{\alpha\beta}} \tag{2.75}$$

Additional assumptions regarding the geometry of common interfaces between two particles make it possible to calculate the number of contacts. Appropriate formulae can be derived in a rather obvious analysis found in Reference 15.

Example 2.14

Figure 2.31 shows characteristic images of microstructures as a function of annealing time. It can be noted that the deformed microstructure contains wavy bands of heavily strained α-Fe (Figure 2.31a). After 15 minutes of annealing at 500°C (Figure 2.31b), small nuclei appear in the form of clusters. The cluster geometry is similar to the geometry of the bands of heavy deformation. During further annealing (Figure 2.31c), the nuclei in clusters grow in size and gradually develop a shape unrelated to the deformation structure. The microstructure in Figure 2.31d shows that no relationship seems to exist between the deformed and the recrystallized structure at the end of the recrystallization process. Figure 2.31e shows fully recrystallized material, which is similar to the microstructure observed in fully recrystallized α-Fe after further annealing that results in grain growth (Figure 2.31f).

From a quantitative metallography point of view the material studied contains two "phases":

1. phase α—recrystallized α-Fe;
2. phase β—heavily deformed α-Fe.

The surface-to-volume ratio of the interfaces have been measured by the intercept method and the following values of C obtained ($C = [2 (S_v)_{\alpha\alpha}/(2 (S_v)_{\alpha\alpha} + (S_v)_{\alpha\beta})]$.

t [h]	Recrystallized phase [%]	C
0.00	0	0
0.25	29.4	0.24
0.50	42.1	0.27
1.00	93.8	0.60
3.00	100	1.0
6.00	100	1.0

The value of C = 1 is typical of a space filling aggregate.

2.24 Curvature of lines on a plane

The shape of linear objects revealed on a plane can be described in terms of their curvature. The curvature describes the rate of change of direction of a curve at a particular point on that curve. A definition of this parameter is given in Figure 2.32.

For any two points, A and B, on a given curve the total curvature, k_{AB}, is the angle between the tangents to the curve at these two points:

$$k_{AB} = \phi_A - \phi_B \tag{2.76}$$

The mean curvature for the segment of a line is defined as the ratio of the total curvature (k_{AB}) to the length of the curved segment, l_{AB}, between the points A and B:

$$(k_{AB})_{mean} = \frac{k_{AB}}{l_{AB}} \tag{2.77}$$

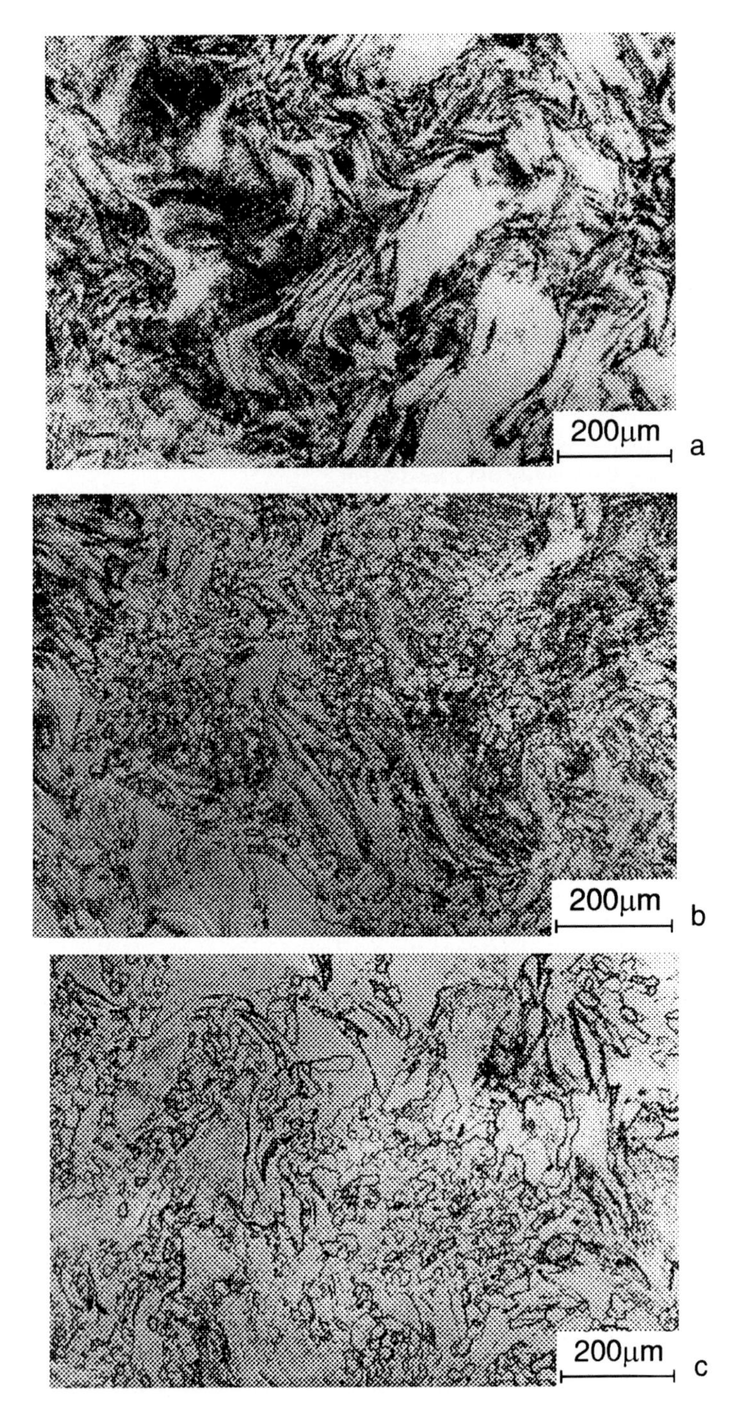

Figure 2.31 Micrographs of heavily deformed strained α-Fe (a) and microstructures representative for material after annealing at 500°C for (b) 15 min, (c) 30 min, (d) 1 h, (e) 3 h, (f) 6 h. (Figure is continued on the facing page.)

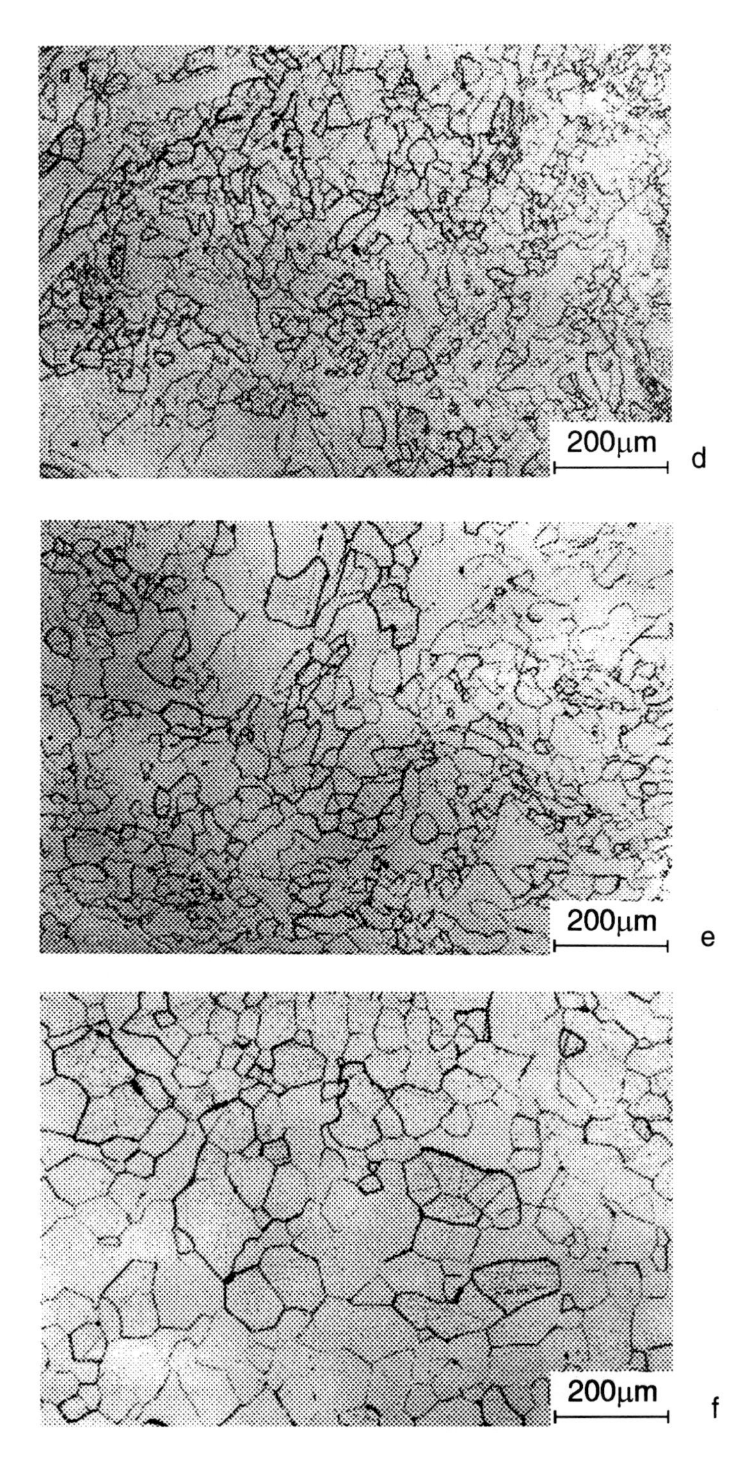

Figure 2.31 (*Continued*)

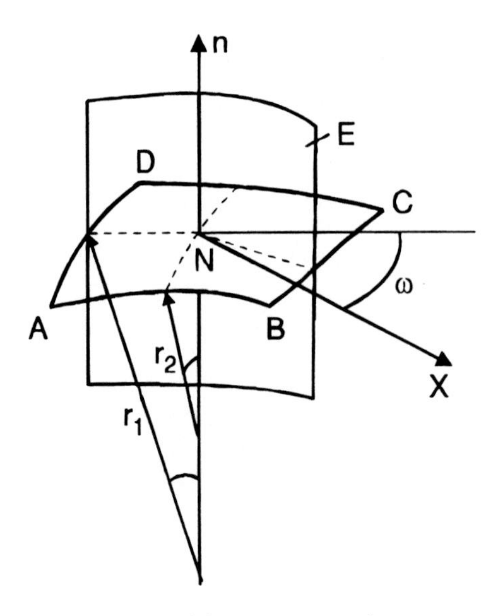

Figure 2.32 Definition of the curvature of a given surface element.

It can be shown that the total and the mean curvature of a straight line is equal to zero while for a circle the mean curvature is constant and equal to 1/R, where R is the circle radius:

straight line $\quad k_{AB}$ = 0 for any segment
circle of radius R $\quad k_{AB}$ = 1/R for any segment

The concept of curvature can be further extended from the one for a segment to the curvature at a point on the line. The curvature, k(A), at the point A is the limiting value of the mean curvatures of the segment AB as B approaches A:

$$k(A) = (k_{AB})_{mean} \quad \text{for } B \rightarrow A \tag{2.78}$$

The value of k(A) depends on the radius, R_A, of the largest circle tangential to the curve at a given point A:

$$k(A) = \frac{1}{R_A} \tag{2.79}$$

The relationship above defines the absolute value of the curvature. Its sign is determined by an additional definition. It is commonly assumed that the curvature is positive if the tangent line is placed outside the area of the particle enclosed by the curve and negative otherwise.

For an open segment of a curve the curvature depends on the angle between the tangent lines at the ends of the segment. For any closed loop the total curvature is equal to 2π. A system of lines revealed in the area A can be described in terms of curvature density, k_A. This parameter is the ratio of the total curvature of lines included in the area A divided by A:

$$k_A = \frac{\sum k^i}{A} \tag{2.80}$$

where k^i gives the curvature of i-th line segment.

DeHoff[16] has shown that the curvature density of the lines in a given area A is related to the number of tangent points of a line sweeping through the area in a perpendicular direction. In this procedure the tangent points are distinguished by a sign plus or minus. If we study a system of closed loops, as frequently happens, the sign at a given tangent point can be defined as "plus" if the figure is locally convex and "minus" otherwise. This means that for a single connected 2-D particle the sign plus distinguishes a convex segment of the circumference and minus a concave one. It has been shown[17] that the number of positive T_A^+ and negative T_A^- tangent points determine the average curvature, k_A, according to the equation:

$$k_A = \pi\left(T_A^+ - T_A^-\right) \tag{2.81}$$

Curvature measurements are frequently carried out for two-dimensional figures which are plane sections through a system of 3-D convex particles. In such a case the sections are also convex and each of them has two positive tangent points. As result of this the last equation reduces to:

$$k_A = 2\pi\, N_A \tag{2.82}$$

where N_A is the density of the particles.

The concept of curvature can be extended so as to cover the curvature of a surface, S, at a given point, say Q. To this end, let us assume that the surface S is cut by a set of planes containing the normal, $\mathbf{n}_Q$, at point, Q. The planes are defined by an angle, γ, to some reference direction, $\mathbf{t}_Q$, perpendicular to $\mathbf{n}_Q$. These planes create a system of curves which are the intersection lines of S and the planes containing $\mathbf{n}_Q$ and $\mathbf{t}_Q$. In the next step each intersection line has its curvature prescribed in the way described previously. As result, a system of curvature values is assigned to the point Q. Among these curvature values there are minimum and maximum curvatures, k_{min} and k_{max}, respectively. These are the so called main curvatures of the surface S at the point Q.

The concept of main curvatures is used to define the mean curvature of a surface at a point. The mean curvature, H, is the mean of these two extreme curvatures:

$$H = \frac{1}{2}(k_{\min} + k_{\max}) \tag{2.83}$$

This value is equal to zero for a plane and has a constant value for a sphere:

flat surface H = O for any point
sphere of radius R H = R for any point

For a surface with a more complicated geometry the curvature varies from point to point, and the total and the mean curvatures, M and M_s, can be computed according to the following equations:

$$M = \int_S H \, ds \tag{2.84}$$

$$M_S = \frac{M}{S} \tag{2.85}$$

respectively, where S is the area of the surface studied.

If the surface microstructural elements are enclosed in a specified volume it is useful to define the density of surface curvature, M_V:

$$M_V = \frac{M_S}{V} \tag{2.86}$$

where V is the volume and M_S the mean curvature of all the surfaces in this volume.

DeHoff[18] and Cahn[19] have proved that the mean curvature of the surfaces enclosed in volume V is related to the mean curvature of the lines of intersections of these surfaces with the plane of observation in the form:

$$M_V = k_A \tag{2.87}$$

This equation can be used to propose a procedure for surface curvature estimation through examination of 2-D sections of a microstructure. It can be performed via counting tangent points with a sweeping line passing through the image of the section. In the case of convex surfaces the number of tangent points is always two and the following simplified equation can be derived:

$$M_V = 2\,N_A \tag{2.88}$$

where N_A is the planar density of the intersections. In this case the interfacial curvature of a system of particles can be estimated by a simple counting procedure on the profiles revealed in the plane of observation. It may be also noted that if the surface-to-volume ratio, S_V, is known then the volume curvature density, M_V, can be turned into the surface curvature density according to the equation:

$$M_S = \frac{M_V}{S_V}. \tag{2.89}$$

Measurements of particle curvature have proved to be useful in studies of the kinetics of phase transformations (see, for example, Reference 20). An example of curvature measurements is given in Chapter 8.

DeHoff[21] has extended the concept of curvature to curvature of lines in space. Such lines are assigned curvature as well as torsion, which is related to the inflection points on the projected images of the lines. Both parameters can be estimated using the technique of a sweeping line applied to projected images of a known thickness. The total curvature per unit volume is given as:

$$k_V = \frac{\pi}{t}\left(T_A\right)_{proj} \tag{2.90}$$

where t is the thickness and $(T_A)_{proj}$ is the density of tangent points. The total torsion per unit volume can be estimated using a similar equation of the form:

$$t_V = \frac{\pi}{t}\left(I_A\right)_{proj} \tag{2.91}$$

where t is the thickness and $(I_A)_{proj}$ is the density of inflection points in the projected image. This concept may prove useful in the characterization of the curvature of dislocation lines.

If the lines studied are edges of some volumetric elements such as particles or grains they can also have dihedral angles assigned to them. These are angles between particle/grain faces joined along a common edge. The average dihedral angle can be estimated by measurements carried out on sections of the microstructure. Once again the technique of a sweeping line is used. However, in this case the points of the emergence of the edges are studied on the cross-section. This situation is considered in more depth in Chapter 5.

References

1. Okazaki, K. and Conrad, H., Grain size distribution in recrystallized alpha-titanium, *Transactions* JIM, 13, 198, 1972.
2. Conrad, H., Swintowski, M., and Mannan, S. L., Effect of cold work on recrystallization behavior and grain-size distribution in titanium, *Metallurgical Transactions*, 16A, 703, 1985.
3. Bystrzycki, J., Przetakiewicz, W., and Kurzydłowski, K. J., Study of annealing twins and island grains in f.c.c. alloy, *Acta Metallurgica et Materialia*, 41, 2639, 1993.
4. Bucki, J. J. and Kurzydłowski, K. J., Measurements of grain volume distribution parameters in polycrystals characterized by a log-normal distribution function, *Scripta Metallurgica et Materialia*, 28, 689, 1993.

5. Kurzydłowski, K. J., McTaggart, K. J., and Tangri, K., On the correlation of grain geometry changes during deformation at various temperatures to the operative deformation modes, *Philosophical Magazine*, 61, 61, 1990.
6. Cochran, W. G., *Sampling Techniques*, Wiley, New York, 1977.
7. Matern, B., Precision of area estimation—a numerical study, *Journal of Microscopy*, 153, 269, 1989.
8. Spiegel, M. R., *Theory and Problems of Statistics, Schaum's outline series*, McGraw-Hill, London, 346, 1992.
9. Lantuejoul, C., Ergodicity and integral range, *Journal of Microscopy*, 161, 387, 1991.
10. Cruz-Orive, L. M. and Weibel, E. R., Sampling design for stereology, *Journal of Microscopy*, 122, 235, 1981.
11. Mathieu, O., Cruz-Orive, L. M., Hoppeler, H., and Weibel, E. R., Measuring error and sampling variation in stereology: comparison of the efficiency of various methods for planar image analysis, *Journal of Microscopy*, 121, 75, 1981.
12. De Hoff, R. T., Aigeltinger, E. H., and Craig, K. R., Experimental determination of the topological properties of three-dimensional microstructures, *Journal of Microscopy*, 95, 69, 1972.
13. Rhines, F. R., Craig, K. R., and Rouse, D. A., Measurement of average grain volume and certain topological parameters by serial section analysis, *Metallurgical Transactions*, 7A, 1729, 1976.
14. Jensen, E. B. and Gundersen, H. J. G., Stereological ratio estimation based on counts from integral test system, *Journal of Microscopy*, 125, 51, 1982.
15. Underwood, E. E., *Quantitative Stereology*, Addison Wesley, Massachusetts, 1970.
16. De Hoff, R. T., The quantitative estimation of mean surface curvature, *Transactions of the Metallurgical Society*, 239, 617, 1967.
17. Rhines, F. N. and De Hoff, R. T., *Quantitative Microscopy*, McGraw-Hill, New York, 1968.
18. De Hoff, R. T., Quantitative serial sectioning analysis—preview, *Journal of Microscopy*, 131, 259, 1983.
19. Cahn, J. W., The significance of average mean curvature and its determination by quantitative metallography, *Transactions of the Metallurgical Society*, 239, 611, 1967.
20. Wiencek, K. and Ryś, J., Coagulation of cementite particles in carbon-steels, *Neue Huette*, 26, 137, 1981.
21. De Hoff, R. T. and Gehl, S. M., Quantitative microscopy of lineal features in three dimensions, *Proceedings of the Fourth International Congress for Stereology*, National Bureau of Standards, Washington, D.C., 29, 1976.

chapter three

Modern stereology

Detailed stereological procedures designed in the past frequently relied on assumptions regarding the shape and/or distribution of the microstructural elements. A number of solutions have been developed for elements of a particular shape. They are reviewed in Reference 1 where references can also be found to the original papers. Some solutions are discussed in Chapters 2, 6, and 8. The basic conjectures involved in these procedures are the following:

a) the shape of microstructural elements, for example, particles, can be approximated to one of a simple convex figure such as a sphere, ellipsoid, rod, or disk;
b) the elements are either randomly distributed and oriented, or the anisotropy/texture of the microstructure is known.

These assumptions proved to be justified in a large number of microstructures successfully studied in the past. However, there are many examples of nonconvex microstructural elements including, for instance, cementite particles in pearlite, the treatment of which is virtually impossible with the procedures developed in the past. Rapid advances in the development of new materials in recent years have resulted in a large number of new systems with little data on their microstructures accumulated so far. In studies of these materials it is difficult to make rationalized assumptions with regard to the shape and anisotropy, and the old procedures are found ever more frequently to be inadequate.

In recent years a number of new procedures have been designed that are free of assumptions with regard to the shape and isotropy of the microstructural elements. These procedures include:

a) the disector;
b) the selector;
c) the fractionator;
d) vertical sectioning;
e) combined point and intercept probing.

These methods have been described in the original papers published mainly in the *Journal of Microscopy* since 1980, although some of the ideas can be traced back to earlier publications. A systematic review of these publications might be an interesting exercise which, however, is not undertaken in the present book. A review of the main results can be found in Reference 2.

Before entering into a more detailed description of the new procedures it is desirable to give their general characteristics with reference to the traditional approach. In making such a comparison it can be shown that the new procedures have provided new solutions to the problem of sampling, rather than a significant advance to the measurement side. These new methods help to select, in a more objective way, fields for observation and microstructural elements for measurement, representative of the microstructure studied.

3.1 Disector—new approach to particle density

The concept of the disector was formally described by Sterio[3] in 1984. The concept is based on the observation that sectioning of a material leads to a biased sampling of larger microstructural elements. More precisely, the individual particle volumes, V_i, or other volumetric microstructural elements are cut by a plane of observation with the probability, $P(V_i)$, proportional to their number, N_i, and height, h_i, in the direction normal to the plane of section:

$$P(V_i) = \frac{N_i h_i}{N_1 h_1 + \cdots + N_k h_k} \tag{3.1}$$

where k is the total number of elements.

As a result, whatever measurements are carried out on the sections of the particles revealed, they produce height-weighted estimates, $E_h(x)$, of the parameter studied, x. This estimate depends not only on the number of the particles showing property x but also on their height (see Figure 3.1 and Example 3.1).

Example 3.1

Figure 3.1 schematically depicts a system of particles which differ in shape and size. There are two subpopulations of particles: gray particles A and black particles B in the proportion 1:1. The particles significantly differ in their height in the direction normal to the plane of section. The height of particles A, h_A, is much smaller then the height of particles B, h_B. It can be seen from Figure 3.1 that the sections of the longer particles (B) appear more frequently compared with the sections of the A particles. A direct counting of the particle sections on the plane of observation yields a biased estimation of the proportion of the two particle types.

In the case of particles with a simple geometry, typically of spherical shape, the effect of a height-dependent probability of particle intersections can be accounted for using the procedures described by Saltykov and others[1], and reviewed in Chapter 2. In this way appropriate corrections can be made to obtain an estimate that is based on the number of particles showing the property

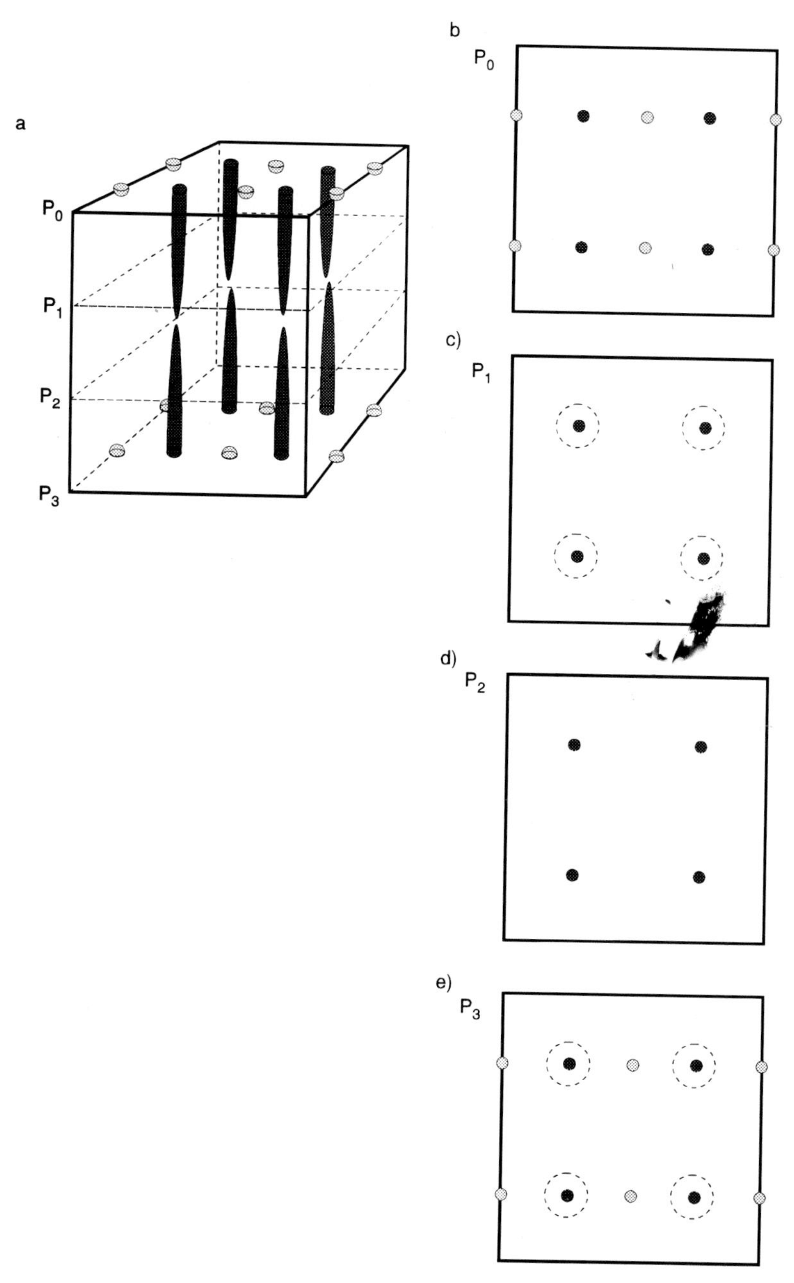

Figure 3.1 A schematic representation of a system of particles studied with the disector (a) and images seen on the sections of the system: (b) to (e). The system consists of equal numbers of two types of particles: gray particles A and black particles B. The A particles are spherical and their diameter is h_A. The B particles are ellipsoids with the longer axis dimension h_B. On the sections of the material the particles appear with a frequency which is systematically different from the frequency of their appearance in the volume of the material (see also Example 3.2).

x, $E_N(x)$. However, for particles of a complex or unknown shape such corrections are not available and the particles should be sampled in a special way. To avoid the effect of particle height, it has been proposed to analyze particles within a certain distance, h, from the section plane. Since this requires two sections to be made the concept is called the disector. This term is used to describe the method as well as the system of two sections and the volume of the material between them.

The disector method is based on measurements in a slice of the material confined between two planes:

a) the upper plane P_0, the so-called "look-up" plane;
b) the lower plane P_1, the observation plane.

These two planes are h apart and the observation area is A as schematically shown in Figure 3.1.

The disector method requires counting the number of particles/elements, N, or selecting for other measurements, **all** and **only** those particles which are placed above the lower plane, P_2, but are entirely below the look-up plane, P_1. These two sections can be used exchangeably and in fact it is recommended to count or study particles on both sections of the disector. Also, a stack of sections is frequently studied with a look-up plane for one disector being the plane of observation for the other one.

There is some similarity between the disector method and serial sectioning. In both cases a system of sections is used and the observations are made on subsequent sections. However, the methods differ significantly in the approach to the data in the form of particle sections revealed on the sections. In the disector method no attempt is made to link sections of the same object and position them one with respect to the other. It is sufficient to know whether or not the object studied is cut by both section planes. This means that from a certain point of view the disector is a simplified version of serial sectioning.

The disector method makes it possible to select particles, and other 3-D features, from some specified volume with the same probability disregarding their size. The particles sampled by the disector can subsequently be used to obtain unbiased estimates of the properties of interest. For instance, the volume density of microstructural elements can be obtained from the equation:

$$N_V = \frac{N}{Ah} \tag{3.2}$$

where N is the number of particles sampled by a disector consisting of two sections of area A placed h apart.

The advantages of the disector method are demonstrated by the following example.

Example 3.2

In Example 3.1 the system of particles studied on individual sections yielded a biased estimate of the particle numbers due to the fact that the bigger particles were more frequently cut by the plane of the section. For example, on the section plane P_0 there are four sections of gray particles and four sections of gray ones while on the section P_1 there are four sections of black particles; none of the gray particles is cut. If the sections of the material are made in the form of a system of sectioning planes according to the concept of the disector, the data obtained are given in Table 3.1. In the disector method only those particles are counted which are not cut by the "look-up plane". On the section P_1 there are no such particles. On the section P_2 there are four black particles meeting this criterion and on P_3, four gray particles. In total, in the studied volume there are four white and four black particles. The results given in Table 3.1 show that the disector provides an unbiased estimate of the number of particles in the material.

Table 3.1 Results of Particle/Particle Section Counting for the System of Particles Shown in Figure 3.1

Section	Number of A/B sections	Number A/B in the disector
P_0	4/4	
P_1	0/4	0/0
P_2	0/4	0/4
P_3	4/4	4/0

3.1.1 *Implementation of the disector method*

Implementation of the disector method is relatively simple for transparent specimens. In this case, each object can be unambiguously identified and by changing the focus of the microscope its sections with the upper and lower surfaces of the slab of material may be studied. As a result, the whole procedure is reduced to counting, or selecting for other measurements, the particles entirely inside the slab or cut by only one of the surfaces. The application of the disector method to studies of nontransparent materials is much more difficult as one needs then to rely on surface observations of the particles, or other 3-D objects, for instance, grains in polycrystals.

Implementation of the disector method for nontransparent materials, at the present stage of the development of this method, practically requires a large number of sections separated by a distance smaller than the size of the smallest particle (object) diameter. In the case of surface observations, using light microscopy or scanning electron microscopy of thick specimens, this condition can be met through serial sectioning. We start the disector procedure by taking a micrograph of the particles revealed in the first section plane, P_0, and subsequently remove a thin layer of the material of thickness t_1 such that:

$$t_1 < \min(h_i).$$

Next, a micrograph of the particles seen in plane, P_2, is taken and the number of sections of the particles not revealed on the preceding one, n_1, is counted. This is repeated when further layers of thickness t_i are removed. After removing k layers the volume density is calculated from the following formula:

$$N_V = \frac{\sum n_i}{tA} \tag{3.3}$$

where i = 1,2,. . . ,k and

$$t = \sum_{0}^{k} t_i \tag{3.4}$$

is the total thickness of the layers removed. It is recommended that t_i be at least a third or fourth of the mean particle height:

$$t_i \approx \frac{E(h_i)}{3} \tag{3.5}$$

This suggests that microstructures that contain particles, or other 3-D elements of interest, whose sizes are in the micrometre range can be studied by a disector method if the surface observations are carried out using scanning electron microscopy. On the other hand, systems of submicrometre particles may require the use of transmission electron microscopy.

Example 3.3

Figure 3.2 shows a series images of a two-phase material sectioned by a system of parallel sections. The images on the subsequent sections have been studied from the point of view of detecting the particles cut by one section and not cut by the subsequent one. The number of these particles are given in Table 3.2. Since the sections were 1 μm apart and the area of observations was 1100 μm^2, the data in Table 3.2 yields the following estimate of the particle density: $N_V = 10^6 mm^{-3}$.

Another way of implementing the disector concept in materials science practice is to combine surface observations of thin specimens with the transmitted images of their microstructures. Examples of such a combination include light microscopy combined with X-ray radiography or scanning electron microscopy combined with transmission electron microscopy. In these cases the surface observations are aimed at counting the number of objects emerging on the surface of the specimens subsequently studied with a technique that provides projected images in the transmitted beam.

One of the difficulties of studies with the disector method is that observations usually need to be carried out on relatively small fields of observation. This is related to the requirement for recognizing all particles (objects) cut by one of the plane sections. The small size of the observation fields calls

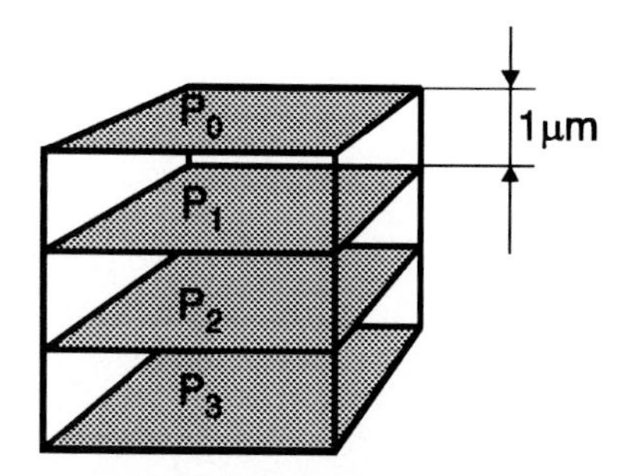

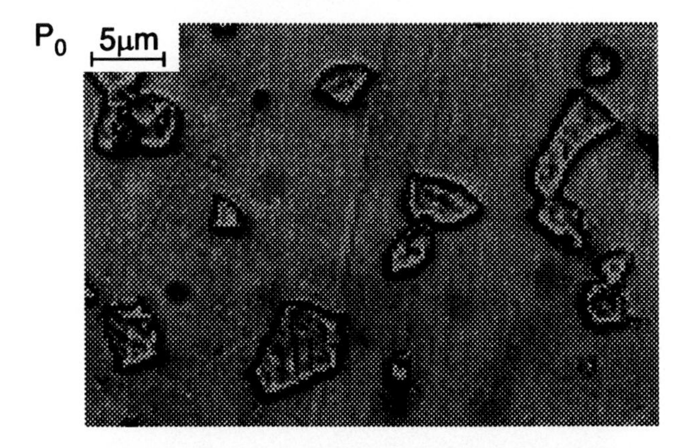

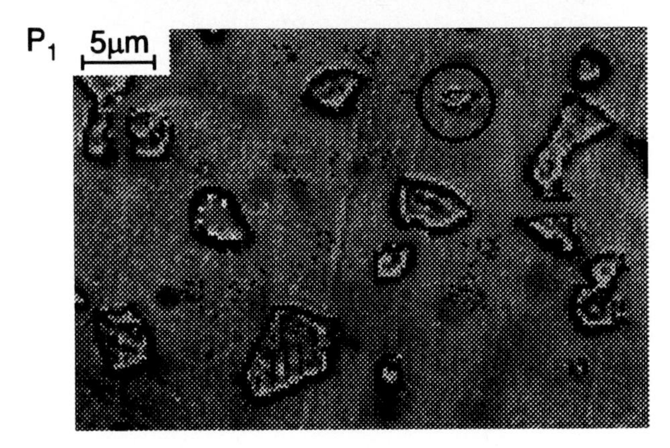

Figure 3.2 A series of micrographs obtained on parallel sections of a two-phase material. (Figure 3.2 is continued on the following page). Circles indicate "new particles", which are cut by the plane of observation, P_{i+1}, but are not cut by the look-up plane, P_i. These particles are the ones sampled with the disecting planes. They are sampled with the same probability, which is not dependent on their size and shape. The microstructures shown yield the following estimate of the particle density:

$$N_V = \frac{N}{A\left(h_1 + h_2 + h_3\right)} = \frac{3}{3000\ \mu m^3} = 10^6\, mm^{-3}$$

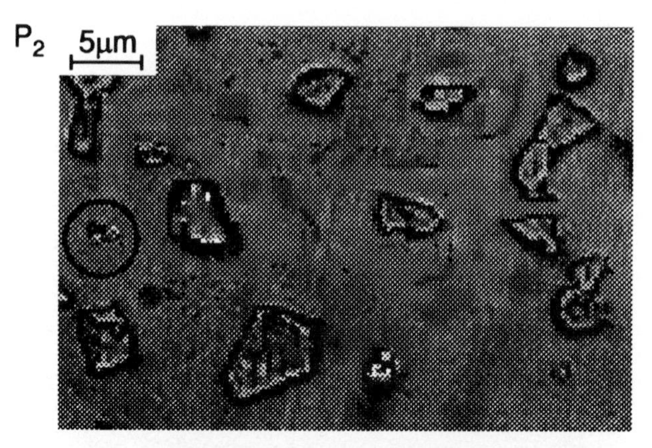

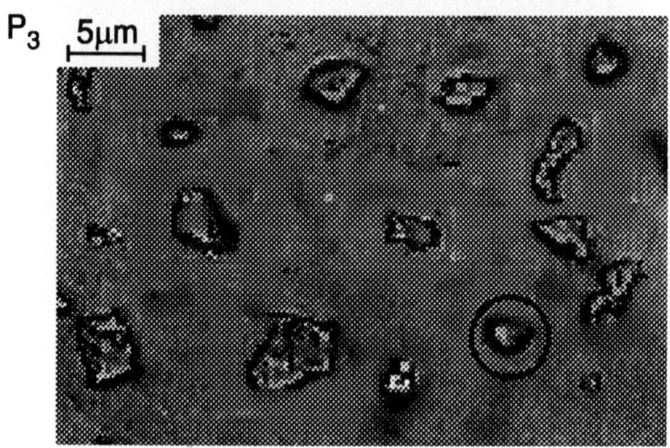

Figure 3.2 (*Continued*)

Table 3.2 Results of Particle/Particle Section Counting for the System of Particles Shown in Figure 3.2

Section	Number of particle sections	Number of particles in the disector		
P_0	17			
P_1	17	1		
P_2	16		1	
P_3	17			1

for special measures to reduce possible bias caused by the intersection of particles by the edges of the observation fields. The observation fields should cover, systematically, some area of the specimen without overlapping and the particles should be counted (selected) in a way that ensures that any particle found in this larger area can be counted (selected) and each of them only once. This problem is solved by one of the following two methods.

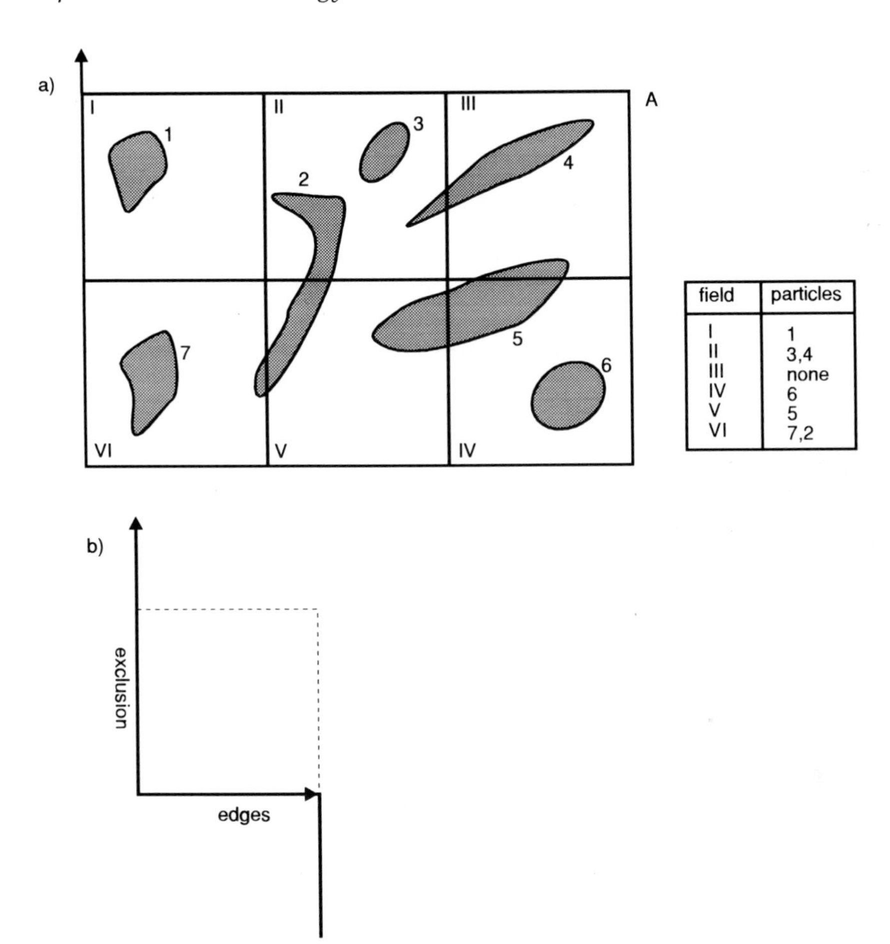

field	particles
I	1
II	3,4
III	none
IV	6
V	5
VI	7,2

Figure 3.3 Schematic explanation of the concept of the exclusion edges and tiling of the observation fields. A bigger section of material, A, was divided into six observation fields. There are seven particles section in A. For each field of observation those sections are counted which are seen in the field and are not cut by exclusion axis shown in (b). Each particle seen in the area of observation is counted as belonging to one and only one field of observation.

The particles can be attributed to being inside the field of observations using the associated point procedure described in Chapter 2. Each particle section has assigned to it one point, or m points in a general case, which are determined on strictly geometrical grounds. This can be, for example, the center of gravity of the particle section or the point of its highest vertical coordinate. Whatever the definition of such a point is, the particle is counted if this point(s) is found inside the frame of the observation field.

The problem of particles cut by the edges of observation fields can also be solved by the tiling method proposed, for example, by Gundersen.[2] Two perpendicular edges of the observation field are taken as exclusion edges, which are further extended as shown in Figure 3.3 (see also Reference 4). The particles are counted which have a point inside the observation field and are not cut by exclusion edges. It should be noted that the definition of the inclusion and exclusion edges makes it possible to cover systematically a larger volume of the material with each of the particles being counted only once. Also, it should be remembered that in the case of nonconvex particles, more than one section can be linked with a particle. In such a case all sections of the same particle are counted as one section.

One of the important parameters in the disector method is the thickness, h, of the material slab. The value of this thickness is usually set in the process of finding a compromise to two conflicting requirements. First, the thickness should be small in order to ensure that all particles inside the disector are recognized. On the other hand, the thickness should be large in terms of the efficiency of the study based on the cost of studying a unit volume of material. In the case of nontransparent materials, the thickness of the disectors can be measured using microhardness indentations. These indentations have a simple geometrical shape and as a result of polishing, which is probably the most convenient way of removing thin layers of material, the indentations decrease in size. With a known geometry of these indentations, changes in their dimensions in the section plane may be converted easily into the thickness of the material removed.

Example 3.4

Figure 3.4a shows an image of an indentation made with a Vickers hardness tester on a surface of a specimen used for disector studies of the density of particles. The specimen was subsequently polished in order to remove a thin layer of the material. Figure 3.4b shows the same indentation after this polishing. The average dimension of the indentation changed from 125 μm to 80 μm. Taking into account the geometry of the indenter, it can be shown that the thickness of the layer removed can be estimated as: $h = 4$ μm (Figure 3.4c).

The disector method, as with other ways of sampling, provides an unbiased description of microstructures only if the examined slices of the material are representative of the microstructure studied. The requirement of the slices to be representative effectively means the slices need to be randomly positioned and oriented. It should be understood that the expression "randomly placed and oriented" is used here in the sense of mathematical statistics and should be executed in a quite formal way. So-called "random choice" carried out by a person examining microstructures is seldom random and in no way satisfactory.

The following important situations can be distinguished:

1. observations are carried out on isotropic uniform random (IUR) sections;
2. observations are made on arbitrarily oriented planes, typically on a stack of parallel disectors;

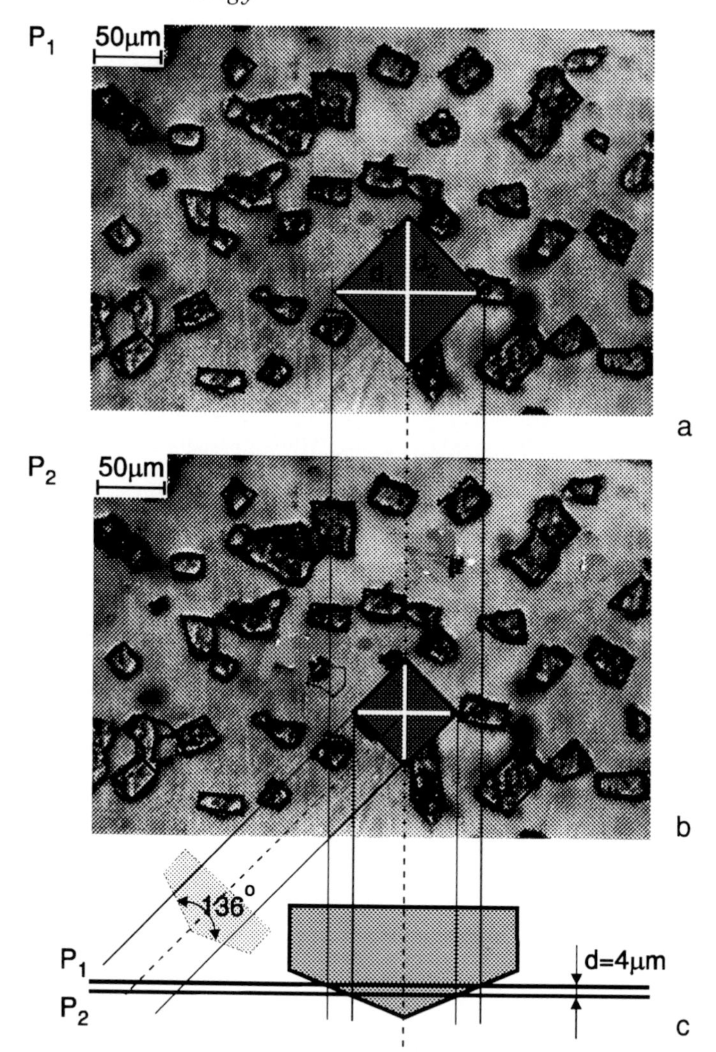

Figure 3.4 Illustration of the method used to estimate thickness of the layer of the material removed during polishing of the sample surface. The diameter of the indentation changes between the surface P_1 and P_2 from 125 μm to 80 μm. A simple geometrical consideration shows that this reduction means that the thickness of the material removed, i.e., the material between these two sections, is 4 μm.

3. observations are conducted on vertical planes that are parallel to an arbitrary direction.

The first case, IUR sections, is a general solution of the problem and such sections can be used in studies of any microstructure. However, it is not a simple task to obtain a system of randomly oriented sections due to the difficulties in sectioning materials used in modern applications. Some simplifications of this method are discussed later.

The second case is rationalized for isotropic microstructures. There are few examples of truly isotropic microstructures among the materials used in technology; however, some materials show relatively low anisotropy and can be approached by this method.

The third situation is again a general solution to the problem. It is discussed in more detail in the subsection on vertical sectioning.

If the number of disectors randomly placed and oriented is equal to m, the mean density of particles may by obtained by averaging over m:

$$N_V = \frac{1}{m} \sum (N_V)_i \tag{3.6}$$

Example 3.5

The system of particles shown in Figure 3.2 was recorded during observation of a two-phase material. A total number of 15 disectors were studied. The results of the measurements are shown in Figure 3.5 in the form of a histogram. The following mean value for these 15 disectors was computed: $N_V = 10.1 \times 10^{-4}\ \mu m^{-3}$. This can be used as an estimate for the material studied.

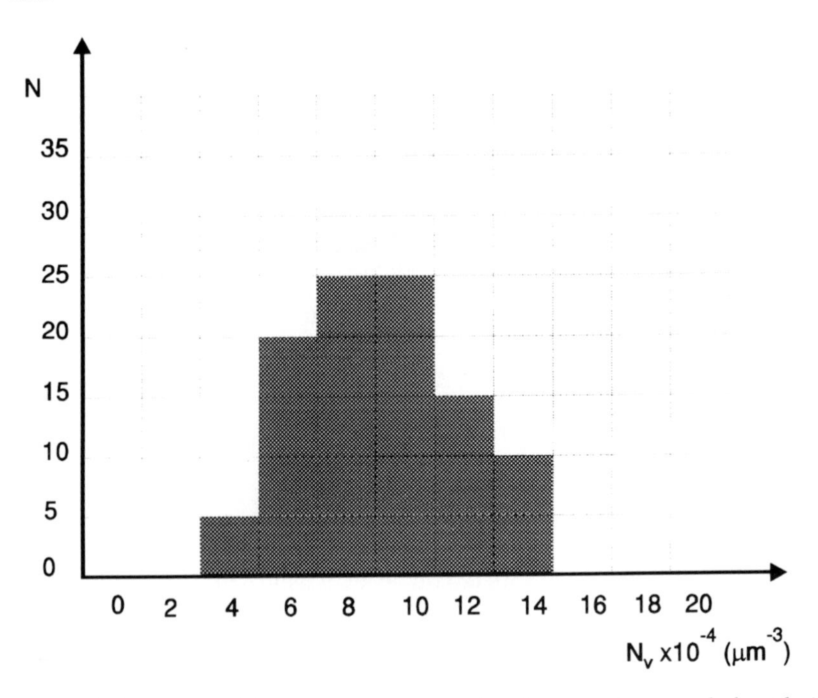

Figure 3.5 Histogram of the particles volume for the particles sampled with 15 disectors in a two-phase material.

3.1.2 *The disector as a method of sampling*

The disector method can be used directly to estimate the number density of particles or other objects. As has been mentioned earlier, the disector also provides a method for unbiased sampling of the particles. It can be used, for instance, to estimate the particle volume distribution function, $f_N(V)$, particle surface distribution, $f_N(S)$, and other characteristics in terms of number distributions. To this end, it is necessary to estimate the volume, V_i, surface, S_i, or other property of the particles sampled with the disector. There are a number of ways of making this estimate.

Example 3.6

The volume of particles sampled by the disector can be estimated from particle section areas, A_i, using the formula:

$$V_i \approx \sum A_i t_i \tag{3.7}$$

(The summation is carried out over all sections of a given particle and t_i is the distance between the sections.) Figure 3.6 shows the images of the particles cut by two parallel sections of a two-phase material. The 17 particles sampled by the disector generated by these two sections have been studied by the serial sectioning method to obtain estimates of the particle volumes. The following mean volume of the particles was computed: $E_N(V) = 47\ \mu m^3$. The histogram of the experimental distribution of $f_N(V)$ function is given in Figure 3.6d.

The above example shows one of the most important advantages of the disector method, which is the possibility to sample particles, or other 3-D objects, in an unbiased way disregarding their shape and size. As a result the disector method is a tool for selecting particles representative of the material studied. If particles selected in this way are further characterized by this or another method the result of these studies can be used rationally to describe the properties of the material. This assertion is the basis for disector-related methods discussed later.

3.1.3 *An example of an application in materials science*

In recent years the advantages of the disector method have been proved by the growing number of successful applications, mainly in the field of biology. Examples of disector method application in the studies of materials are, as of yet, scarce. An interesting example of such an application is the study of grains in polycrystals described by Liu and co-workers.[5] In this case the procedure is based on the definition of three basic patterns in the geometry of grain sections, as schematically shown in Figure 3.7. The appearance of three-sided grain sections marks grains which meet the condition of sampling by the disector: they are hit by a plane of observation n and not by the plane (n – 1). Further, the implementation described by Liu and co-workers makes

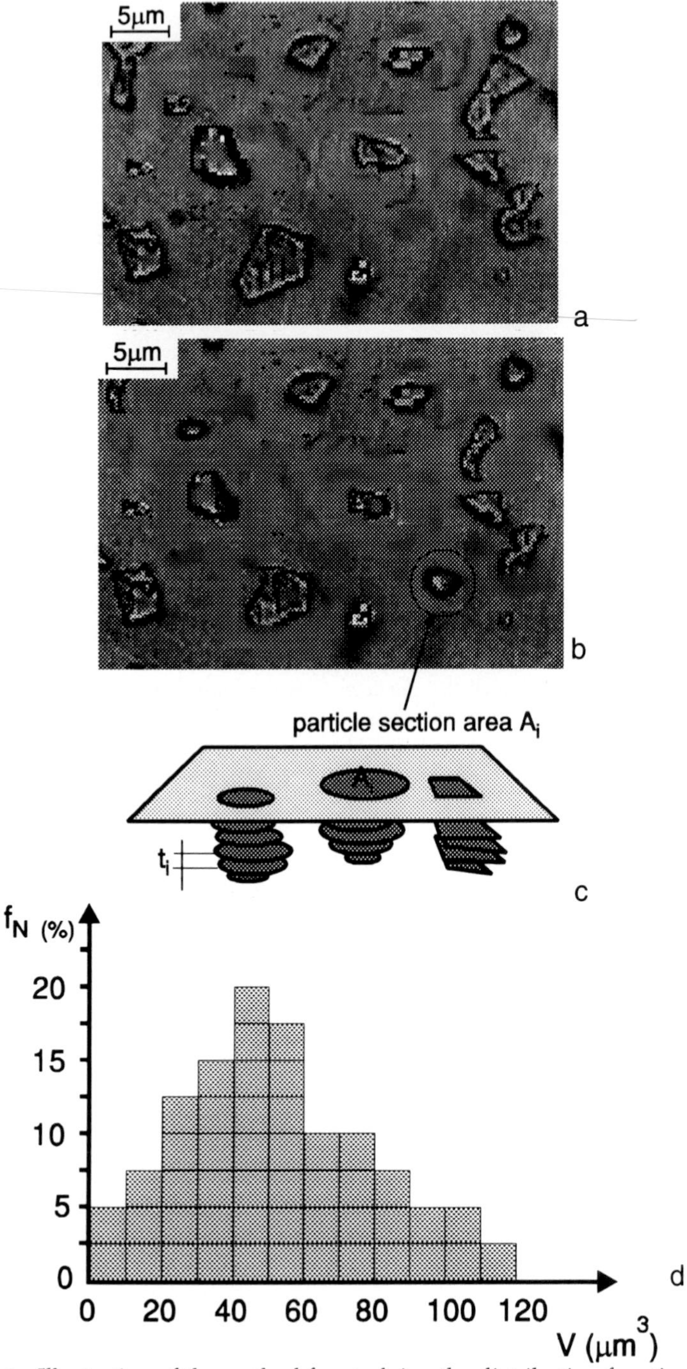

Figure 3.6 Illustration of the method for studying the distribution function of particle volume: (a) and (b) disector sampling of particles; (c) schematic representation of the volume estimation of the sampled particles via serial sectioning; (d) the experimental volume distribution function obtained.

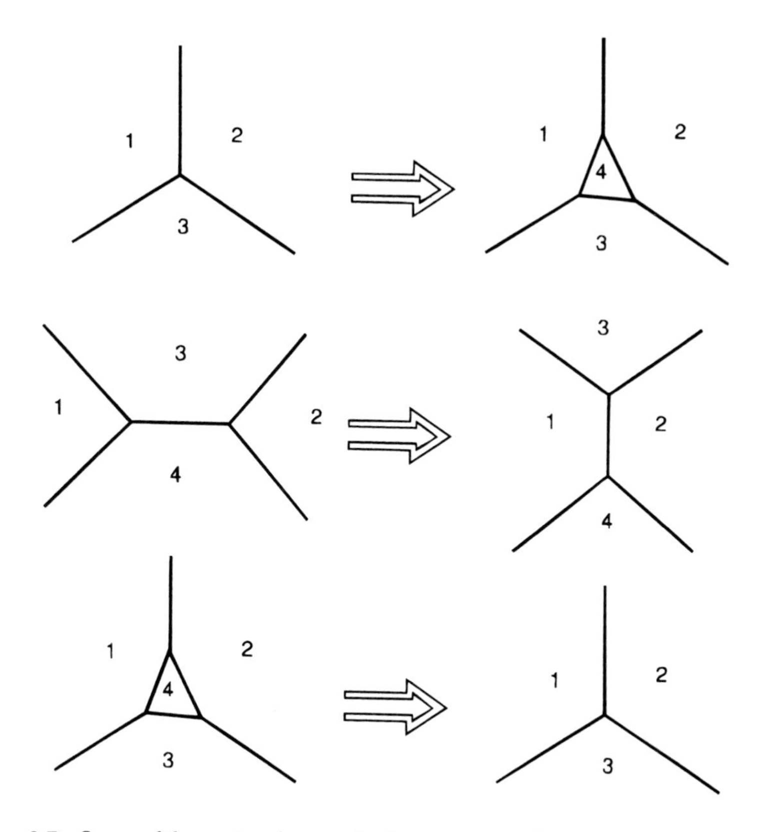

Figure 3.7 Some of the major changes in the geometry of grain boundaries observed during serial sectioning of polycrystals.

it possible to sample grain corners. In this study Liu and co-workers[5] applied the disector method to studies of grain geometry in a steel as a function of annealing conditions. In these studies of grain volume they have observed that the coefficient of variation of the volume, CV(V), is typically in the range from 1.09 to 2.13, which is much larger than the value usually observed in biological materials. They also found that the average number of grain faces, edges, and corners per grain were 14, 36, and 24, respectively, regardless of the variation in the distribution function of grain volume.

3.2 *Selector—particle volume distribution*

The concept of the selector method is a combination of the disector method with point sampled intercepts. It can be viewed as a three-step procedure that consists of:

a) sampling particles, V_i, with a disector from the specimen volume;
b) sampling points, Q_i, in the sections, A_i, of the particles in the observation planes;

c) sampling intercepts, l_i, through the Q_i points to the particle section contours, G_i.

With regard to point a) it is essential in the selector method that the sections are either randomly oriented or vertical (when the microstructure is anisotropic). Parallel sections can be used in the case of isotropic microstructures.

In the selector method the thickness of the disector, and the thickness of the layers removed do not need to be measured. This is due to the fact that the intercepts are used to estimate the volume of individual particles, V_i. To this end, for each particle sampled one calculates the mean value of the intercepts, l_i, raised to the third power:

$$V_i \approx \frac{\pi}{3}\frac{1}{m}\left(l_1^3 + l_2^3 + \cdots + l_m^3\right) \tag{3.8}$$

where m is the total number of intersections with one particle.

In this summation all intercepts for a given particle are taken into account, including the possible double and triple intercepts generated by Q_i points if the particle sections are not concave.

Applying the above equation to all particles sampled with a disector, or disectors, it is possible to estimate the distribution function of particle volume, f(V), or compute an estimate of the mean volume, $E_N(V)$, which is given by:

$$E_N(V) = \frac{V_1 + V_2 + \cdots + V_k}{k} \tag{3.9}$$

and higher-order moments such as the standard deviation, SD(V), which, according to basic relationships, can be computed from the following:

$$SD_N(V) = \sqrt{E_N\left(V^2\right) - \left[E_N(V)^2\right]} \tag{3.10}$$

The selector methods can be used to estimate the volume of individual particles as well as their mean number volume. The mean number volume, can be used in turn to obtain an estimate of the particle density, if the volume fraction V_V is known. The following simple relationship can be used:

$$N_V = \frac{V_V}{E_N(V)} \tag{3.11}$$

The required volume fraction, V_V, may be obtained by any of the relevant methods described in Chapter 2.

Example 3.7

Figure 3.8 shows a set of particles, in a two-phase material, sampled with the disector method. The selector method was applied to make an estimate of the

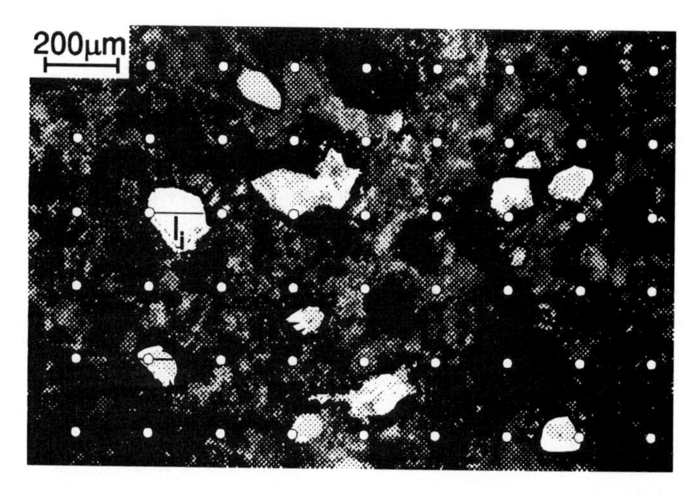

Figure 3.8 Application of the selector and point sampled intercept method to the study of a two-phase microstructure of a concrete. A grid of points is superimposed on the image of the microstructure. Intercepts are drawn through the points that hit the particles of interest. The length of the intercept, l_i, is measured. The mean weighted volume of the particles is estimated using the mean value of l_i^3.

mean volume of these particles. The particles were sampled further with a grid of points, and intercepts were measured through the points. The measurements of the individual intercept lengths yielded the following estimate of the mean volume:

$E_N(V) = 1.33\ mm^3$.

Measurements of the particle section areas yielded the following estimate of the particle volume fraction:

$V_V = 0.08$.

These two values, V_V and E_N, were used to compute an estimate of N_V:

$N_V = 6.01 \times 10^{-2}\ mm^{-3}$.

3.3 *Point sampled intercepts*

An important element of the selector procedure is the point sampling of intercepts. In this procedure, a set of randomly positioned points, P_i, is placed on the image of a section of the microstructure studied. A certain fraction of these points, points indicated as $(P_i)_h$, hit sections of the particles of interest. Through each of these points randomly oriented intercepts are drawn. These intercepts are called point sampled intercepts. It has been shown (see, for example, Reference 6) that the length of the point sampled intercepts, l_i, can be used to estimate the volume weighted mean volume, $E_V(V)$, of particles. For convex particles there is one intercept for each sampled point and the mean volume of particles is estimated using the formula:

$$E_V(V) \approx = \frac{\pi}{3} \frac{l_1^3 + l_2^3 \cdots + l_k^3}{k} \tag{3.12}$$

where k is the number of sampled points.

For nonconvex particles a line can generate a number of intercepts, say m, and then the last equation has to be extended to the following:

$$E_V(V) = E\left[\frac{2\pi}{3}\left[\left(l_o^+\right)^3 + \left(l_o^-\right)^3 + \sum_{i=1}^{m-1}\left|\left(l_i^+\right)^3 - \left(l_i^-\right)^3\right|\right]\right] \tag{3.13}$$

where $l_o^{\pm}$ indicates the distances to the end points $l_i^{\pm}$ of intercepts that do not contain the sampled point.

Example 3.8

Figure 3.8 shows a microstructure typical of a resin concrete. Sections of the filler particles were observed using polarized light illumination. Some of the sections are marked in Figure 3.8.

The mean volume-weighted volume of the particles were studied using the point sampled intercept method, which gives the following estimate:

$E_V(V) = 1.4\ mm^3$.

One of the advantages of the procedure based on the point sampled intercept method is its relative simplicity, which can be implemented by computer-aided image analysis. The measurements are reduced to being made on single sections without the need to make inspections on the look-up planes and the planes of observation—the elements typical of using the disector method. The price paid for this simplicity is that the procedure yields volume-weighted estimates of volume. As discussed before, probing by sectioning results in the bigger elements being chosen more frequently and combined probing by sections and points results in the particles being sampled in proportion to their volume.

3.4 *Second order statistics—methods for estimating the variance of particle volume*

The mean volume of particles, by definition, does not uniquely describe the important property of most particle populations in materials, which is variability in their size. A simple parameter that describes the degree of this variability is the variance of particle volume.

The variance of a random variable, VAR, is an important parameter that can be used to characterize distribution functions. By definition the variance of the particle volume, VAR(V), is given by:

$$VAR(V) = \frac{\left((V_1 - E(V))^2 + (V_2 - E(V))^2 + \cdots + (V_N - E(V))^2\right)}{N} \tag{3.14}$$

The methods that can be used to estimate the mean volume of particles, $E_N(V)$, have been already discussed in the preceding sections. These are the disector and selector methods. On the other hand, point sampled intercepts provide an estimate of the weighted mean volume, $E_V(V)$. Thus, depending on the choice of the experimental method, two variances of particle volume can be obtained: the number variance, $VAR_N(V)$, and the volume-weighted variance, $VAR_V(V)$.

The spread in the size of particles can be described alternatively by the standard deviation, SD(V), and coefficient of variation, CV(V) defined as:

$$SD(V) = \sqrt{VAR(V)} \tag{3.15}$$

$$CV(V) = \frac{SD(V)}{E(V)} \tag{3.16}$$

Following the distinction in the number and volume-weighted means, a population of particles can be described in terms of the number standard deviation, $SD_N(V)$, the volume-weighted standard deviation, $SD_V(V)$ and the number and volume-weighted coefficients of variation, $CV_N(V)$ and $CV_V(V)$, respectively.

Example 3.9

Systems of particles in some materials are described by the distribution functions shown in Figure 3.9. Table 3.3 lists the number and weighted mean values, and coefficients of variations for these three types of populations.

It can be shown that parameters obtained for number distribution functions, such as $E_N(V)$, remain in some relationship with parameters computed for volume-weighted distributions, such as $E_V(V)$. These two mean values remain in the following relationship:

$$CV_N(V) = \sqrt{\frac{E_V(V)}{E_N(V)} - 1} \tag{3.17}$$

The variance of particle volume can be computed directly from the data on the distribution of the particle volumes, which, in turn, can be obtained from the disector or selector methods. However, the variance of a random variable may also be computed from the following general relationship:

$$VAR(X) = E(X^2) - (E(X))^2 \tag{3.18}$$

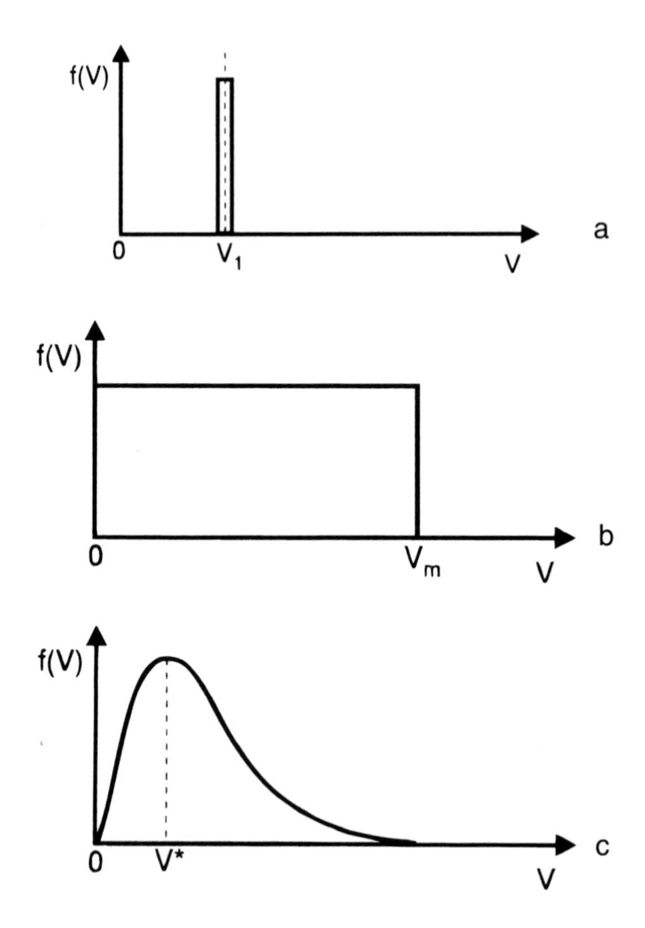

Figure 3.9 Characteristic particle volume distribution functions: (a) monosized particles; (b) uniform volume distribution function; (c) log-normal distribution of particle volume. Values of the basic parameters describing the populations (a)–(c) are given in Table 3.3.

This relationship shows an alternative way to estimate the volume variance: via studies of the mean, E(V), and the mean squared volume, $E(V^2)$.

The volume-weighted variance of the particle volume can be estimated[7] using the method of random triangles. In this method the sections of the particles in a given field of observation have to be sampled with a system of points as shown in Figure 3.10. For each particle section hit by a point from the grid, two more points are randomly selected inside the section. Both the particle section area, A, and the thus formed triangular area, Δ, are measured. The measurements can be carried out with an image analyzer, where special software can be used for random selection of the points inside the particle sections.

Table 3.3 Basic Parameters Defining the Populations of the Particles Described by the Distribution Functions Shown in Figure 3.9

Parameter of the population	Distribution function of particle volume		
	Monosized a)	Uniform b)	Log-normal c)
$E_N(V)$	V_0	$\frac{V_m}{2}$	$e^{V^* + \frac{SD^2}{2}}$
$E_V(V)$	V_0	$\frac{2}{3}V_m$	$e^{V^* + \frac{5SD^2}{6}}$
$CV_N(V)$	0	$\frac{\sqrt{3}}{3} = 0.58$	$\left(\exp\left[SD^2\right] - 1\right)^{\frac{1}{2}}$
$CV_V(V)$	0	$\frac{\sqrt{3}}{4} = 0.43$	$CV_N(V)$

Note: V_m, V^* are explained in Figure 3.9. CV is an additional parameter used to define the spread of the particle volumes.

It has been shown in the literature (for example, Reference 3) that the mean value of V^2 can be estimated from the following formula:

$$E_V\left(V^2\right) = 4\pi \frac{A_1^2\Delta_1 + \ldots . A_N^2\Delta_N}{N} \tag{3.19}$$

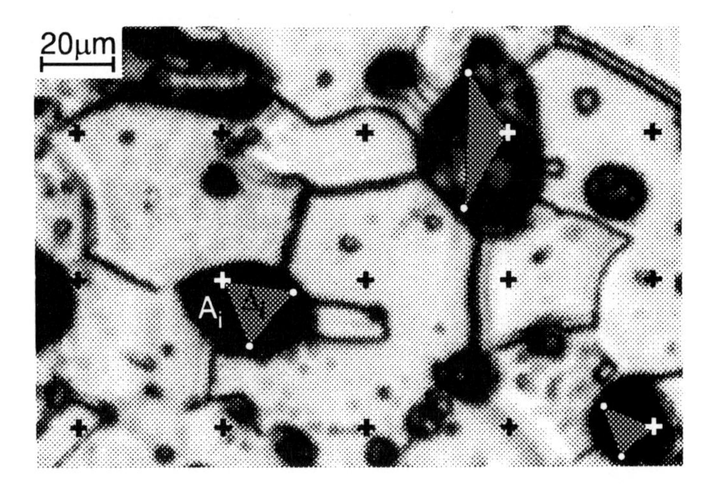

Figure 3.10 Illustration of the method used to estimate the weighted variance of the particle volume. A grid of points is imposed on the image of the microstructure. The area of the sections of the particles hit by the points, A_i, are measured. In each section hit, two points are randomly selected and the area of the thus formed triangle, Δ_i, is computed. The values of A_i and Δ_i are used to estimate the variance of the particle volume.

The estimates of $E_V(V)$ and $E_V(V^2)$ can be further used to obtain an estimate of the volume-weighted volume variance, $VAR_V(V)$, and volume-weighted coefficient of variation $CV_V(V)$.

Example 3.10

The system of particles shown in Figure 3.10 was studied using the method of triangles and point sampled intercepts. The measurements yielded: $E_V(V) = 56.0$ mm^3 and $E_V(V^2) = 3139.1$ mm^6.

The following values of $VAR_V(V)$ and $CV_V(V)$ were obtained:

$$VAR_V(V) = E_V\left(V^2\right) - \left(E_V(V)\right)^2 = 3.1\ mm^6 \tag{3.20}$$

$$CV_V(V) = \frac{\sqrt{VAR_V(V)}}{E_V(V)} \approx 0.3 \tag{3.21}$$

By measuring the mean volume of the particles, $E_N(V)$ with the selector method, it was found that:

$E_N(V) = 51$ mm^2

In summary, it is worth noticing that the values of $E_V(V)$ and $E_V(V^2)$ can be used to estimate the value of $CV_V(V)$. The weighted mean volume can be obtained using the point sampled intercept method already described in this chapter. An estimation of the weighted mean squared volume, which is required for computation of $CV_V(V)$, can be achieved by random sampling with points according to the following steps:

a) point sampling with a system of points, Q_i, superimposed on the sections of the particles;
b) further point sampling with two more points randomly selected inside the intersections selected in a.

For the particle intersections revealed on a cross-section of the specimen this procedure can be used to estimate the volume-weighted variance, $VAR_V(V)$, using the following development:

$$E_V\left(V^2\right) \approx 4\pi E\left(A_o^2 \Delta\right) \tag{3.22}$$

where A_o is the area of the section sampled in the first step a) and Δ is the area of the triangle formed by the two additional random points selected inside the section.

It has been shown recently[8] that the procedure for the estimation of the weighted mean squared volume can be significantly simplified. The following relationship has been suggested:

$$E_v\left(V^2\right) = 4\pi k E\left(A^3\right) \tag{3.23}$$

where:
A is the area of the grains hit by the Q_i points and
k is a constant with a value in the range from 0.071 to 0.083.

The value of k in this relationship depends on the shape of particles; however, it varies over a relatively narrow range. Also, its value can be determined either experimentally or by modelling. It should be noted that the relationship discussed above is based on the measurement of particle section areas, which may be easily effected with an image analysis system.

Using this formula one can derive the following relationships:

$$VAR_V\left(V\right) = 4\pi k E\left(A^3\right) - \frac{\pi^2}{9}\left[E\left(l^3\right)\right]^2 \tag{3.24}$$

and

$$CV_V^2\left(V\right) = \frac{36kE\left(A^3\right)}{\pi\left[E\left(l^3\right)\right]^2} - 1 \tag{3.25}$$

Further simplification has been shown to be possible for the case of particles whose volume is log-normally distributed, in which case the volume-weighted and number-weighted coefficients of variation are equal:

$$if\ V\ \log - normal \rightarrow CV_N\left(V\right) = CV_V\left(V\right) \tag{3.26}$$

Implementation of the procedure based on the use of the above formulae combines standard metallographic measurements such as measurements of intercept length and the section area. All these measurements can be carried out rather easily with the help of an automatic image analyzer.

Example 3.11

The system of particles shown in Figure 3.11a was tested by the point sampled intercept method, l_i, and the areas of the point sampled particles, A_i, were measured. The distribution functions of the intercepts, f(l), and the area, f(A), are given in Figure 3.11b and 3.11c. The following values were computed from these data: $E(A^3) = 123.5 \times 10^6\ \mu m^6$ and $E(l^3) = 5832\ \mu m^3$. Consequently, $CV_V(V)$ can be estimated to be near 1.526. If the distribution function $f_N(V)$ is close to log-normal, this is also a reasonable estimate of $CV_N(V)$.

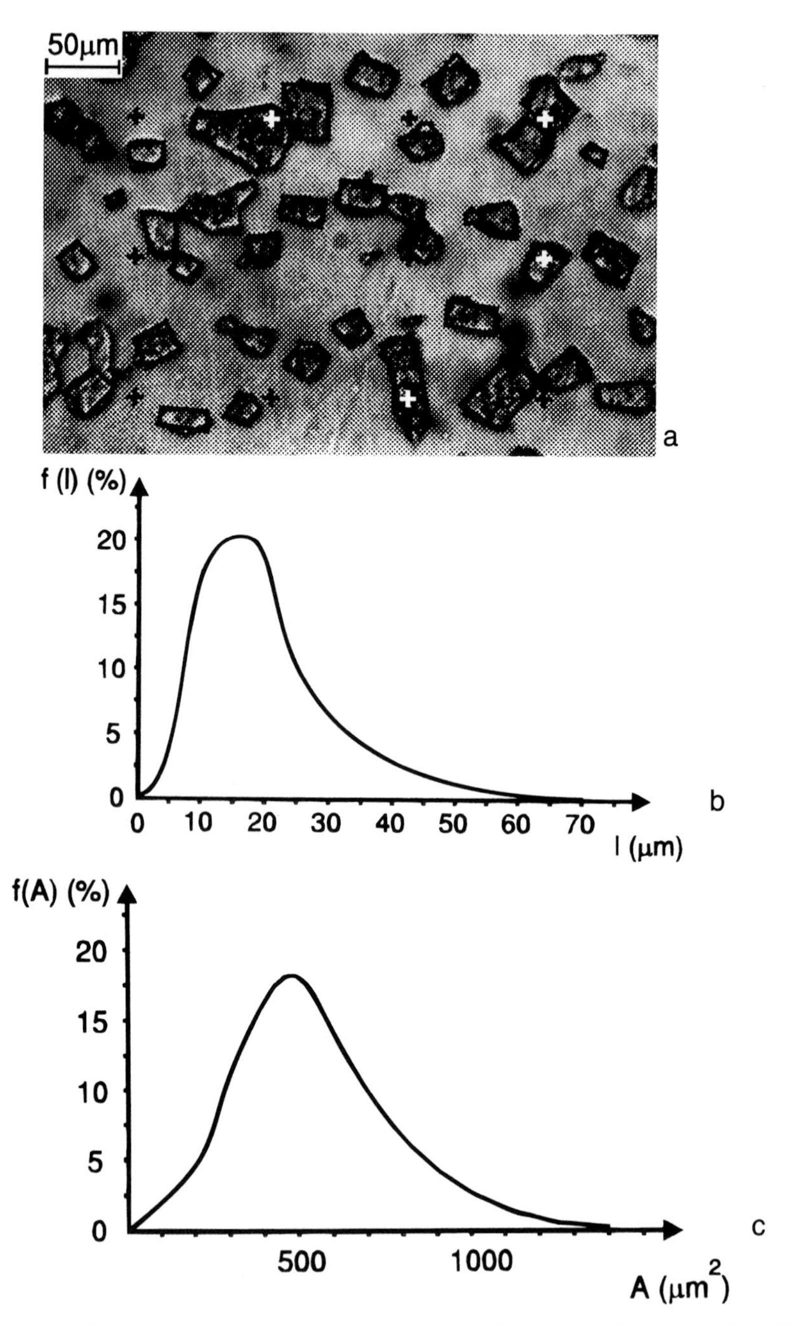

Figure 3.11 Illustration of the method for estimating the weighted mean and coefficient of variation of the particle volume. A grid of points is superimposed on the image of the microstructure. The area of the particle sections hit by the points is measured as well as the length of the intercepts through the points of the grid that are inside the particles. The experimental distribution functions f(l) and f(A) are used to determine the values of $E_V(V)$ and $CV_V(V)$.

3.5 Fractionator

The concept of the fractionator was introduced by Gundersen and described, for example, in Reference 9. It can be used in studies of large specimens that need to be characterized in terms of their particle content or the density of other 3-D elements. The fractionator method is a combination of a disector and a sampling procedure based on a systematic division of the specimen, or an artifact, into K elements and sampling every m-th element of it. It is assumed here that:

a) the specimen or the artifact studied is cut into K elements, which may differ in volume and are of unspecified thickness;
b) despite the subdivision into K samples, the elements of the microstructure studied may be unambiguously identified from their sections.

In practice, the division into K elements might be carried out by slicing the original specimen/artifact into pieces of a thickness that may vary and does not need to be known.

If the number of elements, K, into which the specimen has been divided is large, and this is true for most cases of practical significance, m elements are selected for further observation and measurement. The number m is defined as a ratio of K and it determines the probability, P(K), for any of the K elements to be sampled according to the obvious equation:

$$P(K) = \frac{m}{K} \tag{3.27}$$

One of the important features of this method is the selection of the m elements, which must be of random character. This means there must be an equal probability for any of the K elements to be selected for further measurement. To this end the following practical procedure is recommended:[2]

Step 1: all the K elements are arranged in a sequence of arbitrary order;
Step 2: the elements are numbered from 1 to K;
Step 3: the number of samples for further study, m, is decided on the basis of an estimate of the time and cost of examining a single element resulting from the division of the specimen;
Step 4: the sampling fraction 1/P is computed, with P being the nearest integer to the fraction m/K;
Step 5: a random integer R in the range from 1 to P is selected (this can be done with the help of appropriate tables or computer software);
Step 6: elements numbered R, R + P, R + 2P, R + 3P, ... are sampled for further analysis.

If the aim of the study is to determine the total number of particles in the artifact, the next step is to count the particles in the m elements selected for examination. This is carried out using the disector method (i.e., via examination of the look-up and the observation planes). If the number of such particles is equal to N_m then the total number of particles, N_K, can be estimated using the following relationship:

$$N_K \approx PN_m \tag{3.28}$$

If at the same time the volume of the artifact, V_K, can be measured (which should be relatively easy for larger specimens) the density of the particles can be estimated from the formula:

$$N_V \approx P\frac{N_m}{V_K} \tag{3.29}$$

This procedure may be extended to other 3- and 2-D microstructural features, as well. It can be used for example, to estimate the surface-to-volume ratio of cracks of some critical size, a_c, in a thick piece of material by sectioning the sample into thinner slices of thickness smaller than a_c.

In some cases, it might be found that the m elements sampled with a probability $P_1 = m/K$ prove to be too large to be examined thoroughly with the available instrumentation. In this situation, the whole procedure can be repeated. This means that the m sampled elements are exhaustively cut into smaller pieces and these are sampled with a probability P_2, assumed to the second stage of the investigation. In such an instance the total number of particles, or other features can be estimated from the following equation:

$$N_{total} \approx P_1 P_2 N_{counted} \tag{3.30}$$

This equation can be modified in an obvious way to account for a third and further stages of the division of the artifact into still smaller elements, as might be necessary in particular studies.

The concept of the fractionator has been successfully used in studies of biological specimens. It has attracted less attention among specialists in the field of materials science. This is related to the fact that technical materials are usually characterized by a considerably higher uniformity of their microstructures. In the case of relatively uniform materials, the fractionator is equivalent to random sampling of specimens and averaging over the values measured for the sampled specimens.

3.6 Nucleator

A description of the concept of the nucleator can be found in Reference 9. This is a procedure designed for the measurement of particle volume. It assumes that particles are sampled by disectors and then a point, Q_i, is randomly chosen inside each particle section, A_i, on the observation plane. In particular, these points can be selected from some recognizable subpart of the particles (which are equivalent to nuclei inside biological cells—which is how the name of the procedure can be justified). From the point Q_i a line randomly oriented in space is drawn and the distance from the point to two or more intersections with the particle boundary, l_i, measured. All sections of the disector should be hit at least once.

For each of the Q_i points the mean cubed ray distance is calculated in the following way:

$$l_i^3 = \frac{1}{2}\left(l_+^3 + l_-^3\right) \tag{3.31}$$

where subscripts "+" and "–" are used to differentiate the two intercepts from the point to the boundary. If in any of these directions there are more than one intersection point, the equivalent cubed distance is calculated according to the formula:

$$l^3 = l_1^3 - l_2^3 + l_3^3 - \ldots + \ldots \tag{3.32}$$

where the plus sign stands before the distances to the intersection points where the line passes outwards and the minus sign where it passes inwards.

Gundersen[9] has shown that an unbiased estimator of the volume of a given particle can be obtained in the following way:

$$V_i \approx \frac{4\pi}{3} l_i^3 \tag{3.33}$$

If applied to all the sections sampled with a disector, this formula gives the mean number volume of the particles:

$$E_N(V) = \frac{1}{k}(V_1 + V_2 + \ldots + V_k) \tag{3.34}$$

In applications of the nucleator method it is important to remember that unless the particles are isotropic, the disectors should be randomly oriented and the same applies to the lines passing through the points inside the particles. If applied to the sections revealed on a cross-section (no disector sampling), the nucleator yields the volume weighted mean:

$$E_V(V) = \frac{4\pi}{3k}\left(l_1^3 + l_2^3 + \ldots + l_k^3\right) \tag{3.35}$$

This estimator is very similar to the one based on the intercept through the points. However, it is more efficient from the point of view of the precision of the estimate.

3.7 *Vertical sectioning—area of the surfaces*

The method of vertical sectioning has been devised as an effective solution for the characterization of anisotropic systems, in particular for the measurement of the surface area of 2-D elements such as grain and interphase boundaries. As has been said before, in isotropic systems the surface of selected elements can be determined based on measurements of intercepts with uniformly distributed lines that are drawn on sections of the microstructure. In the final step of this procedure the basic stereological relationship used is:

$$S_V = 2P_L \tag{3.36}$$

where S_V is the surface area in unit volume and P_L is the density of the intersection points of the test lines with the surface studied (see Chapter 2).

This approach fails, however, in the case of anisotropic systems. For such systems the parallel testing lines placed on a system of parallel sections provide biased information on the boundary density. For anisotropic systems of surfaces the testing lines need to be truly randomly oriented in space.

One possible solution for anisotropic systems has been discussed in Chapter 2. The approach described there requires prior knowledge of the type of anisotropy. Based on this, appropriate sections of the microstructure are made and these sections are tested with specifically ordered systems of lines. Subsequently, a modified version of the basic relationship for S_V is used (see Chapter 2).

A more general approach to anisotropic systems is based on the concept of vertical sectioning of the specimen. In the vertical sectioning method,[10] an arbitrary axis, **v**, is chosen and a specimen is examined on a series of sections, C_i, parallel to **v**. These vertical sections can be further distinguished by angles, ϕ_i, which their normals, $\mathbf{n}_i$, make with some reference direction, **r**, perpendicular to **v**.

The concept of vertical sectioning assumes an unrestricted freedom in the choice of the vertical axis, **v**. However, in certain cases a proper choice of this axis simplifies the implementation of this method. It is generally recommended that the axis be aligned with some characteristic direction in the microstructure. The following examples are given to illustrate this point:

1. fracture surface:
   the vertical axis is taken perpendicular to the macroscopic surface of the fracture;
2. cylindrical specimens:
   the vertical axis is taken parallel to the axis of the specimens as it is likely to be the axis of the texture;
3. flat specimens:
   the vertical axis is taken normal to the specimen and possibly the reference direction, **r**, taken parallel to the specimens width.

These situations are illustrated in Figure 3.12.

It can be seen intuitively that, in principle, densely cut vertical sections of a specimen or an artifact provide a full description of the microstructure studied. As the sections rotate around the vertical axis, all microstructural features are revealed systematically and can be included in the measurements. However, the problem arises that the elements of the microstructure are revealed in consecutive sections at a level that depends not only on their

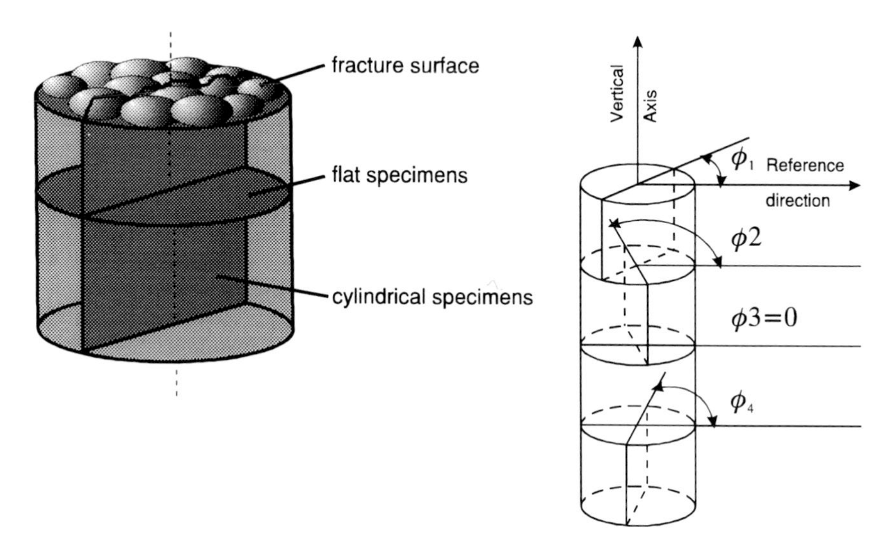

Figure 3.12 Example of the implementation of the principle of vertical sectioning.

volume and surface area but also on their orientation with respect to the vertical axis.

Consider a system of surface elements, S_i, enclosed in a piece of the material of volume, V. The total surface area of these elements can be estimated from the basic relationship if the testing lines, L_i, are piercing the surfaces with equal probability in any possible direction in space. However, this simple concept is difficult to implement as it requires a methodology for dealing with a spatial distribution of test lines, while in practice the lines are drawn on surfaces of the sections through the volume studied. The method of vertical sectioning provides a solution to this problem. This solution is based on the following observations:

1. For each line of the system of randomly oriented lines, L_i, a vertical section can be found that contains the axis **v** and this line.
2. For any vertical section the number of lines from the set L_i making an angle α with the vertical axis is proportional to sin α.

The second comment can be explained using the concept of a stereographic projection. The direction of a line in space passing through the origin is represented by a point on the projection sphere. For a given direction of the vertical axis, the number of directions at an angle α is proportional to sin α as illustrated in Figure 3.13. As a result, 3-D directions appear with different probabilities on the vertical sections. In fact, the length of a test line at an angle α on such a section should be proportional to sin α. This is the property shown by curved lines, which are termed cycloids in mathematics.

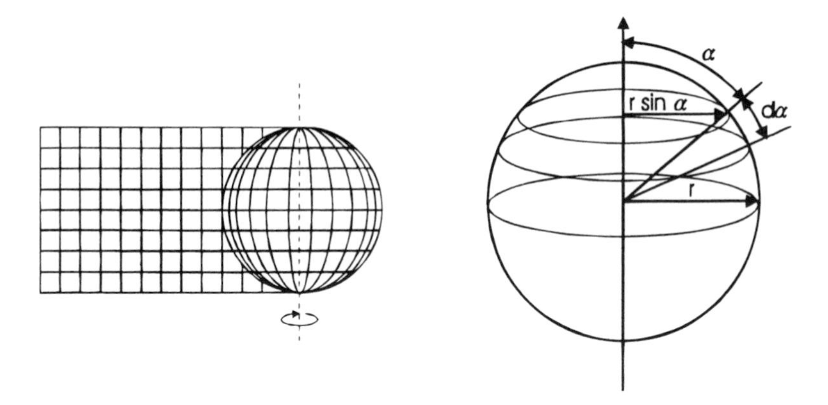

Figure 3.13 Explanation of the relationship between the selecting of random direction in space and on plane sections.

Example 3.12

Figure 3.14 shows a microstructure on which a test set of cycloids and points has been superimposed. The test system is characterized by two parameters: the total length of the cycloids, L, and the number of points, N_P. The number of intersection points of the cycloids with interphase boundaries was measured, as well as the number of the points hitting the particles. For the microstructure shown in Figure 3.14, $S_V \approx 10^{-2}\ \mu m^{-1}$ if the reference space is the volume of material and $S_V \approx 0.9\ \mu m^{-1}$ if the surface area is referred to the volume of particles.

The cycloids are used to determine the number of intersections, I_i, of a given 2-D feature with lines randomly oriented in space. The points, on the other hand, are there to calculate the number of points, N_i, hitting the sections of the reference volume. As a reference volume one may consider the volume of the specimen studied and then all the points within the field of observation are counted. One can also assume the reference volume to be the volume of the particles and accordingly count only the points hitting the sections of the particles. In any case, an estimate of the surface area to volume ratio is given by:

$$S_V \approx \frac{2N_P}{L}\frac{\sum I_i}{\sum N_i} \tag{3.37}$$

where the summation is carried out over all vertical sections.

One of the crucial elements in the implementation of the vertical sectioning concept is proper sampling of the sections. As said before, the procedure starts with a choice of the vertical axis, **v**. A decision for the selection of **v** should be based on the criterion of maximum convenience and no methodological restrictions are imposed. Once the vertical axis is selected, an axis normal to it is chosen. Again, no restrictions are dictated as this axis is used only

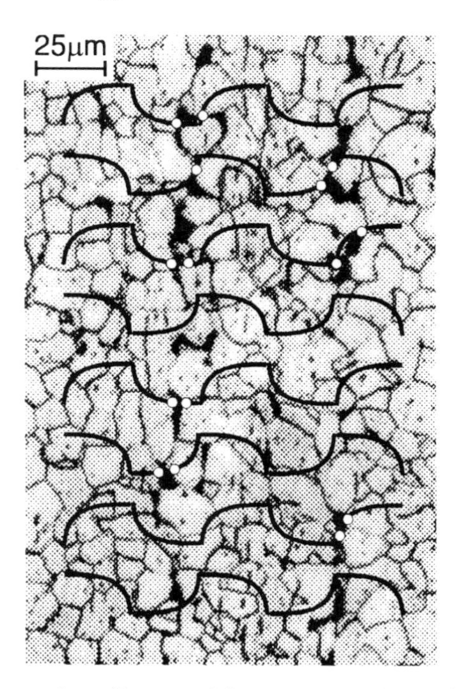

Figure 3.14 A system of test lines used for estimating the surface area in unit volume by the vertical sectioning method.

to register the orientation of the vertical sections to be cut. Then the desired number of different orientations of sections, N_S, has to be specified. In the next step, the initial angle, ϕ_1, is randomly selected using random number tables or a computer-generated random number. The following sections should have orientations given by:

$$\phi_i = \phi_1 + (i-1)\frac{2\pi}{N_S} \tag{3.38}$$

This gives their orientation with respect to the preselected normal axis. For each orientation, one can take more than one section. However, the number of sections for each orientation needs to be constant. It should be pointed out that the vertical axis, **v**, is a vector that defines a direction and as a result is not attached to any specific point in the specimen volume. Therefore, the sections cut do not need to intersect each other although such intersections are not forbidden.

As a short summary, the following steps are executed in the procedure of vertical sectioning:

1. selection of the axis **v**;
2. selection of the auxiliary normal axis;
3. selection of the orientation angles;
4. sectioning and microstructural measurements.

In the last step it is important that:

a) the sections are parallel to the vertical axis;
b) the orientation of the vertical axis is known for each section.

In addition to S_V measurements, vertical sectioning can be used directly to estimate:

1. V_V—volume fraction of particles;
2. $E_V(V)$—volume weighted volume of particles;

and

3. N_V—density of particles.

In estimating the first parameter, a system of points is used while for the second, a system of points and intercepts are employed.

The concept of vertical sectioning is illustrated by the following example.

Example 3.13

A specimen in the form of a rod was sectioned along planes containing the rod axis. It was found that all sections are equivalent and three of them were selected for measurements with cycloids. The test methodology is illustrated in Figure 3.15. It was found that $S_V \approx 0.7\ \mu m^{-1}$.

An efficient method for estimating the mean volume uses a combination of vertical sectioning with point sampled intercepts, as described in Reference 11. In this case, a system of vertical sections is required and on each section a grids of lines, L_i, and points, Q_j, are imposed. The lines are of nonequidistant, sine-weighted orientation. They can be obtained by the simple geometrical construction, as shown in Figure 3.13, or generated by a computer program if the measurements are carried out with the help of an automatic image analysis system.

The mean volume weighted volume of particles, or other 3-D objects, can be estimated from the measurement of the intercepts on L_i line which have as their ends Q_j points falling into the sections of the particles and the intersections with the particle profiles.

The vertical sectioning method can also be used to determine the density of lines in space. As discussed in Chapter 2, the specific length of a line segment in space can be estimated using the equation:

$$L_V = 2N_A \tag{3.39}$$

where N_A is the density of intersections of the line elements with planes taken isotropically, uniformly, and randomly.

The method of L_V measurement in anisotropic systems via vertical sectioning of slices of unknown thickness has been described by Gokhale and co-workers.[12] The method requires cutting vertical slices parallel to the

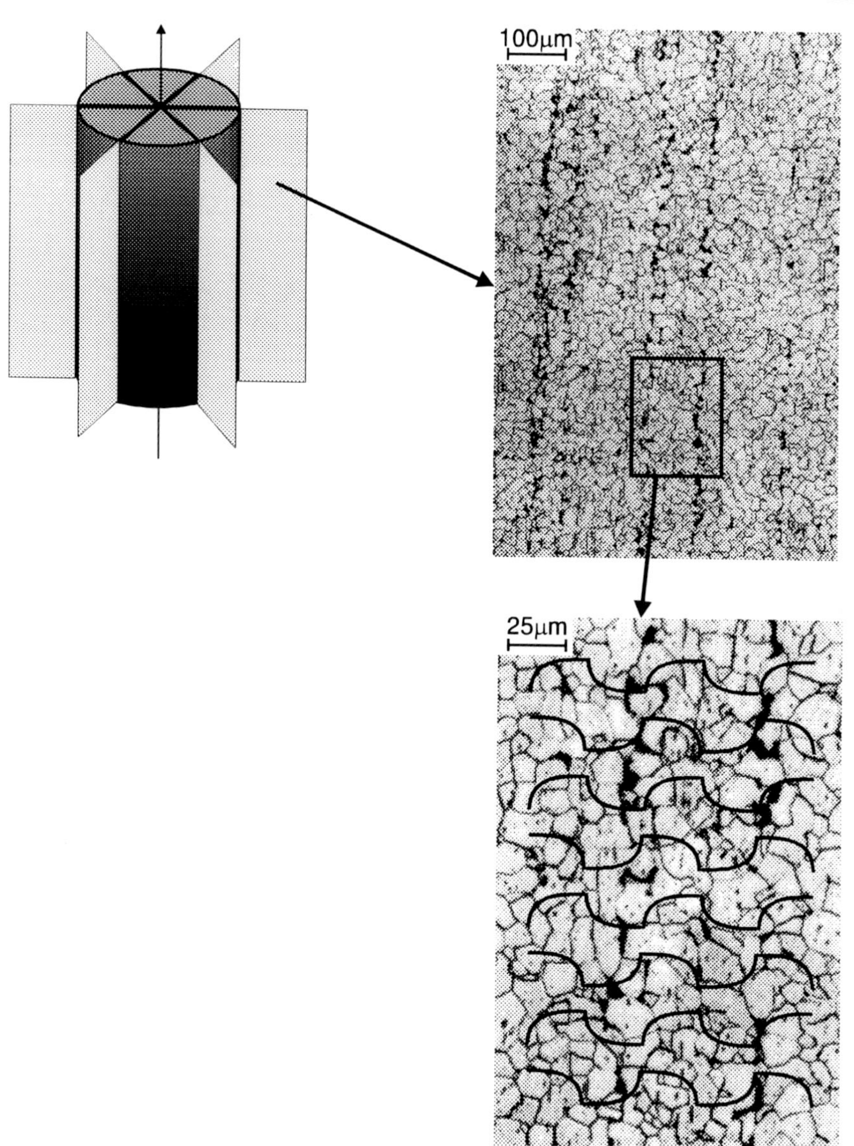

Figure 3.15 Application of the vertical sectioning procedure to studies of the surface area of particles in a rod of a low-alloy steel.

preselected **R** axis. The slices should be oriented randomly or systematically with respect to a reference direction, **U**, perpendicular to **R**. This means that the vectors normal to the slices, $\mathbf{N}_i$, should make angles, α_i, varying from 0 to 180°. The thickness of slices needs to be neither constant nor known.

Observations of the microstructure in the vertical sectioning method are carried out on n images projected along $\mathbf{N}_i$ with the position of the axis **R** identified. The images are tested with a set of lines parallel to **R** and the average density of the intersections determined:

$$(N_L)^{\|}_{proj} = \frac{1}{n}\sum_{i=n}^{i=1} (N_L)_{\alpha_i} \tag{3.40}$$

In the second step, the same images are subjected to cycloid test measurement with the minor axis parallel to **R** in order to estimate the density of intersections:

$$(L_V)^{cycl}_{proj} = \frac{1}{n}\sum_{i=n}^{i=1} (N_L)_{\alpha_i} \tag{3.41}$$

Next the density of intersections with elements is studied using the parallel planes of the slices. This leads to the following estimation of L_V via $E(N_A)$:

$$L_V = \frac{2E\left(N_L^{cycl}\right)_{proj} E(N_A)^V}{E\left(N_L^{|}\right)_{proj}} \tag{3.42}$$

The procedure described can be simplified in the case of measurements that ensure that the thickness of the sections is constant and known. The L_V density is given then as:

$$L_V = \frac{2E\left(N_L^{cycl}\right)_{proj}}{T} \tag{3.43}$$

where T is the thickness of the slices.

3.8 Orientator

The concept of the orientator has been described in Reference 13. It provides a procedure for measurement of the surface and length of 2- and 1-D elements of a microstructure from cross-sections. In principle, such measurements can be based on two fundamental stereological relationships for surface, S_V, and length, L_V, to volume ratios described in Chapter 2. These are Equation 3.37, for a system of surfaces in space, and the following:

$$L_V = P_A \tag{3.44}$$

for a system of lines in space, with P_A standing for the density of intersection points of cross-sections with the linear elements. However, these simple formulae in a general case require isotropic uniform probing with lines or planes. The requirement for isotropic probing can be abandoned only if the microstructure studied is isotropic. In this case any probing can be used and usually parallel lines and/or sections are employed. Unfortunately the microstructures of materials are seldom isotropic or at least their isotropy rarely can be assumed. As a result, parallel lines and/or sections do not provide an unbiased estimation of P_L and P_A, and in turn S_V and L_V.

The orientator is a procedure for isotropic random sampling of section orientations for measurements of P_L and P_A. Consider a specimen of interest placed in some reference system of coordinates X,Y, Z and its section C_1. The orientation of the section is defined by its normal, $\mathbf{n}_1$, and can be described by two angles, θ and Ω, such that:

$$\theta = \arccos(n_{1z}) \tag{3.45}$$

$$\Omega = \arccos\left(\frac{n_{1x}}{n_{1x}^2 + n_{1y}^2}\right) \tag{3.46}$$

where n_{1x}, n_{1y}, n_{1z} are the coordinates of n_1.

The angle θ by analogy with map making is called colatitude, whereas Ω has the meaning of longitude. With these two coordinates any direction of the section normal can be depicted as a point on a hemisphere defined as:

$$0 \le \theta \le \pi \tag{3.47}$$

$$0 \le \Omega \le \pi \tag{3.48}$$

The problem now arises as to how to select randomly a point from such a hemisphere that can be used subsequently to determine a random section of the microstructure studied. The way of sampling such a point should ensure that over a large number of selections the selected points would cover the surface of the hemisphere in a uniform way.

One solution of the problem can be obtained if the surface of the hemisphere is divided into a finite number, N, of small patches of the same area, each representing one direction determined by its center of gravity. The centers can be numbered from 1 to N and then random numbers of m directions can be used in the selected studies.

Another approach is based on the random selection of pairs of numbers (a,b) each from the interval [0,1]. It can be shown that the pairs of such numbers generate pairs of θ, Ω that are uniformly distributed on the hemisphere of directions if the following formulae are used:

$$\Omega = \pi\, a \tag{3.49}$$

and

$$\theta = \arccos(1 - 2b) \tag{3.50}$$

In this way points (a,b) randomly sampled from a unit square ABCD, defined by the following points: A[0,0], B[1,0], C[1,1], and D[0,1] are converted into random points on the hemisphere of all possible orientations for directions in space.

Points from the square ABCD can be sampled according to different schemes. The first point sampled always needs to be of random position. Successive points, however, can be positioned systematically with respect to the first one or obtained by random selection. In most cases, systematic sampling of the subsequent directions seems to be more efficient. In fact a simplified version of the orientator that consists of sampling with orthogonal triplets has been proposed in Reference 14. This method is called "ortrips" from **or**thogonal **trip**let **p**robes.

The ortrips method is implemented through the following steps:

1. assign to the specimen a system of coordinates (it is convenient to use some directions defined with respect to the specimen geometry);
2. randomly sample the first direction, $\mathbf{n}_1$;
3. sample the second direction, $\mathbf{n}_2$, in the plane perpendicular to $\mathbf{n}_1$;
4. determine the orientation of the third direction, $\mathbf{n}_3$, as a vector product of $\mathbf{n}_1$ and $\mathbf{n}_2$;
5. make sections of the specimen normal to the directions $\mathbf{n}_1$, $\mathbf{n}_2$, $\mathbf{n}_3$ and make the measurements on these sections.

An efficient estimate of the density of linear and surface features, in terms of L_V and S_V, can be obtained if the results are averaged over the three sections selected for observation. The theoretical background of the method can be found in Reference 14.

Example 3.14

Figure 3.16 schematically shows a specimen of material in the form of cube cut from a larger block of the material studied. Microstructures typical of the sections parallel to the faces of the cube are shown in Figure 3.16. Measurements of the particle volume fraction, V_V, were carried out on these microstructures. The results of these measurements are given in Table 3.4.

Three other directions of normals to the sections were determined by random selection of numbers in the interval [0,1]. The results of these measurements on these sections are also given in Table 3.4. It may be noted that the results for the randomly selected orientations are characterized by a lower spread.

3.9 *Trisector*

Another method for surface area measurement has been proposed.[12] This method, called by the authors the trisector, combines the principles of vertical sectioning with the orientator. It requires selecting three planes parallel to a common axis, $\mathbf{n}$: C_1, C_2, and C_3 that are mutually at angles of 120° to each other (see Figure 3.17). On these sections measurements are carried out with

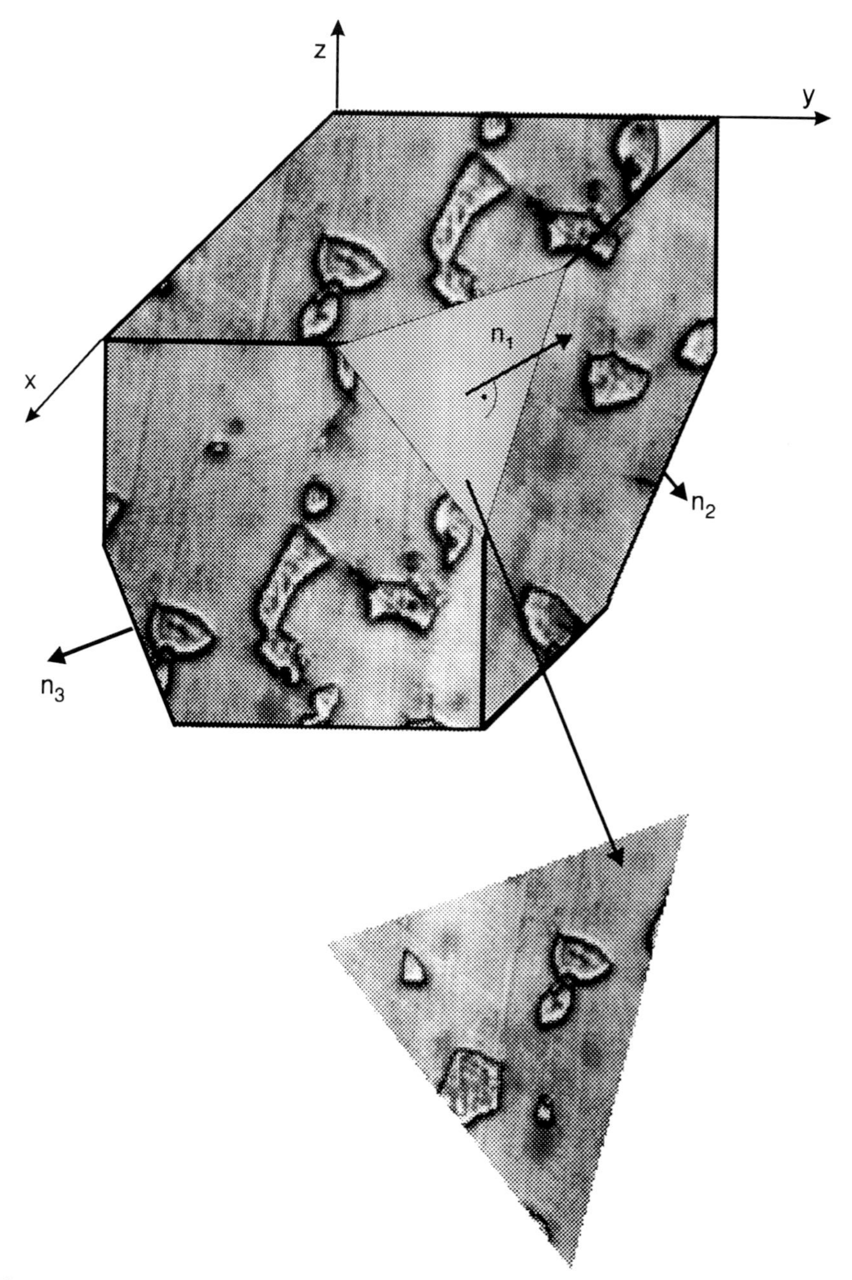

Figure 3.16 Implementation of the orientator method in studies of the microstructure of materials. The results of the measurements are given in Table 3.4.

a system of oriented cycloids in the same way as in the standard vertical sectioning procedure. This means counting the number of intersection points with the system of cycloids of a known length, L, and computing the S_V ratio from the formula:

Table 3.4 The Results of the Measurements Carried Out on the Specimen Depicted in Figure 3.16

	Orientation of the section						
	n_X	n_Y	n_Z		n_1	n_2	n_3
V_V	0.18	0.17	0.19		0.19	0.15	0.20
$E_3(V_V)$		0.18				0.18	
$E_6(V_V)$				0.18			

Note: The symbol $E_3(V_V)$ stands for the mean from the fields perpendicular to the axes X, Y, Z, on the one hand, and to the axes, n_1, n_2, and n_3 on the other; $E_6(V_V)$ gives the global mean.

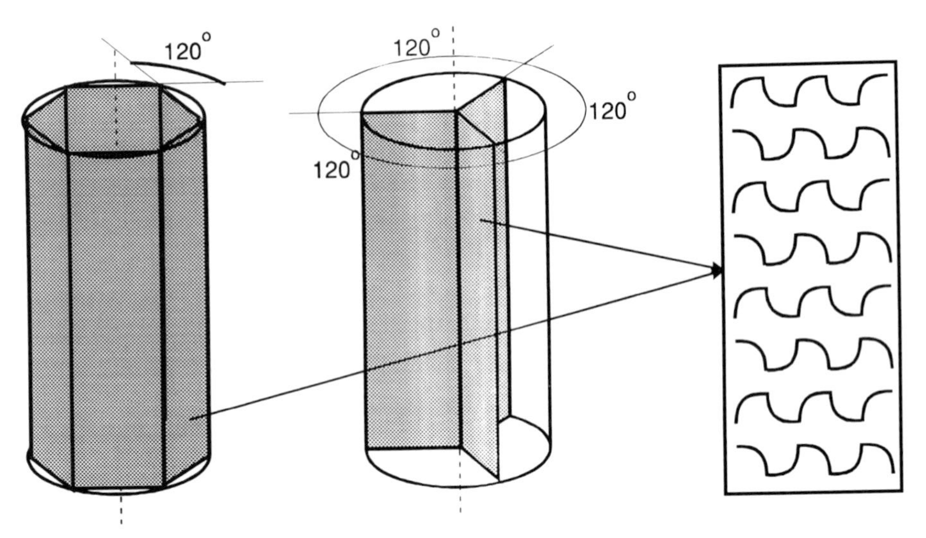

Figure 3.17 Schematic representation of the possible implementation of the trisector method.

$$S_V = \frac{2I}{L} \tag{3.51}$$

where I is the total number of intersection points on all three sections. It has been shown[12] that this procedure yields a satisfactory estimate, with an error smaller than 5%, for any microstructure, assuming that the vertical axis, **n**, is not parallel to the direction of elongation of the features investigated. This result is of importance from the point of view of optimizing the experiment, since cutting an increasing number of sections of different orientations makes the measurements much more time-consuming.

The same authors suggested also[12] that in the case of vertical sections reduced to three sections arranged symmetrically about a selected direction the following formula can be used:

$$S_V \approx 2.012\left(P_L\right)_{cucl} \tag{3.52}$$

where $(P_L)_{cycl}$ is the density of the intersection points per unit length of the cycloids.

References

1. Underwood, E. E., *Quantitative Stereology*, Addison Wesley, Massachusetts, 1970.
2. Gundersen, H. J. G., Stereology of arbitrary particle, *Journal of Microscopy*, 143, 3, 1986.
3. Sterio, D. C., The unbiased estimation of number and sizes of arbitrary particles using the disector, *Journal of Microscopy*, 134, 127, 1984.
4. Gundersen, H. J. G., Notes on the estimation of the numerical density of arbitrary profiles, *Journal of Microscopy*, 111, 219, 1977.
5. Liu, G., Yu, H. and Li, W., Efficient and unbiased evaluation of size and topology of space-filling grains, presented at 6th European Congress for Stereology, Prague, September 7–10, 1993.
6. Jensen, E. B. and Gundersen, H. J. G., Fundamental stereological formulae based on isotropically oriented probes through fixed points with application to particle analysis, *Journal of Microscopy*, 153, 249, 1989.
7. Jensen, E. B. and Gundersen, H. J. G., The stereological estimation of moments of particle volume, *Journal of Applied Probability*, 22, 82, 1985.
8. Jensen, E. B. and Sorensen, F. B., A note on stereological estimation of the volume-weighted second moment of particle volume, *Journal of Microscopy*, 164, 21, 1991.
9. Gundersen, H. J. G., The nucleator, *Journal of Microscopy*, 151, 3, 1988.
10. Baddeley, A. J., Gundersen, H. J. G., and Cruz-Orive, L. M., Estimation of surface from vertical sections, *Journal of Microscopy*, 142, 259, 1986.
11. Sorensen, F. B., Stereological estimation of the mean and variance of nuclear volume from vertical sections, *Journal of Microscopy*, 162, 203, 1991.
12. Gokhale, A., Drury, W. J., and Whited, B., Quantitative microstructural analysis of anisotropic materials, *Materials Characterization*, 31, 11, 1993.
13. Mattfeldt, T., Mall, G., and Gharehbaghi, H., Estimation of surface area and length with the orientator, *Journal of Microscopy*, 159, 301, 1990.
14. Mattfeldt, T., Mobius, H. J., and Mall, G., Orthogonal triplet probes: an efficient method of estimation of length and surface of objects with unknown orientation in space, *Journal of Microscopy*, 139, 279, 1985.

chapter four

Computer-aided characterization of the microstructures of materials

Microstructures of materials are studied experimentally using various methods of imaging to resolve specific microstructural elements. These methods include transmission microscopy through thin slices and reflected microscopy on surfaces and polished sections. A large range of wavelengths are employed, from acoustic to electron via visible light. In all cases, attention is focused on images of the microstructure and these, in many cases, are found to be of considerable complexity. More precisely, images from the microstructure of materials are distinguished by:

a) geometrical complexity of the features of interest, which differ in their dimensions (point, line, surface, and volumetric elements) and assume complicated shapes;
b) diversity in size and shape of the elements of interest, which, as a result, can only be properly described if a proper statistical methodology is available;
c) the presence of specific errors caused by imperfections in the imaging techniques.

In brief, characterization of microstructures usually requires an analysis of a large number of complicated images with each of them containing the microstructural elements required as well as some artifacts produced by the imaging technique. This justifies the need for automation and explains the reasons for computer-aided procedures in the characterization of materials.

The present chapter can be viewed as consisting of two parts. In the first part, attention is focused on computer-aided analysis of microstructural images. The text explains how images of the microstructure can be made accessible to computerized analysis and the main procedures to carry out such an analysis.

The second part deals with the problem of data analysis. This problem is logically related to the subject of the first part of the text because the methods of automatic image analysis are responsible for creating a great deal of data, which can again only be effectively processed with the use of computers.

4.1 Image processing

The term image processing is used in the present context to describe operations that are performed on images of microstructures in order to correct them or to make more accessible to quantitative analysis. Among all the possible methods that can be used for such purposes, including manually made corrections, of special interest are in this case computer-aided methods that are based on the concept of digital images. In such a case the process of image processing can be viewed as consisting of five steps:

1. image acquisition;
2. digital processing;
3. threshold operations;
4. mathematical morphological operations;
5. measurements.

After the above-mentioned steps of image processing have been performed, data analysis is undertaken on the data obtained in the measurement steps.

4.1.1 Image acquisition

The term image acquisition covers all steps involved in transforming the image of a microstructure into a set of 2-D pictures that provide information on the form and position of the microstructural features. Each material and microstructural element may require an individually designed imaging technique and these techniques are not discussed here—they are the subjects of a number of monographs devoted to separate experimental imaging methods such as: electron and light microscopy, ultrasonic microscopy, radiography, etc. (see, for example, References 1 to 3).

Whatever the imaging technique, it produces a 2-D image that can be recorded either on a film or displayed on a monitor. An image contains specific patterns of points and regions, which differ in their intensity or color. From this point of view, the image is a function of two spatial coordinates, for instance (x,y), which ascribes to each point on the plane a specific value of point image intensity.

Two examples of microstructural images are shown in Figure 4.1; one of these was obtained by light and the other by scanning electron microscopy. These examples illustrate that the images are characterized by spatial distributions of image intensity, in this case variations in the gray level. This means that each point on the image has its specific value of intensity leading to spatial intensity profiles. Intensity profiles for the lines indicated on Figure 4.1 are depicted in Figure 4.2 The images are also characterized by a distribution of intensity levels. The intensity distribution functions for the images from Figure 4.1 are shown in Figure 4.3.

One word of caution here concerns the question: When is an image an image? Essentially, to be an image of the microstructure in the true sense, it

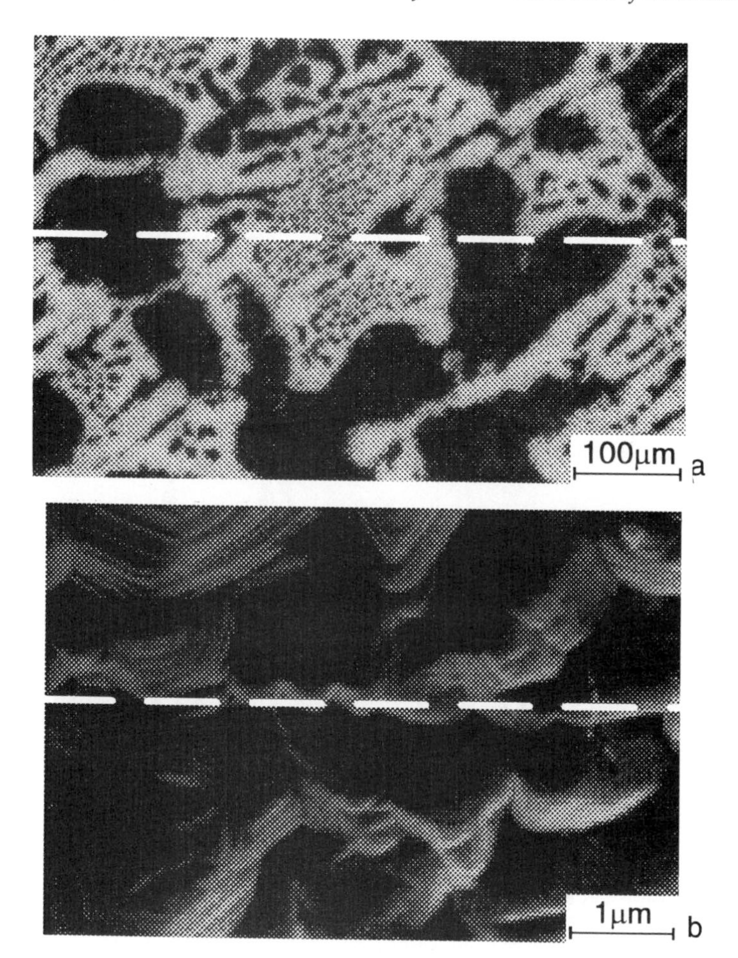

Figure 4.1 Examples of microstructures obtained by: optical microscopy (a) and scanning electron microscopy (SEM) (b). These images, although clearly different, are both characterized by spatial distributions of intensity.

must contain information that can be discriminated on the basis of microstructural features that are of interest. In other words, the intensity distribution on a point-by-point basis must carry useful information on the distribution of atoms, particles, defects, or energy distribution.

4.1.2 Digital processing

The human brain seems to perceive images registered by the eyes in a continuous form in the sense that the objects are seen as elements of continuous space (or a set of zero-sized points). For an image to be "understandable" by a computer it has to be discrete (digital), not continuous (analog). This discretization has to be conducted on both spatial and intensity (or energy) level distributions. The spatial discretization occurs usually by dividing the image

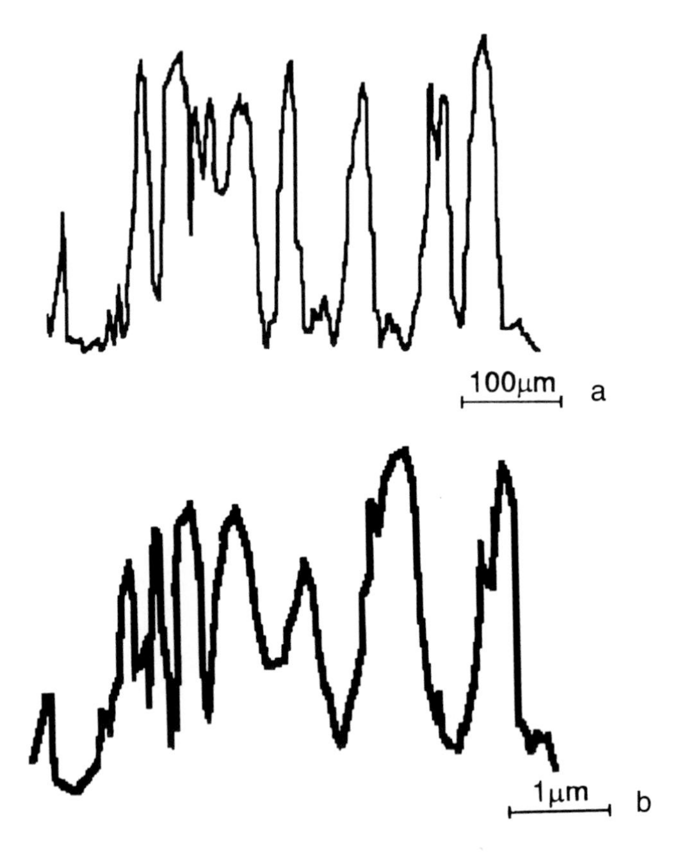

Figure 4.2 Line density profiles for images from Figure 4.1 taken on lines marked on Figure 4.1.a and Figure 4.1.b, respectively. These density profiles show the distribution of intensities along the selected lines.

"plane" into a regular rectangular or hexagonal grid of points with finite dimension (picture elements). These points are usually called pixels or pels. Energy level/intensity discretization usually occurs by means of an A/D (analog-to-digital) converter. Typical values of pixel size are expressed as a ratio of their dimensions to that of the image frame. In fact, in most cases numbers are given which define the number of pixels along the horizontal and vertical edges of the image. These numbers vary from 256 × 256 or 512 × 512 in basic level image analyzers up to 8192 × 4096 or more in sophisticated systems dedicated to special tasks (as, for example, satellite weather prediction systems). It should be noted that these numbers are powers of 2 ($256 = 2^8$, $512 = 2^9$, etc.). This is related to the binary system used by computers. The number of discrete intensity/energy levels that may be ascribed to each pixel vary from 16 or 64 levels in simple systems up to 256 levels or 256 levels for each basic color in advanced color image analysis systems and even more in dedicated systems. (A color image can be seen as the superposition of three images in three basic colors—red, green and blue.)

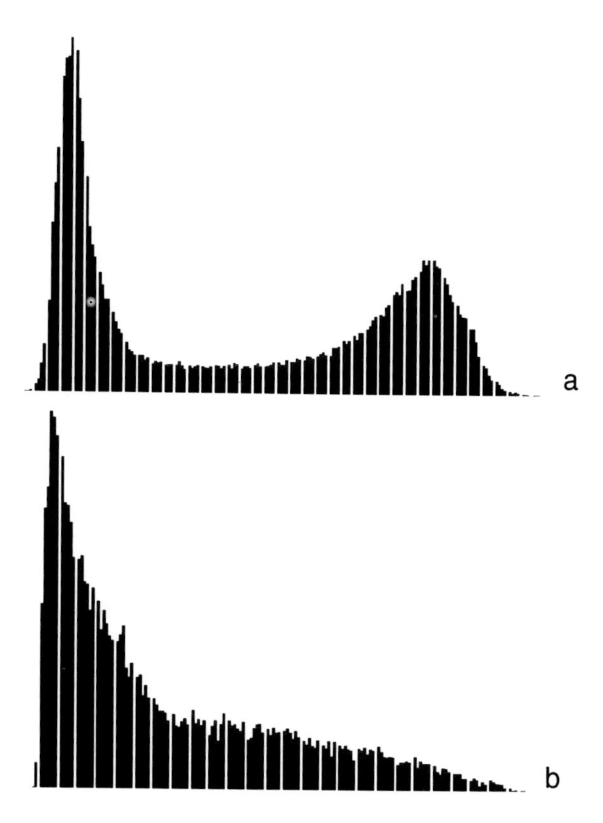

Figure 4.3 Intensity histograms for images from Figure 4.1. These histograms show distribution of probability that any point selected on the image has a specified intensity.

The concept of a digital image is strictly related to the technical conditions of recording the images. Two-dimensional images of the microstructure can be viewed as sets of different color/gray level dots filling the picture. Such a representation in many applications is convenient for recording images on films or using video equipment. Consider the image shown in Figure 4.4. This image can be divided into a finite number of elements (pixels) in the way shown schematically in Figure 4.5. The number of pixels, N_p, depends on the number of rows, N_r, and columns, N_c, assumed in the division of a given image. The pixels forming the image are distinguished by different gray levels. If the intensity of the gray tone is measured on a scale ranging from 0 to N_b, then the image analyzed can be represented as a matrix with a size $N_r \times N_c$:

$$\begin{matrix} p_{11} & p_{12} & \cdots & p_{1N_c} \\ p_{21} & p_{22} & \cdots & p_{2N_c} \\ \cdots & \cdots & \cdots & \cdots \\ p_{1N_r} & \cdots & \cdots & p_{N_rN_c} \end{matrix} \tag{4.1}$$

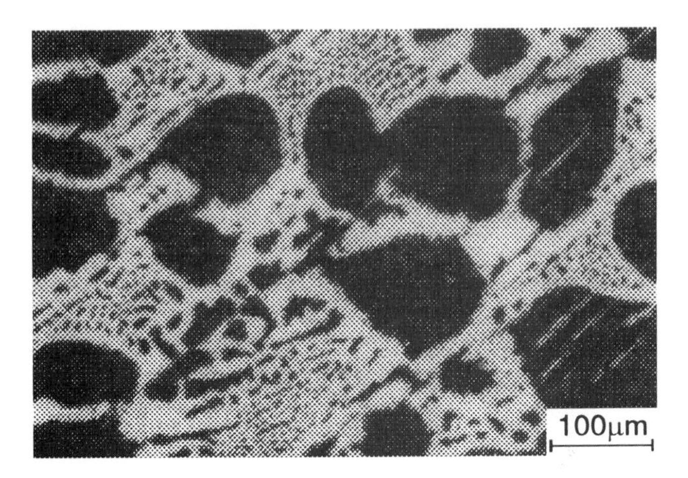

Figure 4.4 Example of a light microscopy image of a eutectic alloy used in the discussion on digital and analog images.

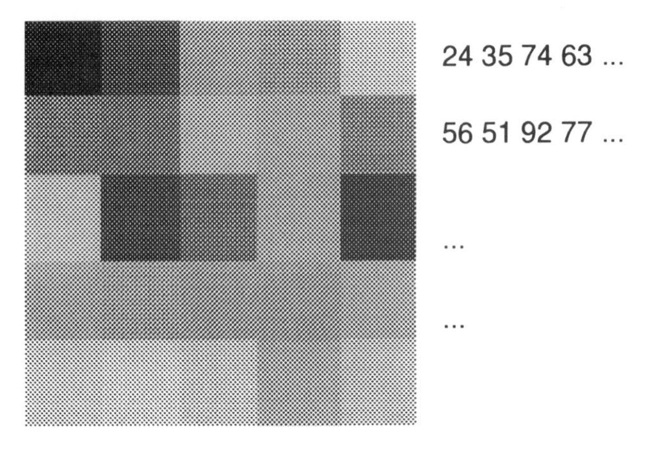

Figure 4.5 Example of the digitization of a small area from Figure 4.4 shown as a zoom image with visible square-shaped pixels and as a numerical representation in the form of a matrix.

with elements, p_{ij}, which are of integer value in the range, 0 to N_b. The value zero in this case indicates a black pixel and N_b a white one. The numbers between distinguish intermediate gray levels.

The pixels of the image have prescribed geometrical dimensions. If the total area of image analyzed is $A \times B$ then the pixels are quadrilaterals of a size $p_r \times p_c$ such that:

$$p_r = \frac{A}{N_r} \qquad p_c = \frac{B}{N_c} \tag{4.2}$$

This means that any point in the image has specified coordinates with respect to the frame and can be addressed to an appropriate element of the microstructure studied.

Digitization of an image into the form of a matrix is illustrated by an example in Figure 4.5. It should be noted that digitization results in a specific distortion of the image. Features in the microstructural image are changed into sequences of square-like pixels and lose their apparent smoothness. By contrast such a representation gives an important advantage in allowing the possibility of applying the rules of algebra to the analysis of the digitized image. The digitized image can be subjected to all matrix algebra operations where a large number of these operations have a clear physical interpretation.

The digitized image can be decomposed into a sum of images:

$$\begin{bmatrix} P_{11} & P_{12} \\ P_{21} & P_{22} \end{bmatrix} = \begin{bmatrix} P_{11} & 0 \\ 0 & 0 \end{bmatrix} + \begin{bmatrix} 0 & P_{21} \\ P_{12} & P_{22} \end{bmatrix} \tag{4.3}$$

where P_{ij} are smaller sized images and 0, as an element of the image matrix, indicates a completely black area. In a similar way the images can be superimposed by the operation of adding two or more matrices:

$$\begin{matrix} im\,1 & & im\,2 & & im\,3 \\ \left[A_{ij}\right] & + & \left[B_{ij}\right] & = & \left[A_{ij}+B_{ij}\right] \end{matrix} \tag{4.4}$$

Example 4.1

An image shown in Figure 4.4 is digitized into a 256 × 256 matrix. Figure 4.5 shows a part of such a matrix and its visualization. Figure 4.6 presents the distribution of gray levels (intensities) of image pixels. The figures that follow (Figures 4.7 and 4.8) show images produced by simple algebraic operations of adding and subtracting a constant from the matrix defining the image.

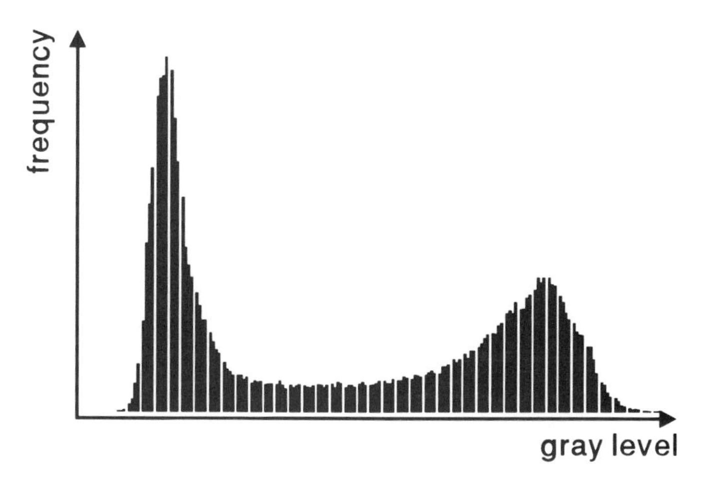

Figure 4.6 Histogram of intensities of pixels from Figure 4.4 showing the distribution of probability that a selected pixel has a given value (intensity).

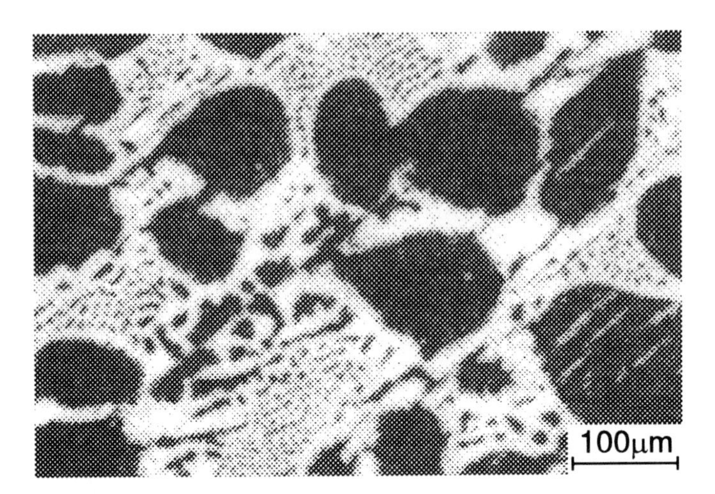

Figure 4.7 Example of results of a simple algebraic operation on a digital image. This image was produced from Figure 4.4 by adding a constant. One can easily observe an effect of lightening the image.

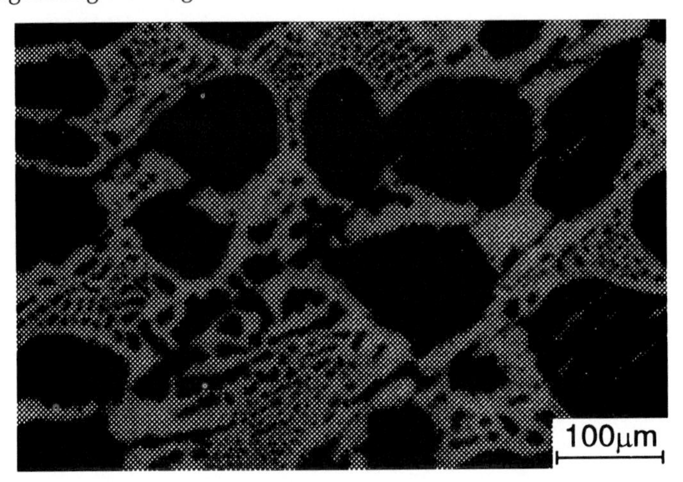

Figure 4.8 Example of results of a simple algebraic operation on a digital image, produced by subtracting a constant from the image shown in Figure 4.4. One can observe an effect of darkening of the resulting image.

4.1.3 *Threshold operations*

The pixels forming a digitized image of a microstructure differ in their color or gray level. These differences can be used to discriminate parts of the image representing microstructural elements if the rule that leads to contrast between these features and others in the image is known. Consider a specimen of a two-phase material prepared for light microscopy observations via polishing and subsequent etching. Polishing and etching of such a material often leads to significant differences in the intensity of light reflected from the parts of the area studied that are occupied by different phases. As a result, the particles revealed on a cross-section appear to have a gray level distinguishing them from the background.

Differences in the gray level of various phase constituents can be used to detect particles and conduct simple measurements on their geometrical features. Let us assume that phase 1, under specified conditions of specimen preparation and illumination, is lighter than a certain critical gray level, N_b^*, which is the limit of background brightness from the other phase. Thus, the pixels of the background can be found from the condition:

$$\left(p_{ij}\right)_{backgrd} \leq N_b^* \tag{4.5}$$

while the pixels representing the particles of phase 1 must satisfy the condition:

$$\left(p_{ij}\right)_{phase} > N_b^* \tag{4.6}$$

This means that the total area of the particles can be estimated by counting the appropriate pixels. Formally this may be stated in the following way:

$$\left(A_A\right)_{phase1} = \frac{1}{N_r N_c} \sum_{i=1}^{i=N_r} \sum_{i=1}^{i=N_c} p_{ij}^*\left(N_b^*\right) \tag{4.7}$$

where

$$p_{ij}^*\left(N_b^*\right) = \begin{pmatrix} 1 & for\ p_{ij} \geq N_b^* \\ 0 & otherwise \end{pmatrix} \tag{4.8}$$

In more complicated cases the pixels forming the image of a given constituent can be detected using more sophisticated methods. Such methods may employ a lower threshold gray level value as well as an upper limit of particle brightness. In this case, the pixels, p_{ij}, representing a given phase can found from the following:

$$N_w^* \leq p_{ij} \leq N_b^* \tag{4.9}$$

The threshold values, N_w^*, N_b^* can be determined by the operator of the image analysis system based on his/her judgement as to what extent the values used make it possible to distinguish the particles from the rest of the image. This choice can also be made more objective with computer-aided decision rules.[4]

The operations discussed turn an image characterized by a distribution of intensity levels into a binary image, which consists of points with intensity 0 or 1. An example of the formation of a binary image is shown in Figures 4.9 and 4.10. This operation is a special case of more general transformations of image intensity.

Consider an image that is described by an intensity function $I_o(x,y)$. This function, for any pixel position defines the local intensity as a discrete

Figure 4.9 Example of an analog image of a single-phase microstructure with grain boundaries being the most visible element of microstructure.

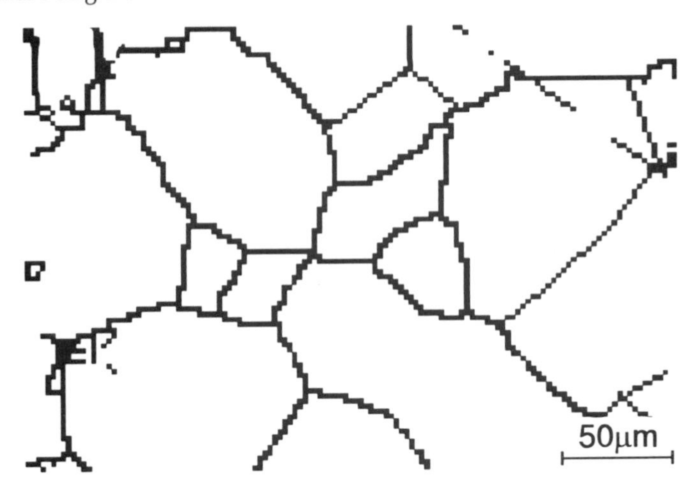

Figure 4.10 Binary image obtained from Figure 4.9 by means of threshold operation with a specific threshold value. The resulting image consists of only black and white pixels, because every pixel of higher intensity than the threshold value has been assigned white (1), while pixels of intensities lower than threshold value have been assigned black (0).

number in the range from, I_{min}, usually 0, to I_{max}. A single variable function, F, is given, which, for a given level of intensity, I, ascribes a new, corrected intensity, I_c:

$$F: \quad I_c = F(I) \tag{4.10}$$

This function should be of monotonic character and should preserve or narrow the range of intensities:

$$F(I_{\min}) \geq I_{\min} \qquad F(I_{\max}) \leq I_{\max} \tag{4.11}$$

For the case of turning images into binary images, the function F is step-like, and can be formulated in the following way:

$$F_{binary} = \begin{cases} 0 \; for \; I < I_o \\ 1 \; for \; I \geq I_o \end{cases} \tag{4.12}$$

where I_o is the threshold intensity. Figure 4.11 shows schematically a series of images produced from an initial image using different values of the threshold intensity I_o.

Other examples of the function F include both linear and nonlinear operations, which lead to a better discrimination of the features. Examples of such operations are given below.

Example 4.2

Pixels forming an image of a microstructure are characterized by distributions of gray levels and Figure 4.12a shows an example of the gray level histogram detected for the single-phase microstructure exemplified in Figure 4.9. Figure 4.12b shows a transformed image of the microstructure from Figure 4.9 that was obtained by the following transformation:

$$I_c = \begin{cases} 1 \quad for \; I \geq I^* \\ \dfrac{I}{I^*} \; for \; I < I^* \end{cases} \tag{4.13}$$

Example 4.3

By examining a series of micrographs it has been established that the elements of the image that are of interest are made up of pixels with their intensities in the range $I_1 > I_{min} = 0$, $I_2 < I_{max}$. Figure 4.13 shows the initial image of the microstructure in which the intensity of the pixels vary from 0 to I_{max}. Figure 4.14 presents the same image after an operation of enhancement using the operation defined by the following function:

$$F(I) = \begin{cases} 0 & for \; I < I_1 \\ \alpha(I - I_1) & for \; I_1 < I < I_2 \\ I_{\max} & for \; I > I_2 \end{cases} \tag{4.14}$$

with α defined as:

$$\alpha = \frac{I_{\max}}{I_2 - I_1} \tag{4.15}$$

Example 4.4

Figure 4.15 shows a series of images obtained from the initial image (Figure 4.13) by operations on the image intensity with the following functions:

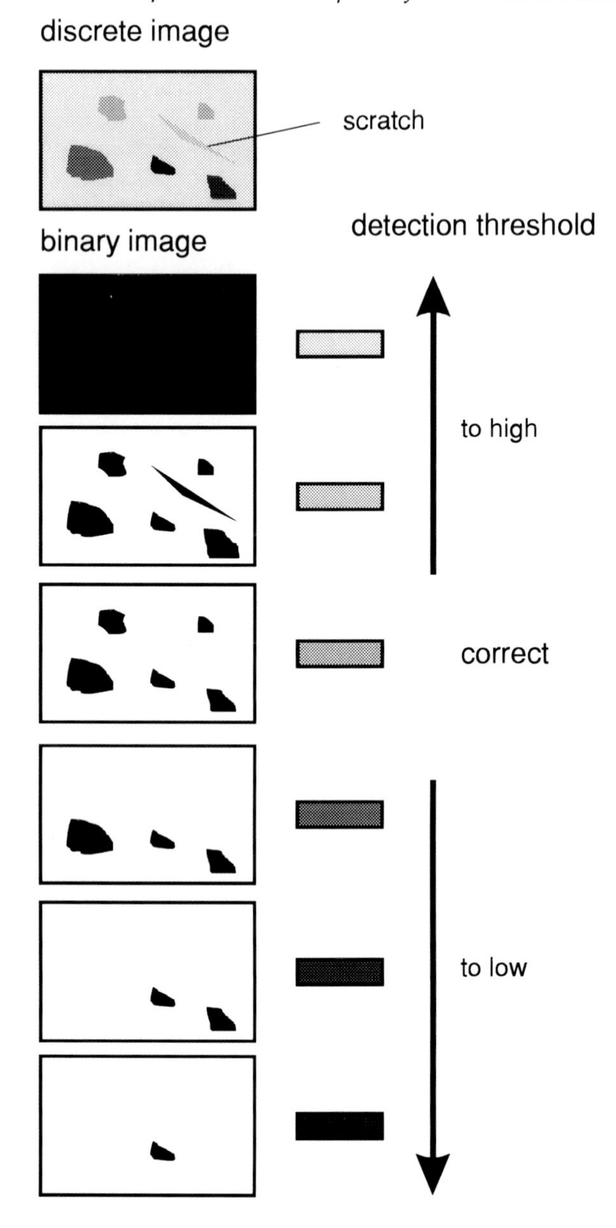

Figure 4.11 Schematic example of the influence of too low and too high threshold values on the appearance of a binary image obtained from a digital one by means of a threshold operation.

a) $F(I) = A \log(I)$;
b) $F(I) = A\,I$.

Operations on the image intensity are carried out for various purposes. In some cases the main objective is an enhancement of the discrimination leading

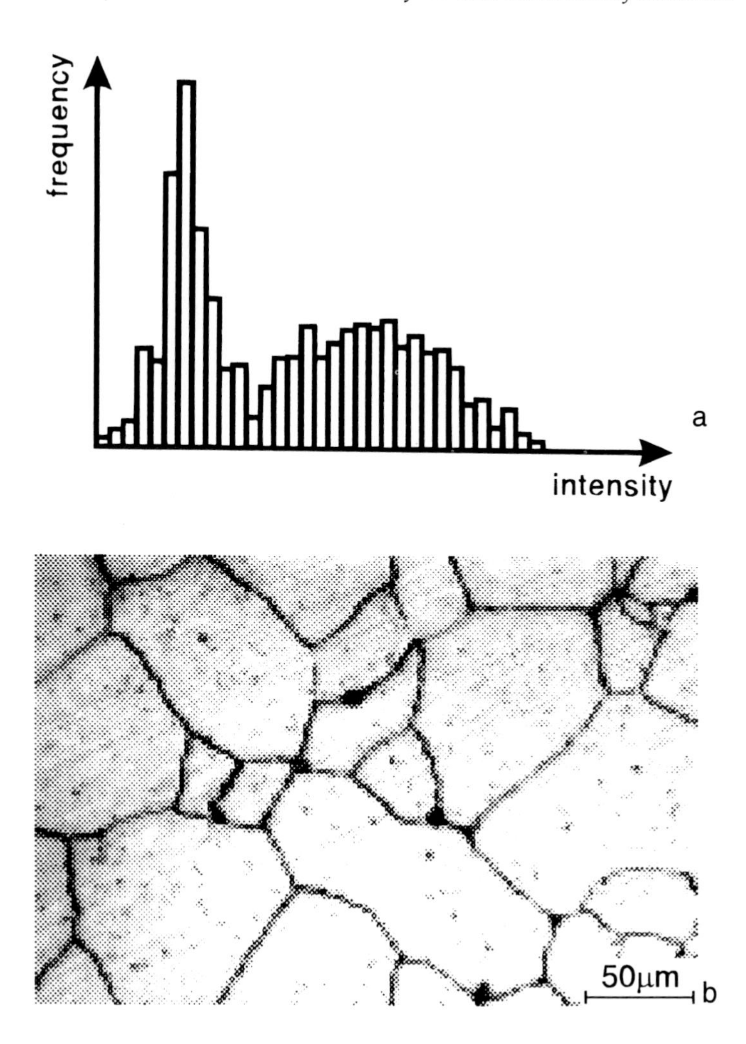

Figure 4.12 Histogram showing the distribution of pixel intensities on the digital image obtained from Figure 4.9 (a) and digital image obtained from Figure 4.9 after transformation by Equation 4.13, i.e., by extending the intensity scale in the range of lower intensities (b).

to a better visualization of the microstructural elements. However, the main reason for these operations is the need to select elements pertinent to the analysis. The assumption is made that these elements are imaged into the pixels, which are distinguished by their intensity or color. Figure 4.16 shows an example where such an assumption is justified. By setting a threshold value, I_o, the image can be turned into binary form indicating the position of elements of interest. A commercial steel sample with a TiN particle SEM image is shown in Figure 4.16a, the Ti X-ray emission image is shown in Figure 4.16b and the binary image of the particle is shown in Figure 4.16c.

Figure 4.13 Example of the image of a microstructure with intensities of pixels falling generally in a limited range, I_1–I_2.

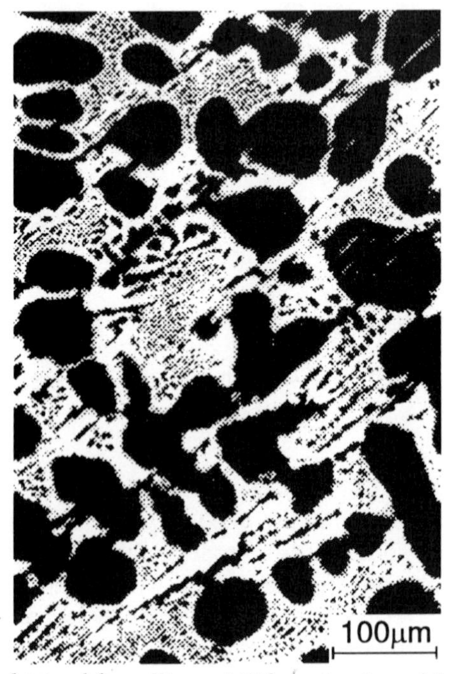

Figure 4.14 Image obtained from Figure 4.13 by extension of the intensity range according to Equations 4.14 and 4.15. The resulting image shows better resolution in terms of intensity differences between neighboring pixels than the original one (Figure 4.13).

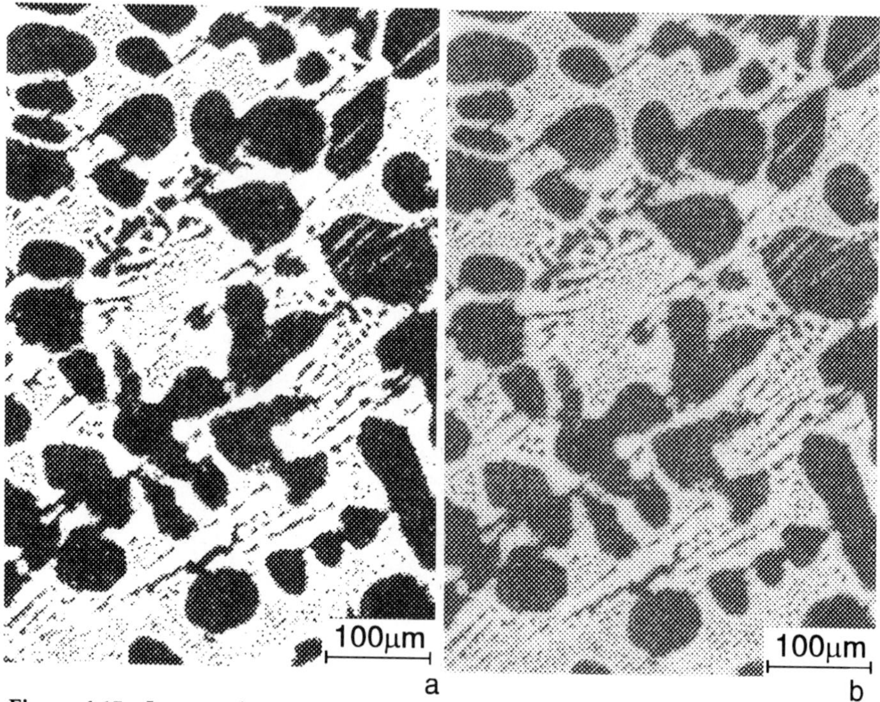

Figure 4.15 Images obtained from the image shown in Figure 4.13 by means of intensity functions defined in Example 4.4.

4.1.4 *The applications of mathematical morphology*

One of the basic problems faced in the analysis of images of a microstructure is the large amount of information included in the image. On the other hand, a great deal of this information forms a peculiar "background" that is irrelevant from the point of view of a quantitative description of specific microstructural features, which should strictly be subordinated to the relations under consideration.

In order to reduce the amount of information to an adequate level three measures are undertaken:

1. reducing the observation area;
2. transforming a digital image into a binary one;
3. transforming the image by means of mathematical morphological operations.

These three possibilities are illustrated in Figure 4.17.

Reducing the observation area is an effective way of reducing the information included in the image. However, the information is reduced in a nonselective way, i.e., relevant information is reduced to the same extent as irrelevant information. Therefore, this method is not of any great importance. It is mainly applied because of the limited resolution of devices used for the image analysis.

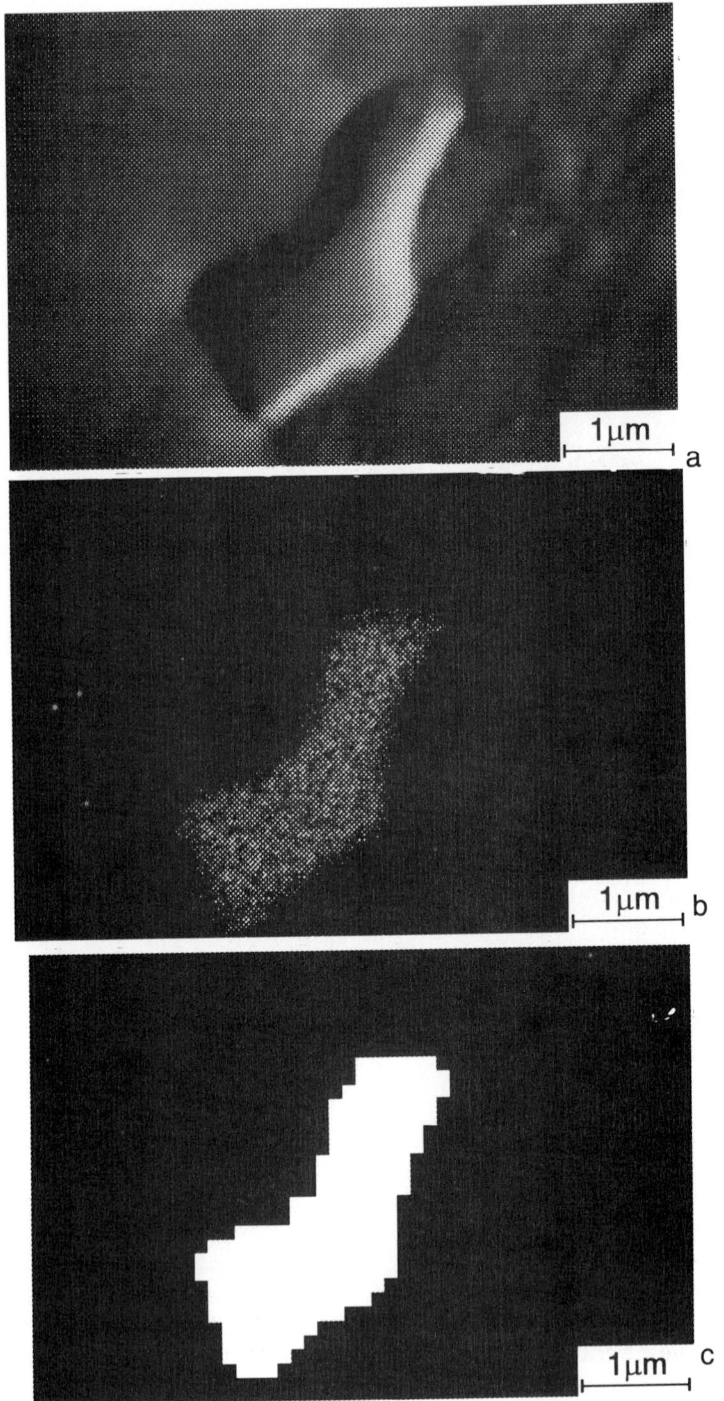

Figure 4.16 Example of the use of different imaging modes to study specific properties of microstructural elements. TiN particle in steel : SEM image (a), X-ray image (b), binary image after erosion (c).

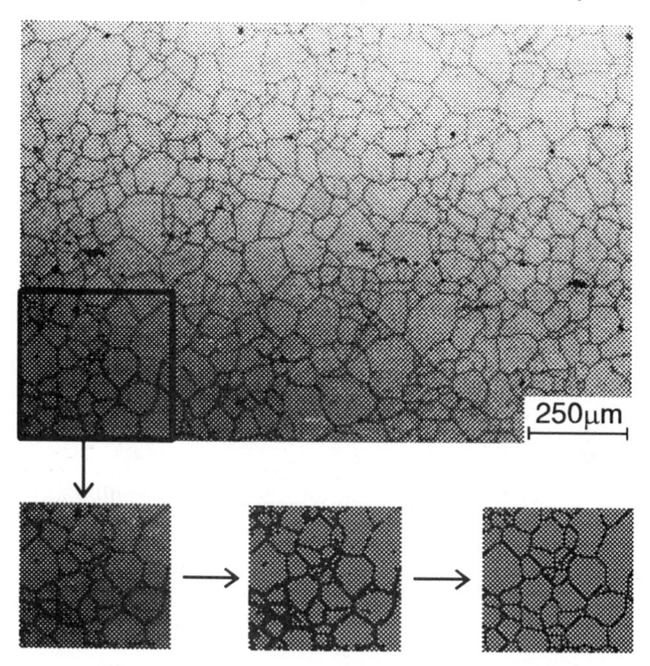

Figure 4.17 Application of three steps for information reduction in image analysis. By reducing the observation area, relevant information is reduced to the same extent as irrelevant, but one can better deal with resolution limits of image analysis devices. Transformation of a digital image into a binary one emphasizes certain elements of the microstructure. Using the appropriate mathematical morphological methods one can separate the microstructural elements chosen and even enhance the apparent image quality.

Transforming a digital image into a binary one is a powerful method of focusing the attention on preselected elements of the microstructure as discussed above. In a binary image all the relevant pixels are ascribed a white color or the number 1 while the pixels of the background are zeros. Binary images may be obtained directly from the initial images by a simple operation of discriminating pixels by brightness in a certain range of specified brightness level (the threshold operation). However, this approach is effective only if all pixels from the elements studied show consistent brightness clearly different from the other pixels in the image. This condition is not very often met in practice and the binary images obtained in this way may contain a number of artifacts. This is exemplified by the images shown in Figure 4.18. The grain boundary microstructure of Figure 4.18a is transformed into a segmented system of lines in Figure 4.18b despite the fact that the boundaries form a continuous network.

Simple discrimination techniques do not utilize additional information on the geometrical features of the elements studied. For example, the segmented binary image of the grain boundaries in Figure 4.18a could have been corrected if it were recognized that the lines defining the traces of grain boundaries cannot end inside the image frame. This and other information is used

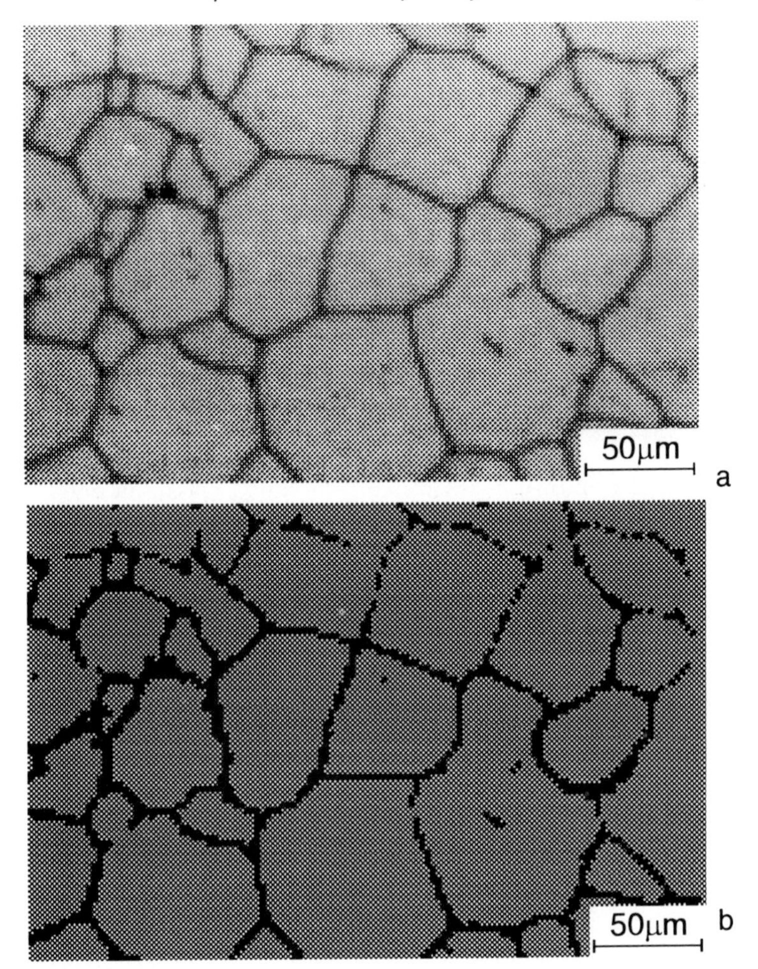

Figure 4.18 Image obtained after the first two steps of information reduction presented in Figure 4.17. Without applying proper mathematical morphological methods after obtaining the binary image the network of grain boundaries becomes a set of segmented lines.

effectively in the case of the third of the above-mentioned approaches—mathematical morphological operations. Mathematical morphology is a discipline that is focused on transformations to and the properties of images. A review of basic mathematical morphological operations performed on binary images and their applications in quantitative microscopy is the subject of the following part of the text.

4.1.4.1. Mathematical morphological operations

Operations of mathematical morphology are based on the concept of a structural element. Structural elements are usually characteristic sets of some small number of pixels arranged according to a certain pattern. The elements, designated by the symbol E can be described in terms of their shape (E_{bar}, E_{square}, etc.), size, E^s, and of a characteristic reference point, E(P). An example of a structural element is a vertical bar consisting of four pixels with the lowest one being the reference point. Such an element, shown in Figure 4.19, if placed at point P on the image can be described shortly as $E^4_{bar}(P)$.

In the operations of mathematical morphology, E elements are placed step-by-step at all pixels, P_i, of the initial binary image, P, and a comparison is made of the surroundings of points, P_i, with the structural element E. Depending on the result of this comparison, the brightness of the pixels is changed and the image, P, transformed into a new one, P′:

$$P \rightarrow P' \tag{4.16}$$

The transformed image contains a reduced amount of information, with the information on certain aspects being more accessible to quantitative description.

This concept can be explained more formally by assuming that a structural element, E^S, is placed at a pixel, P_i, and forms an image, $E(P_i)$, of the same size as the initial image, P. With this notation the process of comparing the surroundings of a given pixel to the structural element E is equivalent to studies of the relationship between the common part of the images P and $E(P_i)$, $C_{PE}(P_i)$, on one hand, and $E(P_i)$ on the other (see Figure 4.20). Three cases of the relationship between these images are of special interest:

a) $C_{PE}(P_i)$ equals $E(P_i)$;
b) $C_{PE}(P_i)$ differs from $E(P_i)$;
c) $C_{PE}(P_i)$ is included in $E(P_i)$.

For each of the cases a) to c) the brightness of the pixel P_i is changed according to some specified rules. For example, one may decide that:

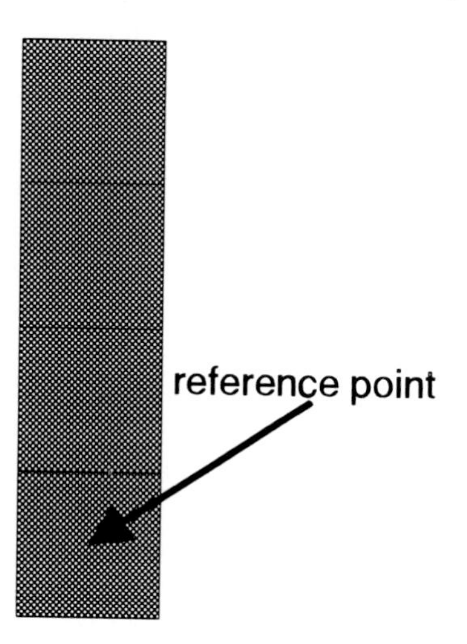

Figure 4.19 Structural element E^4_{bar} having the insertion point P, bar shape, and size of four.

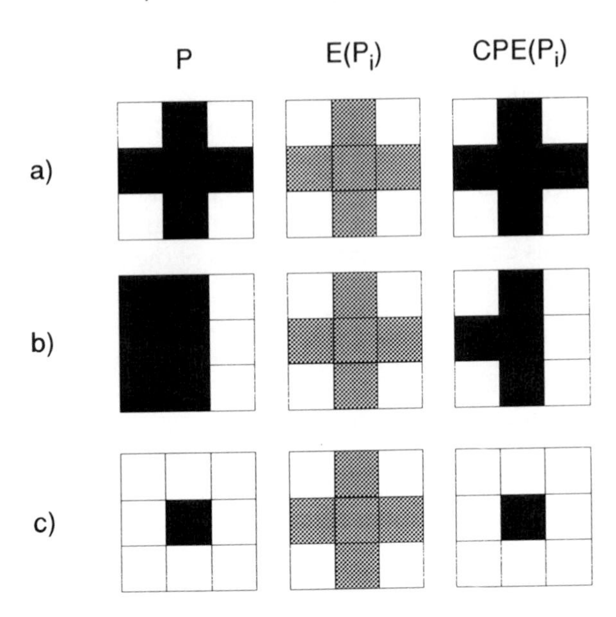

Figure 4.20 The rules for changing the intensity of a pixel in a mathematical morphological operation according to the result of comparison between the common part of the images, P, and $E(P_i)$, $C_{PE}(P_i)$, on the one hand, and $E(P_i)$ on the other.

If $C_{PE}(P_i) = E(P_i)$ then $P_i = P_i$
and
$P_i = 0$ otherwise.

This concept is schematically explained in Figure 4.20. In this case the structural element, E, is a "cross" containing five pixels. In all cases this element is placed in the central pixel of the image P and the common part of the image and the element C_{PE} is obtained. It can be noted that for (A) $C_{PE} = E$.

The process of logical analysis of the relationship between the structural element and elements of the image studied is called an operation of mathematical morphology. Operations of mathematical morphology result in thoughtful transformations of the digital images of microstructures. For instance, an erosion operation, denoted by $\ominus$, is a transformation that excludes all the image points, P, where the surroundings are different from the structural element E as shown in Figure 4.21.

The example shown in Figure 4.22 illustrates how by a proper design of a structural element, combined with an operation on the image, one can obtain a transformed image that images information on a specific aspect of the initial image. By using a vertical bar element and eroding the image the letter E has been transformed into three small bars indicating the vertical bar structure of the image analyzed.

In order to achieve a required result in more complex situations, not one but many of the mathematical morphological operations should be applied. The sequence of their application requires a knowledge of the operation characteristics. The following directions are worth taking into consideration.

structural element	binary image	image after erosion
	0 0 0 0 0 0 0 0 0	0 0 0 0 0 0 0
1 1 1	0 1 1 1 1 1 1 0 0	0 0 0 0 0 0 0
E=1 1 1	O=0 1 1 1 1 1 1 0 0	O⊖E=0 0 0 1 1 0 0
1 1 1	0 0 1 1 1 1 1 1 0	0 0 0 0 0 0 0
	0 0 0 0 0 0 0 0 0	0 0 0 0 0 0 0

Figure 4.21 The schematic concept of the erosion operation, which excludes (assigns 0 value) all the image points, P, where the surroundings (common part of the image and structural element) are different from the structural element E.

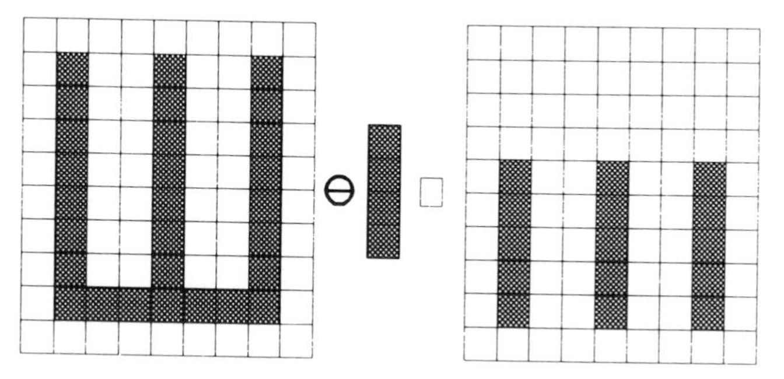

Figure 4.22 Example of the use of a proper structural element, E_{bar}^4, (shown in Figure 4.19) in an erosion operation to visualize the vertical nature of the image.

1. Mathematical morphological operations should be preceded by information reduction, which is achieved by means of the two first measures mentioned above, i.e., transformation of a digital image into a binary one and reducing the observation area (but only in such a way that does not result in a loss of relevant information).
2. The simplest operations, if possible, should go first. It should be pointed out that the transformations of mathematical morphology result in distortions of the initial images and the transformed images can be used only for specific purposes.

The operations of mathematical morphology do not, or should not, change the characteristics studied. They make it more accessible for a quantitative description, especially a computer-aided one.

4.1.4.2. Structural elements

As has been mentioned, a structural element is associated with a reference point, a shape, and a size. Its size is denoted by a single number that determines the scale of the structural element. It is frequently called a radius of the element (although structural elements are usually not circular; for example, the size of a bar-shaped structural element is characterized by the length of the segment). The shape of structural elements used in given studies should be carefully designed in accordance with their purpose. The shape should be adjusted to take into account features pertaining to the digital image. In general practice two types of elements are most common:

—square elements E_s;
—hexagonal elements E_h.

From a theoretical point of view a structural element with a circular shape and with a reference point at its the center is frequently adopted (Figure 4.23a). However, in the real systems encountered in image analysis a discrete image (digital or binary) is employed, which makes applying a structural element of this type impossible. A structural element of hexagonal shape with a side equal to the distance between adjoining pixels and a reference point in its center (Figure 4.23a) is applied in the hexagonal grid method. Another structural element used to approximate circles has a square shape with its side equal to triple the pixel size and a reference point in its center, and is applied in the case of a square grid (Figure 4.23a). Structural elements of other shapes like crosses or octagons (Figure 4.23a) are also applied using a square grid.

The size of a structural element is characterized by a linear dimension, such as its radius. In image analyzers the size/radius of a structural element is described in terms of the number and size of the pixels covering the linear dimension chosen. Therefore, the size of an applied structural element is

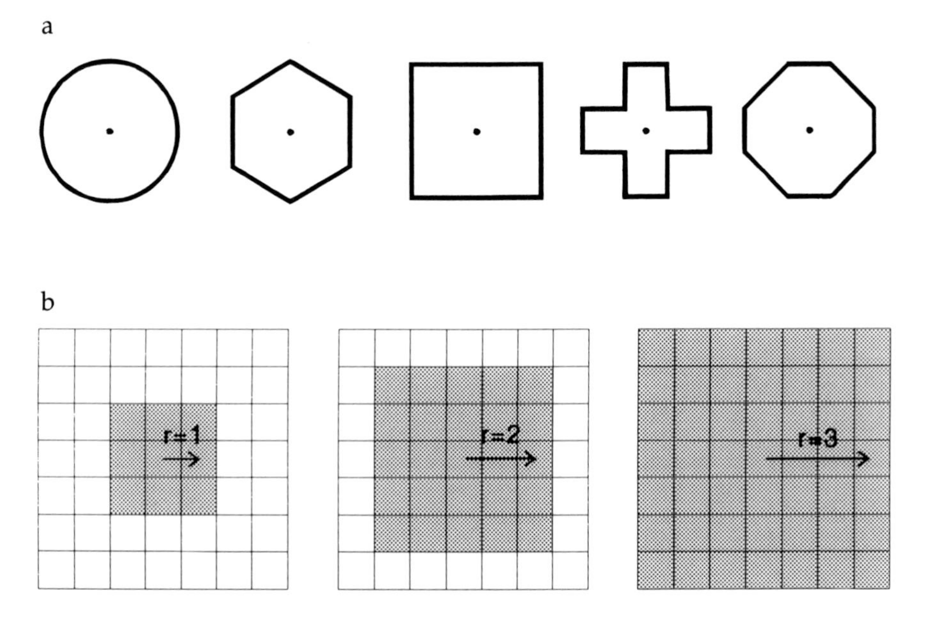

Figure 4.23 Examples of some popular structural elements (a) and square elements of different size (b).

expressed in nonnegative integer numbers that define the number of pixels added to the central point in a given characteristic direction. (For instance the 3×3 pixels square element has a radius equal to 1, because it has 1 pixel added to the central pixel in the specific direction—it could be to the left, right, top, or bottom.) Square elements of radius equal to 1,2, and 3 are shown in Figure 4.23b.

4.1.4.3. Classification of mathematical morphological operations

Mathematical morphological operations can be divided into three classes according to the degree of complexity and their means of realization:

I. Simple operations realized by a single comparison of the surroundings of all image points with a structural element

$$P \to P' \tag{4.17}$$

II. Complex operations realized as a strictly defined sequence of simple operations

$$P \to P' \to P'' \cdots \to P'^{n} \tag{4.18}$$

III. Iterative operations realized as a sequence of simple and/or complex operations, which are completed at the moment of achieving convergence to a certain extent (usually after achieving the point at which repetition of a given operation does not bring a change in the image)

$$P'^{i} = P'^{i+1} \tag{4.19}$$

or with the operation being repeated until the differences between the images are negligibly small:

$$P'^{i} \simeq P'^{i+1} \tag{4.20}$$

4.1.4.3.1. Basic operation of class I

4.1.4.3.1.1. Erosion

Definition:

In Euclidean space, an erosion operation of image O, denoted by $\ominus$, is defined as follows:

$$O \ominus E = P_i \hat{\in} O \qquad C_{OE}(P_i) \cap E \neq E \Rightarrow P_i = 0 \tag{4.21}$$

Characteristics:

Erosion eliminates "islands" smaller than a structural element and diminishes bigger areas by a "thin strip" as wide as the structural element. Erosion eliminates all the points of the O image, the surroundings of which are different from the structural element, E.

Properties:

Erosion is an antiextensive operation, i.e., it transforms some points of an object (brightness 1) into background points (brightness 0), whereas it does not transform any background points into object points.

Comment:

Erosion is an operation preserving a set-containing relation:

$$O \subset O' \Rightarrow (O \ominus E) \subset (O' \ominus E) \tag{4.22}$$

Iterativeness is another important property of erosion:

$$O \ominus nE = O \underbrace{\ominus E \ominus E \cdots \ominus E \ominus E}_{n\ times} \tag{4.23}$$

where n is the radius of the structural element. The property allows one to perform erosion operations by a structural element of any radius in a simple way, providing the method of the erosion operation by a structural element of unit radius is known.

Example 4.5

As an example, erosion can be used to eliminate from a real image small artifacts due to etch pits, dust, etc. An example of an erosion application is shown in Figure 4.24.

4.1.4.3.1.2. Dilation

Definition:

In Euclidean space the dilation operation, denoted by ⊕, is defined as follows:

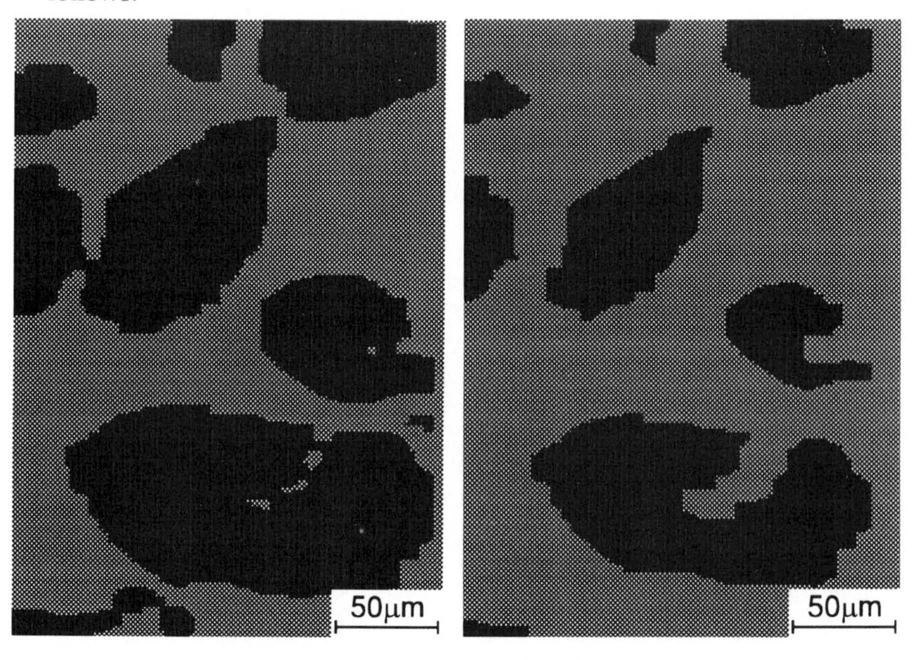

Figure 4.24 Example of an erosion operation.

$$O \oplus E = P_i \hat{\in} O \quad C_{OP}(P_i) \cap E \neq \varnothing \Rightarrow P_i = 1 \tag{4.24}$$

Characteristics:

Dilation transforms all the pixels of the background image, with brightness 0, into object pixels if there is at least one object pixel in the common part of structural element, E, and the image. Dilation "fills" all "holes" and "hollows" not bigger than a structural element and expands all object areas by "a strip" as wide as the structural element.

Properties:

Dilation is an extensive operation, i.e., it transforms some of the background pixels (brightness 0) into the object pixels (brightness 1), whereas object points are never transformed into background points.

Comment:

Dilation is an operation preserving a set-containing relation, i.e.:

$$O \subset O' \Rightarrow (O \oplus E) \subset (O' \oplus E) \tag{4.25}$$

Dilation, similarly to erosion, has an iterativeness property:

$$O \oplus nE = \underbrace{O \oplus E \oplus E \cdots \oplus E \oplus E}_{n\ times} \tag{4.26}$$

where n is the radius of a structural element. This property allows one to perform a dilation operation by a structural element of any radius in a simple way, providing that the method of the dilatation operation by a structural element of unit radius is known.

Example 4.6

An example of dilation application is shown in Figure 4.25.

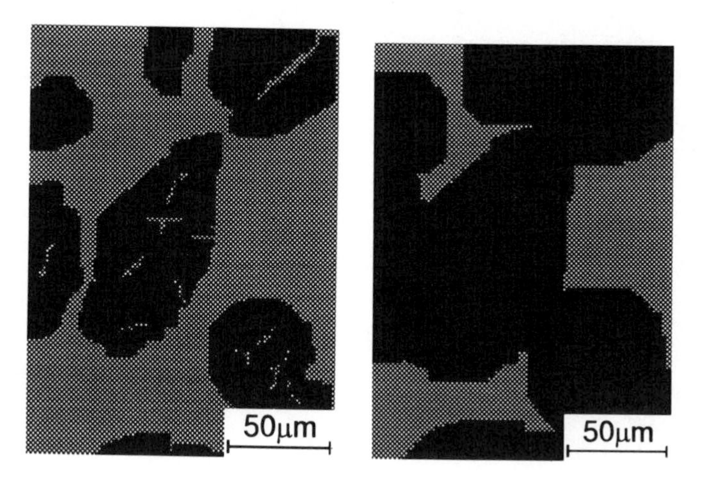

Figure 4.25 Example of a dilation operation.

4.1.4.3.1.3. *Negation*

Definition:

In Euclidean space a negation operation, denoted by "–", is defined as follows:

$$-O = P_i \hat{\in} O \quad \begin{cases} P_i = 0 \Rightarrow P_i = 1 \\ P_i = 1 \Rightarrow P_i = 0 \end{cases} \tag{4.27}$$

Characteristics:

Negation transforms all the pixels of the image of brightness 1 into pixels of the background, with a brightness 0, and the background pixels into the image pixels.

Properties:

Negation is an operation reversing a set-containing relation, i.e.:

$$O \subset O' \Rightarrow -O \supset -O' \tag{4.28}$$

Comment:

Negation is an operation realized by means of a point structural element (element of zero radius). The negation operation is similar to the well-known effect of brightness reversal during photographic printing of black and white negatives.

Example 4.7

An example of the application of the negation operation is shown in Figure 4.26.

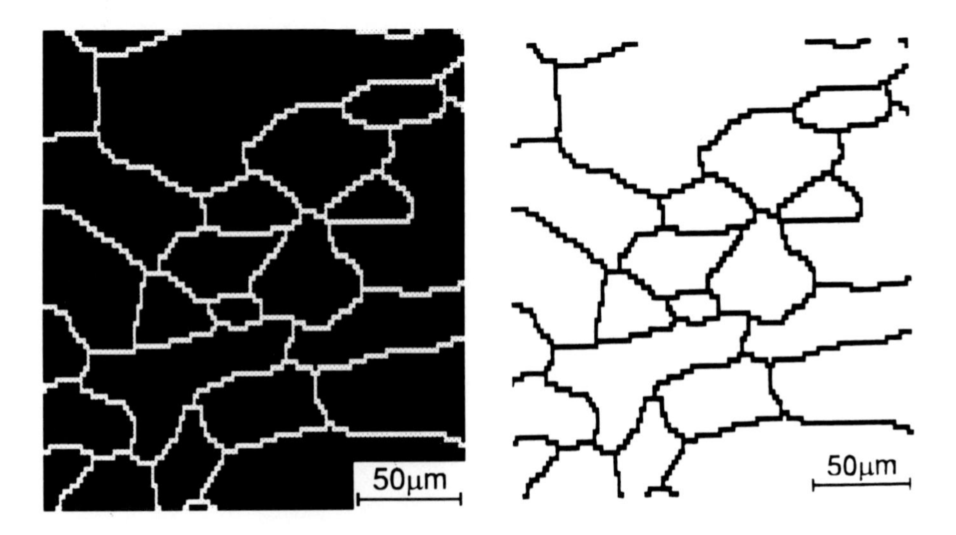

Figure 4.26 Example of a negation operation.

4.1.4.3.2. *Basic operations of class II*

4.1.4.3.2.1. *Opening*

Definition:

In Euclidean space the opening operation, denoted by ○, is simply a combination of erosion and dilation by the same structural element E:

$$O \circ E \equiv \left(O \ominus E\right) \oplus E \tag{4.29}$$

Characteristics:

The opening operation eliminates all "parts" of the image that form "islands" or "peninsulas" of a size smaller than a structural element, E.

Properties:

Opening, similarly to erosion, is an antiextensive operation, i.e., it transforms some of the object points (brightness 1) into background points (brightness 0), whereas none of the background points are transformed into object points.

Comment:

Opening is an operation preserving a set-containing relation, i.e.:

$$O \subset O' \Rightarrow \left(O \circ E\right) \subset \left(O' \circ E\right) \tag{4.30}$$

Opening does not alter a convex set of points, whose size is larger than the structural element. The application of successive opening operations with a structural element of growing size together with the procedure of counting features, may be a simple method of defining the size distribution of particles of simple convex shapes.

Example 4.8

An application of opening is shown in Figure 4.27.

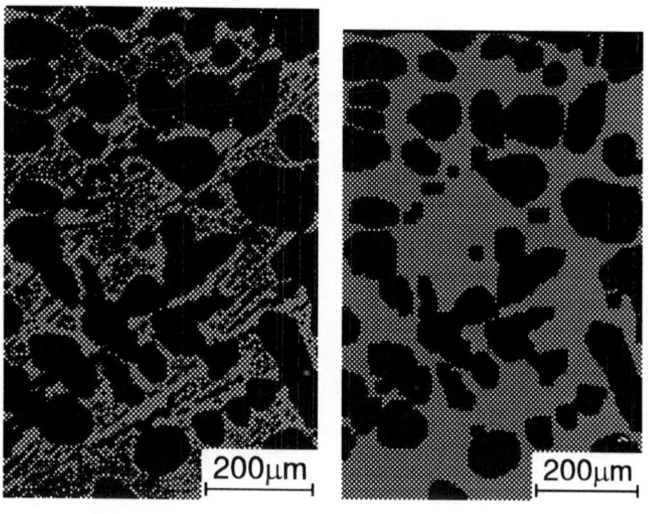

Figure 4.27 Example of an opening operation.

4.1.4.3.2.2. Closing

Definition:

In Euclidean space the closing operation, denoted by ●, is a simple combination of dilation and erosion by the same structural element E:

$$O \bullet E \equiv (O \oplus E) \ominus E \tag{4.31}$$

Properties:

Closing, similarly to dilation, is an extensive operation, i.e., it transforms some background points, (brightness 0), into object points (brightness 1) whereas none of the object points is transformed into background points.

Comment:

Closing is an operation preserving a set-containing relation:

$$O \subset O' \Rightarrow (O \bullet E) \subset (O' \bullet E) \tag{4.32}$$

Closing, as well as opening, is not an iterative operation. It does not change a convex set of any size.

Example 4.9

An example of a closing application is shown in Figure 4.28.

4.1.4.3.3. Basic operations of class III

4.1.4.3.3.1. Thinning

Definition:

In Euclidean space, a thinning operation, denoted by ⊙, is a superposition of k operations employing structural elements, E_j, generated as a result of rotation of element E, by assigned angles $\alpha_1, \alpha_2, \ldots, \alpha_k$.

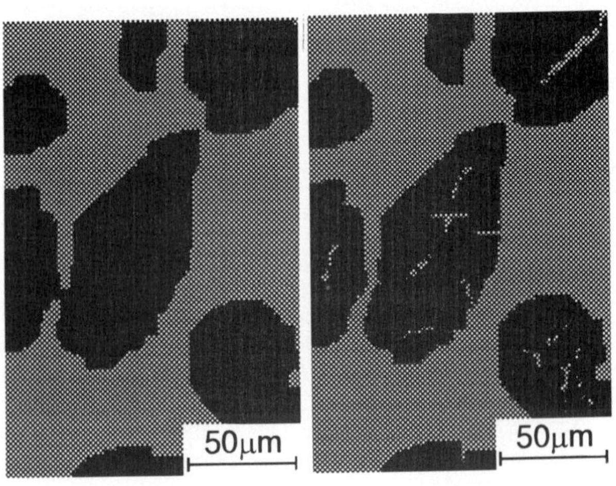

Figure 4.28 Example of a closing operation.

$$\odot : \hat{j}\left\{P_i \hat{\in} O \quad C_{OE}(P_i) \cap E_j = E_j \Rightarrow P_i = 0\right\} \quad j = 1, 2 \ldots k$$
$$E_j - E \text{ rotated by } \alpha_j \tag{4.33}$$

The operations are repeated until the whole cycle j = 1 . . . k does not change the image. A typical structural element used in a square grid for a thinning operation is given as follows:

$$\begin{array}{ccccc} \textit{initial} & \textit{after } \alpha_1 & \alpha_2 & \alpha_3 & \alpha_4 \\ \textit{element} & \textit{rotation} & & & \\ E & E_1 & E_2 & E_3 & E_4 \\ 1\,1\,1 & 1\quad 0 & 0\,0\,0 & 0\quad 1 & 1\,1\,1 \\ 1 & 1\,1\,0 & 1 & 0\,1\,1 & 1 \quad \textit{etc.} \\ 0\,0\,0 & 1\quad 0 & 1\,1\,1 & 0\quad 1 & 0\,0\,0 \end{array} \tag{4.34}$$

Characteristics:

A thinning operation transforms "thick" lines forming closed figures in the image into a line grid of unit width.

Properties:

Thinning, similarly to erosion, is an antiextensive operation, i.e., it transforms some object points, brightness 1, into background points, brightness 0, whereas background points are not transformed into object points.

Comment:

Thinning is neither an operation preserving a set-containing relation nor is it an iterative operation.

Example 4.10

An application of a thinning operation is shown in Figure 4.29.

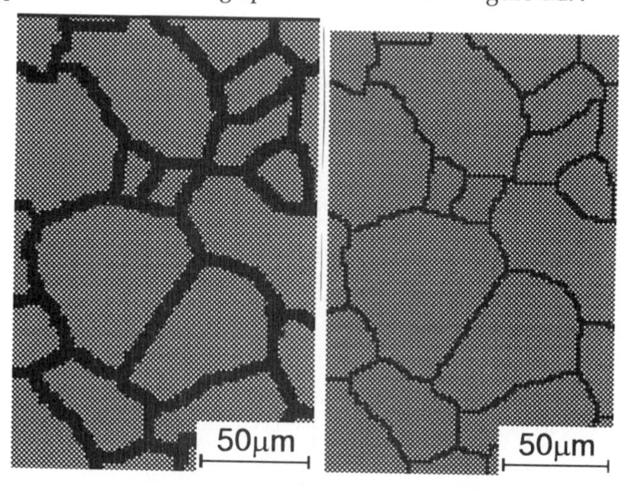

Figure 4.29 Example of a thinning operation.

4.1.4.3.3.2. *Ultimate erosion*

Definition:

In Euclidean space, an ultimate erosion operation, denoted by ∅, is defined as follows:

$$O \varnothing E = P_i \hat{\in} O \quad (C_{OE}(P_i) \cap E \neq E) \wedge ((C_{OE}(P_i) - P_i) \neq \varnothing) \Rightarrow P_i = 0 \quad (4.35)$$

The operation is repeated until repetition does not change the image, i.e:

$$O \varnothing E = O \quad (4.36)$$

Characteristics:

An ultimate erosion transforms continuous convex areas (subsets) into single points. Nonconvex areas are transformed into several points.

Properties:

An ultimate erosion is an antiextensive operation, i.e., it transforms some object points into background points whereas background points are not transformed into object points.

Comment:

An ultimate erosion operation neither preserves a set-containing relation nor is it an iterative operation.

Example 4.11

An application of an ultimate erosion operation is shown in Figure 4.30.

4.1.4.3.3.3. *Dilation without touching*

Definition:

In Euclidean space, a dilation without touching operation, denoted by ⊗, is defined as follows:

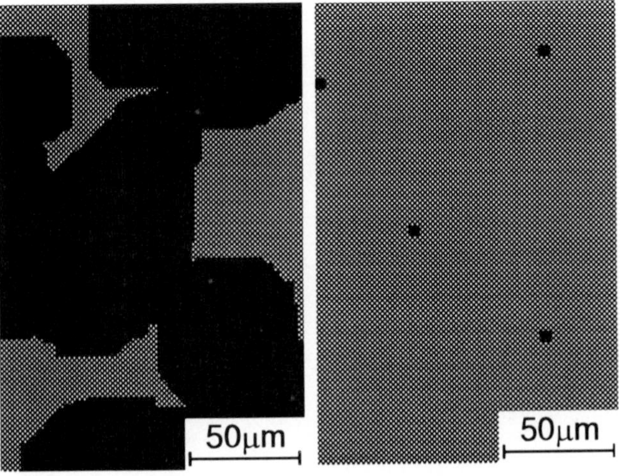

Figure 4.30 Example of an ultimate erosion operation.

$$O \otimes E = P_i \hat{\in} O \quad (C_{OE}(P_i) \cap E \neq \varnothing) \wedge (E - C_{OE}(P_i)) \neq \varnothing) \Rightarrow P_i = 1 \quad (4.37)$$

The operation is repeated until repetition does not change the image, i.e.:

$$O \otimes E = O \quad (4.38)$$

Characteristics:

Dilation without touching places a condition on the dilated image: the background of the image must never be thinned to less than a single pixel width.

Properties:

Dilation without touching is an extensive operation, i.e., it transforms some background points into object points whereas object points are not transformed into background points.

Comment:

Dilation without touching preserves a set-containing relation. It is not an iterative operation. Dilation without touching is sometimes also referred to as exoskeletonization or conditional thickening. Dilation without touching is often preceded by an ultimative erosion operation. This coupling of operations works as a tool for the reconstruction of a continuous network of grain boundaries, for instance.

Example 4.12

An application of a dilation without touching is shown in Figure 4.31.

4.1.4.3.3.4. *Skeletonization*

Definition:

In Euclidean space, a skeletonization operation, denoted by ◇, is defined as follows:

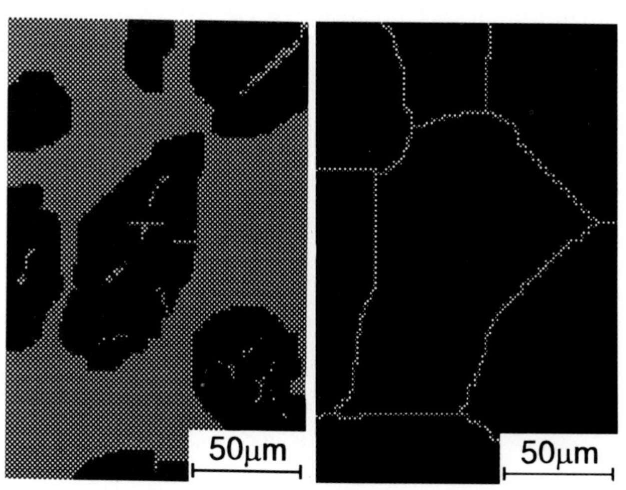

Figure 4.31 Example of a dilation without touching operation.

$$O \diamond E = P_i \hat{\in} O \quad (C_{OE}(P_i) \cap E \neq E) \wedge ((C_{OE}(P_i) - P_i) \neq \varnothing) \Rightarrow P_i = 0 \tag{4.39}$$

The operation is repeated until a repetition does not change the image, i.e.:

$$O \diamond E = O \tag{4.40}$$

Characteristics:
Images produced by skeletonization consist of pixels, which in the initial features had at least two nearest boundary points (the vast majority of pixels have only one nearest boundary point). The resulting feature is called a skeleton. For an initial circular feature, the skeleton is its central point, for a square feature the skeleton becomes a cross, and for an ellipse a line skeleton results. In practice skeletons of features are often more complicated than they should be, because of problems due to the digital nature of the image.

Properties:
Skeletonization is an antiextensive operation, i.e., it transforms some object points into background points whereas background points are not transformed into object points.

Comment:
Skeletonization neither preserves a set-containing relation nor is it an iterative operation.

Example 4.13

An application of a skeletonization operation is shown in Figure 4.32.

4.1.5 *Measurements*

Digitization and morphological operations make it possible to effect efficiently a variety of measurements on the images of the microstructural elements. The typically measured parameters are:

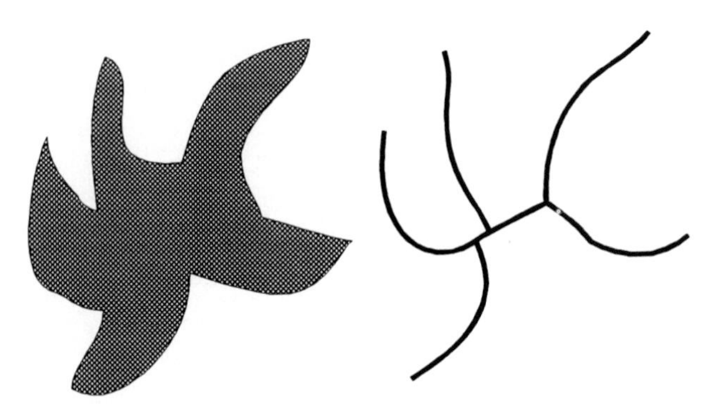

Figure 4.32 Example of a skeletonization operation.

a) number of objects inside the observation field;
b) length of linear elements;
c) area of planar elements.

The latter two parameters can be measured by counting the number of pixels forming the image of a given element. The number of objects can be obtained using any operation that reduces an object to a representative point. In all these cases the measurements are executed as sequences of logical and algebraic operations on the matrixes representing images of the microstructures. There are a number of algorithms that can be used to increase the speed of the measurements; however, formulation of the basic procedures does not require extensive experience in mathematical programming.

Measurements of intercept length are one of the basic operations carried out on single- and multiphase microstructures. In the case of a two-phase microstructure, an algorithm for horizontal intercept length measurements requires the following operations:

1. formation of a digital image matrix, A_{ij};
2. a threshold operation resulting in a binary image in which the particles studied are white and the background is black: if $A_{ij} < p$ then $A_{ij} = 0$, otherwise $A_{ij} = 1$;
3. measurements of horizontal intercepts along a line a specified distance from the edges of the observations field:
   3.1. select the row index i;
   3.2. starting from $j = 1$,
   3.3. find j^+ such that $A_{ij} = 1$; find the lowest j^- such that: $j^- > j^+$ and $A_{ij} = 0$; compute the distance $l_k = (j^- - j^+)$;
   3.4. repeat 3.3. until a whole row of pixels is analyzed;

These operations are illustrated in Figure 4.33.

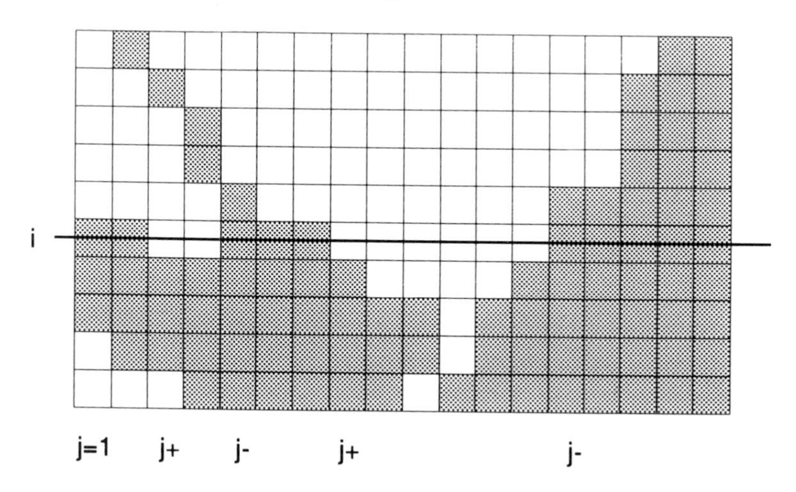

Figure 4.33 Schematic sequence of operations during measurements of intercept length in a two-phase microstructure.

Further geometrical parameters that can be measured on microstructural elements are shown in Figure 4.34. In the case of 2-D elements they include the following parameters:

—center of gravity;
—area;
—equivalent circle diameter;
—perimeter;
—maximum chord;
—orientation of maximum chord;
—width;
—Feret diameters;
—number of neighbors;
—curvature;
—Cauchy perimeter;
—shape factors.

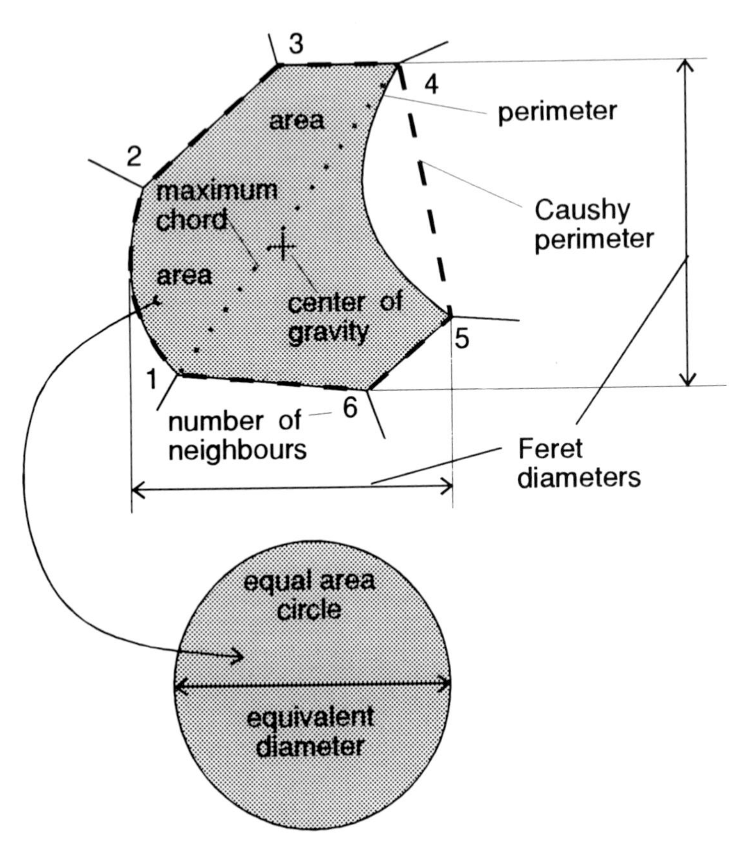

Figure 4.34 Schematic representation of main geometrical parameters of a 2-D microstructural element (grain or particle).

4.1.5.1. Fractal aspects of image digitization

The division of an image into a set of finite-size elements has important implications in the problem of the definition and measurement of the size of elements. It can be shown, for example, that the measured size of a feature depends on the size of the pixels. This effect is especially pronounced in the case of features of a complicated shape, distinguished by a high fraction of pixels on their perimeter. In particular, the measurement of the length of a curve would yield a series of values, $l_1 < l_2 < \ldots < l_n$, for a varying size of pixel employed, $d_1 > d_2 > \ldots > d_n$. In this situation the results of some measurements, such as the length of a curve, can be properly interpreted only if the size of the pixel is known. However, the results of these measurements can be made size-of-pixel independent using the concept of fractals (see Chapter 1).

The fractal dimension of a line is a dimensionless measure of its roughness. It is independent of (and can be measured without defining or establishing) the size of the pixel used. This concept in the present context can be introduced in the following way. Consider the profiles shown in Figure 4.35, which could be images of a particle section, a segment of a trace of a grain boundary on a section, or a profile of a fracture surface. Measuring the line profile using a varying precision of measurement (i.e., different lengths of yardsticks) gives a perimeter length which is a function of this precision. Just to make this concept clear, consider a very simple example. Equip a selection of people (students) with an array of different "yardsticks"—say one 100 m in length, one 1 m long, one 1 cm long, etc. Now ask them to measure the perimeter of your favorite country. It should be evident to you that the person measuring with 1 cm "yardstick" will find the perimeter much longer than the person using the "yardstick" which is 1 m, etc. (see also Figure 4.36).

The relationship between the "sampling distance" and resulting perimeter length of a feature has been suggested by Richardson to be interpreted in terms of the following relationship:[5]

$$L(r) = Mr^{1-D} \tag{4.41}$$

where: L(r) is the length of perimeter measured at a "sampling distance" r,

M is a constant,

D is the fractal dimension (the term fractal dimension was introduced by Mandelbrot[6]).

The values of D for a particular feature are usually found to be constant for a wide range of r values. Ploting log(L(r)) against log(r), in this range, one obtains a line of slope (1 – D).

The fractal dimension can be used to compare roughness of lines surrounding features of different dimensions, regardless of differences in the overall shape of the features compared.

Flook[7] has shown that one of the mathematical morphological operations, closing, is equivalent to increasing the sampling distance for perimeter

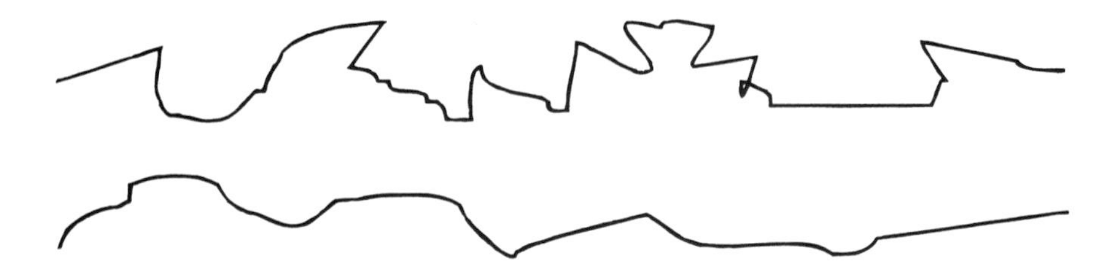

Figure 4.35 Examples of different profiles of brittle/ductile fracture surfaces. The length of the profile lines depends on the precision of the measurements.

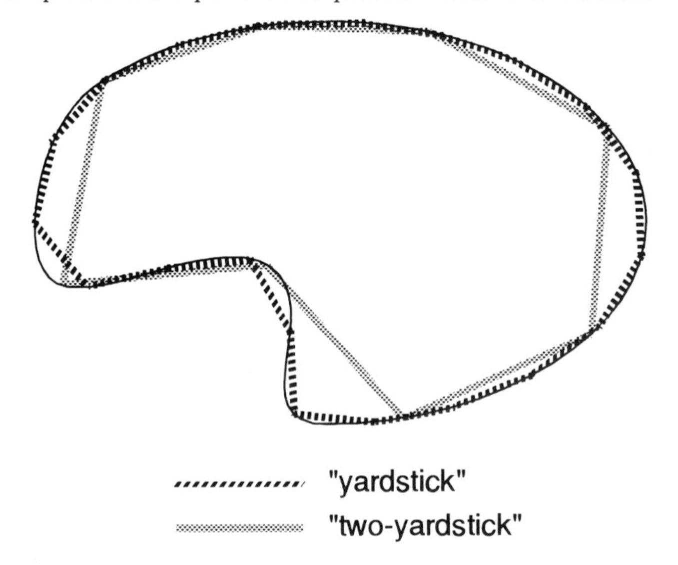

Figure 4.36 Example of measurements of the perimeter of any feature with different "yardsticks". General rule is that the longer the "yardstick" the shorter is the measured perimeter. This is caused by approximating the length of any line by the distance between line ends.

measurements. Taking this into account one can derive the fractal dimension by plotting log(perimeter) against log(diameter of the structural element).

Example 4.14

Two fracture profiles shown in Figure 4.35 are representative of a mixed mode brittle/ductile fracture of α-iron samples. The length of the profiles has been measured by means of an image analysis system with different sampling distances varying from 1 pixel to 20 pixels. The log-log plot of the measured line length against sampling distance for both fracture profiles are shown in Figure 4.37. One can see that in both cases there is quite a good fit between the experimental points and a regression line, but the slopes are substantially different. This observation leads to the conclusion that the nature of the fracture process strongly influences the fractal dimension of the fracture surface.

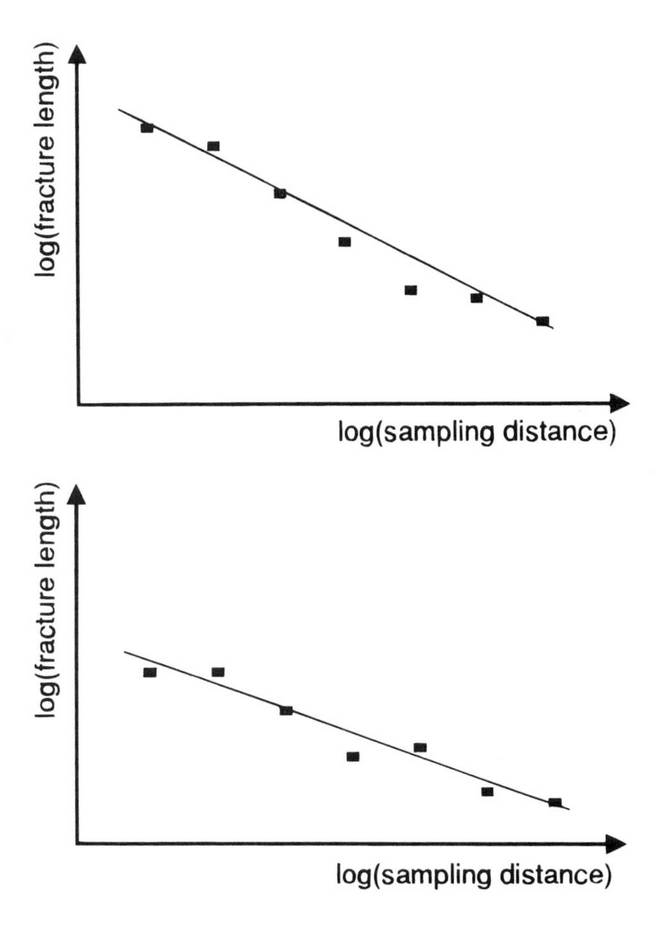

Figure 4.37 Log-log plots of profile length vs. sampling distance for the two fracture profiles shown in Figure 4.35 showing good agreement with Equation 4.41 and differences between the fracture profiles shown in Figure 4.35.

4.2 Computer-aided procedures of data analysis

A modern experimental approach often demands a precise description of the morphology and distribution of particles, grains, and other features forming the microstructure of a material. In the case of polycrystals, for example, such a description is required to decide whether the images of random sections of grains differ among themselves or in relation to a certain model structure.[8–13] In systems of second-phase particles, a description of their positions might help to verify a hypothesis for their preferential location one to another or to some other features in the microstructure. In the past, microscopical practice has been limited, in the main, to calculating "average parameters" such as the mean grain size; whilst the description of shape, size, and position of individual microstructural elements has mostly been of a qualitative character. However, more recently, computerized image analysis systems have become widely available that: (a) permit measurements on a large number of features in the microstructure and (b) can be used to measure various complicated parameters describing position, size, and shape, which were not accessible sensibly by manual procedures. These new computerized techniques have resulted in extensive data being collected from the microstructures studied and also have created a demand for special procedures of data analysis.

Populations of particles, grains, and other microstructural features are characterized by a diversity of geometrical parameters defining their position, size, and shape. Moreover, the populations are frequently highly inhomogeneous and can be viewed as consisting of a number of subpopulations distinguished by a consistently lower spread of the measured parameters. In such situations, it is often more appropriate to distinguish and describe each of the subpopulations than to characterize the overall population by any average property.

Two-dimensional images of grain boundaries revealed in a cross-section of a polycrystal are an example of the variety of geometrical forms observed in the microstructures of materials. The traces of grain boundaries emerging on the surface of observation divide the plane into 2-D grains that differ in size (area), degree of elongation, number of sides, etc. These differences are exemplified by the microstructure shown in Figure 4.38. The question then arises as to what degree the images characteristic of one material microstructure are similar to some model or reference structure (such as a system of regular, constant-size polyhedra) and how to interpret differences in the distribution of the geometrical features among materials that have been processed differently. The same question is also frequently asked in the observations of particle sections in studies of multiphase materials.

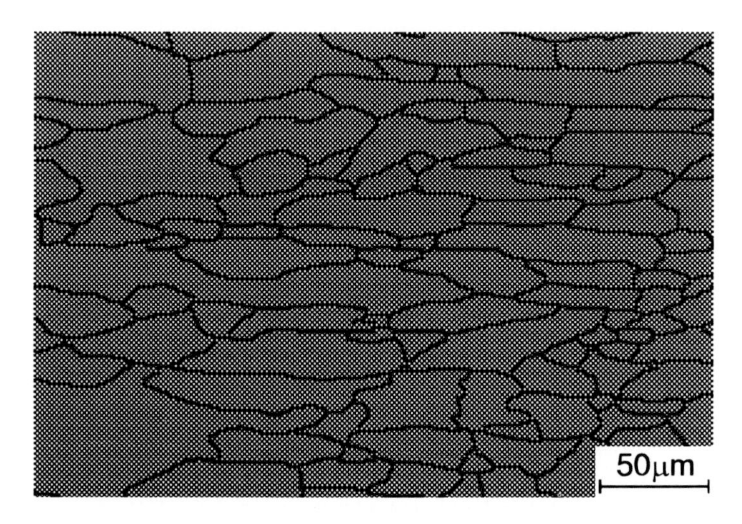

Figure 4.38 Example of the microstructure of a deformed 316L steel showing grains of different size, degree of elongation, number of sides, etc.

Considerable developments in the methods of computer analysis of metallographic images have made it possible to measure many of the parameters describing microstructural features. Typical measurable parameters include:

- x_i the position of the feature (1, 2, or 3 coordinates and possibly time, t);
- A_i its area;
- $(d_2)_i$ its equivalent diameter (defined as the diameter of a circle, having the same area as the element under examination);
- p_i its perimeter;
- n_i its number of sides;
- d_{max} its maximum dimension (chord);
- F_i its Feret diameter.

Other parameters that can be used have been described in the preceding sections of this chapter.

Any of the parameters listed, say z, can be studied separately in the form of a distribution function, f(z). However, it has been shown that in a number of applications it makes more sense to have them analyzed all together[10,13,14] as points in multidimensional space $\mathbf{R}^k$ where k is the number of parameters. This concept is illustrated in Figure 4.39 for the grains shown on Figure 4.38 using two parameters: the grain area, A, and the number of sides, n. This space combines two parameters which define the geometry of grains. Another example of such a space would be a 4-D space (d_2, p, n, d_{max}) in which a point defines

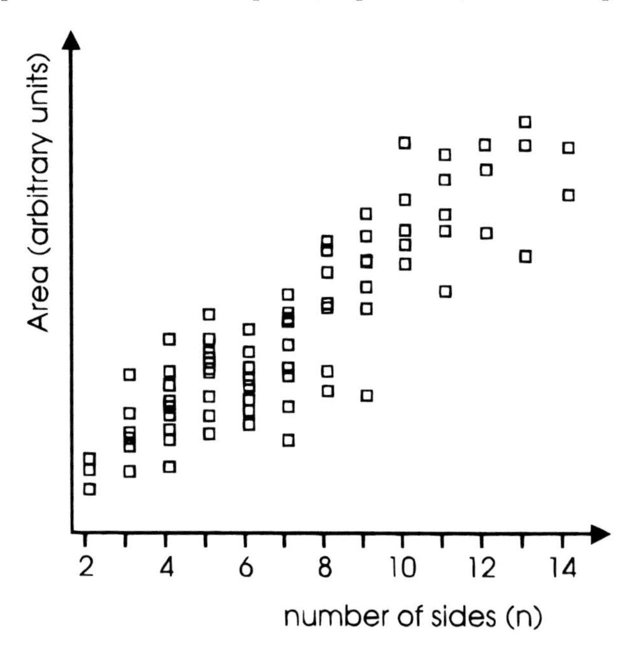

Figure 4.39 Grain area vs. number of sides for the microstructure shown in Figure 4.38.

equivalent diameter, perimeter, number of sides and the maximum chord of the microstructural element, for example, a grain seen on a section of a polycrystal. In some cases the descriptive space can be reduced to the space of positions, $\mathbf{R}^3$ (plus possibly time, t). In a more general case it may combine both spatial coordinates and other factors describing the microstructural features. An example of such a space is discussed in Example 4.15.

Example 4.15

Figure 4.40 shows a microstructure revealed on a perpendicular section through a nitrided steel specimen. Measurements of the particle section area, A, and particle distance from the surface, z, were carried out for randomly selected sections of the particles. The results of the measurements are plotted in Figure 4.41 in the coordinate system (A, z).

The features of a grain, particle, or other element in a space of parameters is represented by a point; while the population of these elements in the microstructure is represented by a set of points. In general, these sets of points are inhomogeneous and local clusters of points are observed. Such clusters indicate the existence of patterns in the factors describing the elements studied. This may be of a great importance in an understanding of the processes taking place in the microstructure and its properties.

A description of the population of microstructural elements, (grains, particles, etc.) in a multidimensional space of characteristic parameters, as well as its classification, may be made by using an appropriate function defining "distance" between points representing the properties of individual elements. Such a function defines a metric in the multidimensional space of parameters. It enables us to separate characteristic groups, e.g., subpopulations of grains, in an objective way using the methods of cluster analysis.

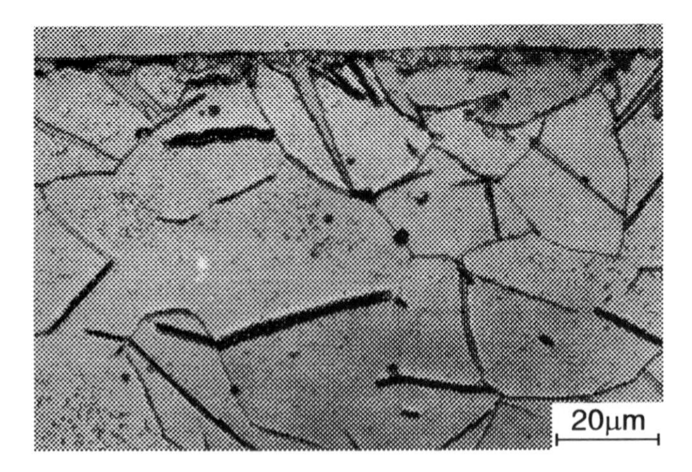

Figure 4.40 Example of the microstructure of a nitrided sample of steel showing changes of microstructure with distance from the sample surface.

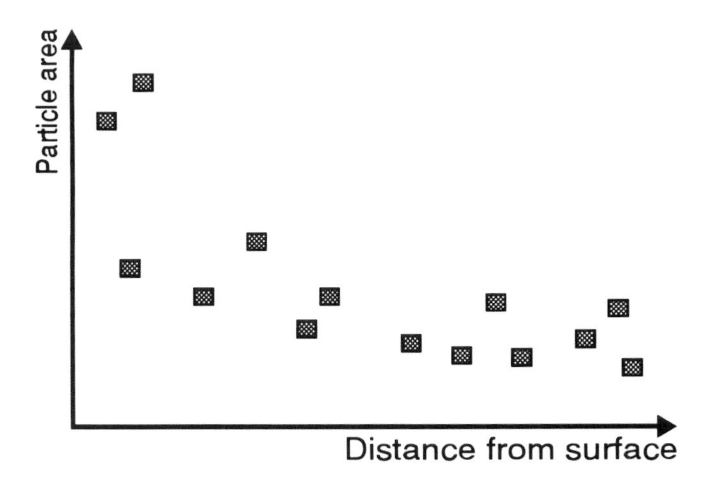

Figure 4.41 Particle area vs. distance from the surface for the nitrided steel sample in Figure 4.40.

Consider a population of microstructural elements with each of them being described by L parameters. The observed values of the parameters of the first element are marked by x_{1i}, of the second element by x_{2i}, and the values of the parameters for element k are x_{ki} (i = 1,2, . . ., L). The description of the population studied is given in the form of a matrix, X_{ij}:

$$X_{ij} = \begin{matrix} x_{11} & x_{12} & \cdots & x_{1L} \\ x_{21} & x_{22} & \cdots & x_{2L} \\ \cdots & \cdots & \cdots & \cdots \\ x_{k1} & x_{k2} & \cdots & x_{kL} \end{matrix} \tag{4.42}$$

In this matrix the rows X_i (i = 1,2, . . ., k) define the parameters measured for a given microstructural element and the columns X^m (m = 1,2 . . ., l) define the values of a given parameter measured for the elements studied.

Since the parameters forming the matrix X_{ij} have defined physical meanings, they also are expected to be measured in specific units. In order to avoid complications arising from the fact that the parameters may have assigned to them different units, it is desirable to make the matrix dimensionless. This can be done by dividing the matrix elements by their respective units:

$$X_{ij} = \frac{X_{ij}}{\left[unit\ X_{ij}\right]} \tag{4.43}$$

or by normalizing them with respect to the mean values of the parameters:

$$X_{ij} = X_{ij} \frac{1}{E_j\left(X_{ij}\right)} \tag{4.44}$$

where $E_j(X_{ij})$ stands for the mean value of the "j" parameter and is defined as:

$$E_j\left(X_{ij}\right) = \frac{1}{k}\sum_{i=1}^{i=k} X_{ij} \tag{4.45}$$

This is illustrated by the following examples.

Example 4.16

A population of grain sections revealed in a cross-section of a polycrystal (see Figure 4.38) can be described by the following parameters: grain section area, A, grain section perimeter, p, number of edges, n, and maximum chord length, d_{max}. This leads to the following matrix:

$$X_{ij} = \begin{matrix} A_1 & p_1 & n_1 & \left(d_{max}\right)_1 \\ A_2 & p_2 & n_2 & \left(d_{max}\right)_2 \\ \ldots & \ldots & \ldots & \ldots \\ A_k & p_k & n_k & \left(d_{max}\right)_k \end{matrix} \tag{4.46}$$

Each row of the matrix X_{ij} describes the geometry of one particular grain. The first column lists all measured grain section areas, the second—grain section perimeters, etc. The following mean values were obtained for the microstructure exemplified by the micrograph shown in Figure 4.38:

$E(A) = 867\ (\mu m)^2$;
$E(p) = 105\ (\mu m)$;
$E(n) = 6$ (this is true for an infinite system of grain sections);
$E(d_{max}) = 45\ (\mu m)$.

These values have been used to normalize the matrix given by Equation 4.46 into X^*_{ij}, such that:

$$X^*_{1j} = \frac{X_{1j}}{E(A)} \tag{4.47}$$

$$X^*_{2j} = \frac{X_{2j}}{E(p)} \tag{4.48}$$

$$X^*_{3j} = \frac{X_{3j}}{6} \tag{4.49}$$

$$X^*_{4j} = \frac{X_{4j}}{E\left(d_{max}\right)} \tag{4.50}$$

The X^*_{ij} matrix contains dimensionless numbers.

Example 4.17

The population of particles observed on a cross-section of a two-phase material shown in Figure 4.40 can be described by the following matrix:

$$Y_{ij} = \begin{pmatrix} z_1 & A_1 \\ z_2 & A_2 \\ \dots & \dots \\ z_k & A_k \end{pmatrix} \tag{4.51}$$

where z is the distance from a surface of the specimen and A is the area of particle section. The normalized form of this matrix can be obtained by dividing elements of the first column, Y_{1j}, by the mean distance, $E(z_i)$, and the elements of the second column, Y_{2j}, by the mean area, E(A). For the microstructure shown in Figure 4.40, E(A) = 21 μm^2 and E(z) = 530 μm.

The microstructural features represented by the values of the parameters describing their properties can be considered to be points in a multidimensional space. Any two points in such a space can have a distance between them assigned. The distance between the feature/point, X_i, and element/point, X_j, is marked with the symbol, $d(X_i,X_j)$, or in abbreviated form,—d_{ij}. For the sake of conformity with the theory of measurements, it should have the following features:

a) $d_{ij} \geq 0$,
b) $d_{ij} = 0$ for i = j,
c) $d_{ij} = d_{ji}$,
d) $d_{ij} + d_{ik} \geq d_{jk}$.

The definition of distance between features is otherwise arbitrary and depends on the feature type and the purposes for which the analysis of their groupings is meant to serve. Such a definition reflects the assumed structure of the space in which we place the features.

One possible approach to defining this distance is to assume that the points representing the features studied are placed in an orthogonal space represented by a rectangular coordinate system. Such a space is distinguished by a perpendicularity condition. It requires that for any feature say X_q, the parameters x_{qj} can be changed one at a time independently of all the others. This condition is met in the case of 2-D space (A, n) where the points define grain area, A, and number of sides, n. This is due to the fact that by changing the shape of the grains, for any value of A, the number of sides may vary from two upwards. An example of the situation where this condition is not fulfilled is the space (d_2, A) since a change in the area of grains leads to direct changes in their equivalent diameters.

In the case of orthogonal space it is possible to employ Minkowski's definition of distance, which is given by the following formula (see, for example, Reference, 15):

$$d_{ij} = \left(\sum_{k=1}^{p} \left| x_{ik} - x_{jk} \right|^{m} \right)^{1/m} \tag{4.52}$$

where m = 1, 2, ...

For the particular case of Euclidean distance (m = 2), this becomes:

$$d_{ij} = \sqrt{\sum_{k=1}^{p} \left(x_{ik} - x_{jk} \right)^2} \tag{4.53}$$

The definition of this distance is quite simple, yet it disregards differences in the range of numerical values among particular parameters employed to describe the geometry of the features of interest. Some parameters are expressed by numbers differing by orders of magnitude, as for instance, the area of grains, A, while others are expressed by numbers in a rather narrow range as, for instance, the number of sides, n. In this case the parameters with low values will have no influence upon the value of the calculated distance. On the other hand, in terms of grain morphology, they are as important as other parameters. It is desirable, therefore, to employ a generalized Euclidean distance:

$$d_{ij} = \sqrt{\sum_{k=1}^{p} w_k^2 \left(x_{ik} - x_{jk} \right)^2} \tag{4.54}$$

where w_k denotes the weighting assigned to the parameter k. By appropriate choice of this weighting, the analysis can be focused on some specific parameters of interest. In the general case the weighting can be taken as:

$$w_k = \frac{1}{E_k \left(x_{jk} \right)} \tag{4.55}$$

where $E_k(x_{jk})$ stands for the mean value of the x_k parameter over the N elements studied.

In a situation where the orthogonality of the space defined is uncertain one can apply a normalized (standardized) Euclidean distance:

$$d_{ij} = \sqrt{\frac{\sum_{k=1}^{p} \left(x_{ik} - x_{jk} \right)^2}{\sqrt{\sum_{k=1}^{p} x_{ik}^2 \sum_{k=1}^{p} x_{jk}^2}}} \tag{4.56}$$

Another option is to use a self-normalizing distance given by the following equation:

$$d_{ij} = \sum_{k=1}^{p} \frac{|x_{ik} - x_{jk}|}{|x_{ik} + x_{jk}|} \tag{4.57}$$

However, it should be stressed that the choice of a proper distance formula is not only a matter of mathematical consideration. The choice is also dictated by the interpretation of the physical parameters of interest.

Among the parameters selected for the analysis of grain geometry listed earlier, the four: A, p, n, and d_{max} are linearly independent. This means that none of them can be computed based on the values of the other three. If they are varied over some restricted range, these parameters can be assumed to form an orthogonal space in the sense that any three of them can be changed with some restrictions, while the fourth is kept constant. Such changes will involve appropriate modifications of the grain geometry. For example, grain perimeter can be increased whilst the grain area is held constant by making the boundaries more undulated.

A definition of the matrix in parameter space is essential for studies of the spacial "structure" and makes it possible to group the features into subpopulations, which contain features with similar parameters. The differences among features belonging to the same group should be as small as possible, while the differences between groups should be large. The grouping of features must satisfy two main conditions:

1. the division must be comprehensive (exhaustive) (i.e., each feature must belong to a given group);
2. the division must be disjunctive (i.e., each feature must belong to only one group).

There are a number of techniques that can be used to define clusters in populations of different types. It has been found that, from the point of view of morphology, the most useful is the nonhierarchical method of grouping, which uses the matrix of distances between features.

Consider a set of N features, for example, grains revealed in a section of a polycrystal, with each feature, X_i, being characterized by parameters, x_{ij}. The distance matrix, d_{ij}, is an N × N symmetric matrix that defines the distances between all the points representing the features in parameter space:

$$d_{ij} = \begin{matrix} 0 & d_{12} & \dots & d_{1N} \\ d_{21} & 0 & \dots & d_{2N} \\ \dots & \dots & \dots & \dots \\ d_{N1} & d_{N2} & \dots & 0 \end{matrix} \tag{4.58}$$

with

$$d_{km} = d[(X_{k1}, X_{k2}, \dots, X_{kL}), (X_{m1}, X_{m2}, \dots, X_{mL})] \tag{4.59}$$

The matrix, d_{ij}, contains nonnegative elements. Moreover, the only elements equal to zero are on the main diagonal ($d_{ii} = 0$) since no two identical features are to be found in a stochastic population.

4.2.1 Cluster analysis

An algorithm of a cluster analysis method employing the concept of the distance matrix has been described in detail in Reference 15. It is presented schematically in Figure 4.42. The idea is that the points in feature space are analyzed from the point of view of their distances from their neighbors and grouped into sets in which the points are close together. Without an *a priori* knowledge of the potential cluster positions, it is convenient to base the grouping of features using as a criterion the nearest neighbor distance. It is assumed in this case that a cluster, C, is formed by points/features such that:

a) for any point, X_i, in the cluster there is another point, X_j, in the cluster that is placed at a distance smaller than a certain critical value, R;
b) each point outside the cluster is at a distance larger than R from any point in the cluster.

The results of grouping based on nearest neighbor criteria obviously depend on the value of the parameter R. There are two strategies that can be adopted in the process of selecting proper values of R. First, it may come out of a physical analysis of the processes taking place in the microstructure, which leads to some "interaction range" R. For example, it can be assumed that two particles interact if they are placed at a distance smaller than R_o, which can be expressed in direct values or as a fraction of their equivalent diameters.

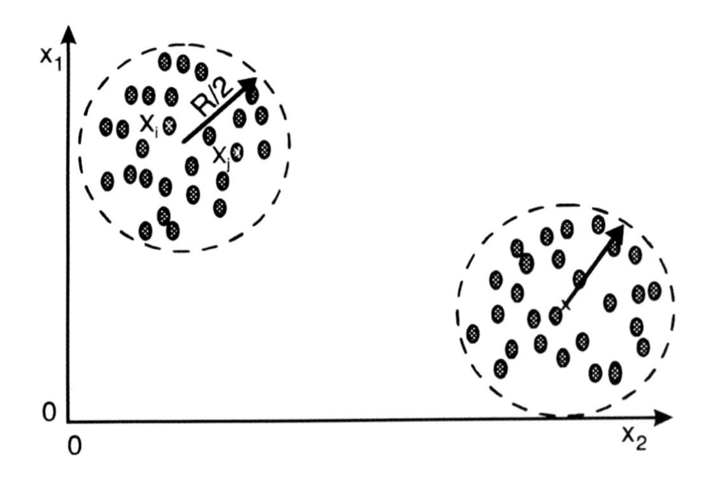

Figure 4.42 Schematic representation of the idea of cluster analysis. Points in feature space are analyzed from the point of view of their distances from their neighbors and grouped into sets in which the points are close together.

Second, the value of R can be obtained by a trial-and-error method. Some suggested starting values to be used in this case are listed below:

1. $R = d$;
2. $R = d - 0.5s$;
3. $R = d + 0.5s$;
4. $R = d + s$;
5. $R = \max_i\{\min_j\{d_{ij}\}\}$

where:

$$d = \left(\frac{1}{n}\right)\sum_{i=1}^{n} \min\left[\left(1-\delta_{ij}\right)d_{ij}\right] \tag{4.60}$$

$$s^2 = \frac{1}{n}\sum_{i=1}^{n}\left(\min\left[\left(1-\delta_{ij}\right)d_{ij}\right]-d\right) \tag{4.61}$$

($\max_i$ and $\min_i$ stand for maximum and minimum nonzero, respectively, of the features in rows and columns of the distances matrix d_{ij}. The Kronecker delta has been used here to account for the fact that $d_{ii} = 0$, i.e., zero is equal the distance of a feature to itself.)

Example 4.18

Different values of the R parameter listed above have been used in Figure 4.43 to evaluate clusters in the 2-D space, Y_{ij}, consisting of normalized diameter dn and shape factor α.

This example shows that it is essential in this approach to set up criteria for defining an acceptable division of the population into clusters. As the value of R changes, so does the number of clusters N_c. $N_c(R)$ may vary from 1 to N, where N is the number of points in the population. The value $N_c(R) = 1$ is obtained for large radii and $N_c(R) = N$ for radii smaller than the smallest distance between the points in the population studied. These two extreme types of population division are of a little value from the point of view of the present analysis. Grouping the features into clusters has a physical significance if the number of subpopulations defined is not too large with respect to the number of features grouped and the densities of points inside the clusters representing the subpopulations are therefore significant.

In cluster analysis the optimum division is to be found and this requires some measure of the effectiveness of the division. One possible measure is based on indices which characterize the structure and the location of clusters, such as parameters Q_A, Q_B, and Q_C proposed in Reference 11 (these Q parameters are defined below).

Consider a division of N features in an L-dimensional space of parameters into a system of clusters C_i ($i = 1,2,\ldots, N_c$) containing, respectively, n_1, n_2, $\ldots n_c$ features. The position of each of the clusters in the spatial parameter can

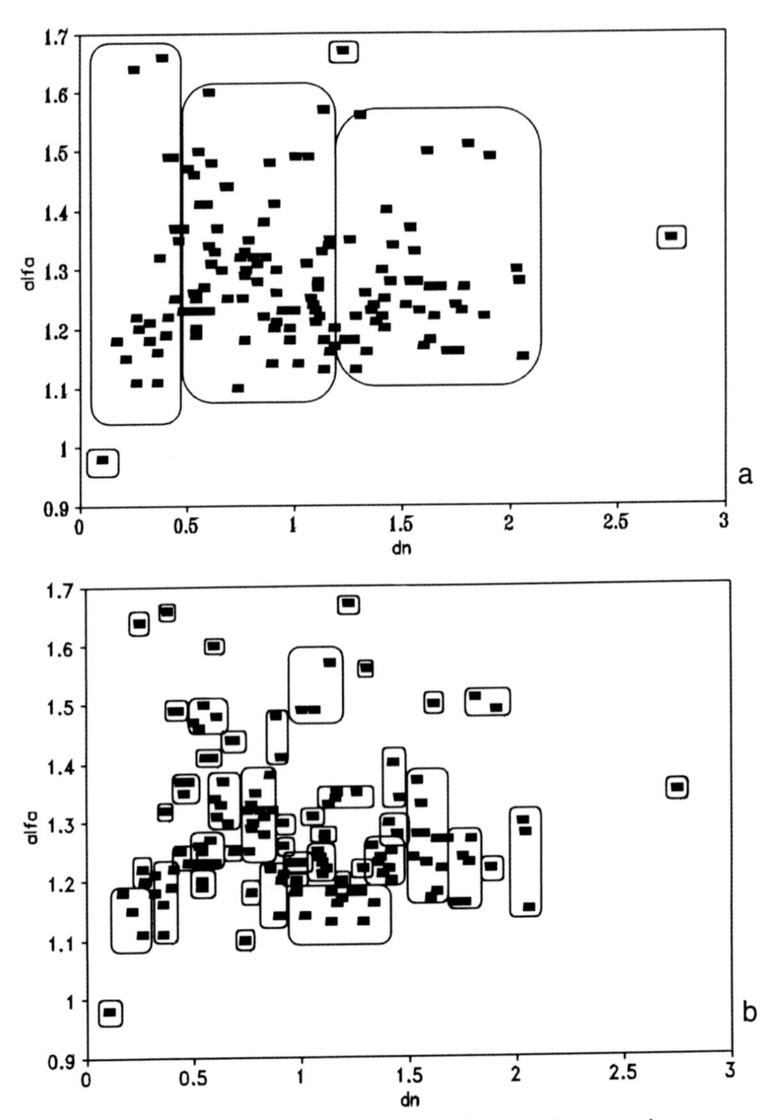

Figure 4.43 Results of changes in the cluster radius for the same feature space: (a) space Y with marked clusters for R = d – 0.5s, (b) space Y with marked clusters for R = d + s.

be defined by its center, S_i. The coordinates of the cluster center are defined as the mean values of the parameters computed over the features that have been categorized as belonging to a given cluster. This means that:

$$S_i = E_{C_i}(X_i) \tag{4.62}$$

The parameter Q_A is defined as a sum of average distances between features divided by the number of features, n_i, in the clusters:

$$Q_A = \sum_{i=1}^{N_C} \frac{1}{n_i} \sum_{j=1}^{n_i} d^*\left(S_i, X_j\right) \tag{4.63}$$

This parameter measures differences in the parameters of features within the clusters.

The in-cluster structures are also described by a parameter Q_B denoting a sum of distances between the features and the cluster centers, which is calculated for all the clusters in the following manner:

$$Q_B = \sum_{i=1}^{N_C} \sum_{j=1}^{n_i} d^*\left(X_j, S_i\right) \tag{4.64}$$

The parameter Q_C, denotes the sum of the mean arithmetic distances between the features within the clusters and it characterizes the degree of feature dispersion:

$$Q_C = \sum_{i=1}^{N_C} \frac{2}{n_i\left(n_i - 1\right)} \sum_{j=1}^{n_i} d^*\left(S_i, X_j\right). \tag{4.65}$$

In these formulae $d^*(S_i,X_j)$ stands for the distance between the center of a given cluster, S_i, and the points/features that belong to that cluster.

Another useful parameter, M, that can be used to define the distance between clusters, is calculated according to the formula:

$$M = \frac{1}{n} \sum_{i,j=1}^{n} d_{ij} - Q_A. \tag{4.66}$$

In terms of the parameters introduced, efficient divisions of the population into clusters yield low values of the Q parameters and large values of M.

In addition to the parameters already discussed, an index, J, can be employed which defines the number of clusters. It is computed in the following way:

$$J = \frac{1 - \dfrac{N_c}{N}}{1 - \dfrac{1}{N}} \tag{4.67}$$

In this form, the influence of the number of features examined upon the number of the clusters observed is accounted for. Different variants of the cluster analysis method for a set of data points in 2-D space are discussed in the following example, which gives cluster analysis of geometrical features of grains.

Example 4.19

The advantages of cluster analysis can be exemplified by the case of studies of the microstructures of polycrystalline materials. In order to characterize the geometry of grains in a polycrystalline metal the following parameters have been chosen:

a) $A/E(A)$ standardized grain section area representing the ratio of grain section area to the mean grain section area;
b) $d_{max}/(A)^{1/2}$ a shape factor sensitive to grain elongation;
c) $p/(A)^{1/2}$ a shape factor sensitive to grain boundary undulations.

The values of these parameters may be obtained from measurements on cross-sections of polycrystals using a computerized image analysis instrument. It should be noted that among the parameters proposed, parameter (c) is related to perimeter measurements and as such is more resolution dependent. The measured values of $p/(A)^{1/2}$ cannot be taken as a "true parameter" of a given feature. In order to simplify interpretation of the results, it is recommended that in studies of a number of images the measurements should be carried out with a constant resolution limit. A Ni-2% Mn alloy was used in a study of the changes in the geometry of grains. The material had been recrystallized at different temperatures ranging from 0.3 to 0.8 T_m (T_m is the melting point in degrees Kelvin) for annealing times from minutes to a few hours leading to various patterns of grain geometry reflecting recrystallization and grain growth. Various sections of the polycrystals were initially examined from the point of view of possible texture in the grain shape. No indication of shape texture was found, and in further analysis of grains on randomly selected longitudinal sections they were sampled systematically. Quantitative measurements were effected with the use of a computerized image analysis system. A typical binary image of the grain boundaries in the material is given in Figure 4.44 and the results of the measurements are shown in Figure 4.45.

The application of cluster analysis to the results of measurements of parameters describing the geometry of grain sections proved that these grain sections in annealed polycrystals show certain patterns in the distribution of their geometrical parameters, which change as a function of thermomechanical history. This rationalizes the use of this method for the description of subtle changes in the microstructure of polycrystalline materials. The results obtained for a series of specimens after different stages of grain growth show that grain growth brings about a gradual increase of homogeneity (similarity) of size and shape of grains. Further applications may include studies of abnormal grain growth and microstructural evolution during phase transformations.

4.2.2 *Space tessellations*

The populations of microstructural features can also be studied using the concept of space tessellations. The Dirichlet tessellation method is a geometrical technique for partitioning a system of points, Q_i, in n-dimensional space into a set of convex polyhedra, V_i, each of which is associated with and contains one of the points. In applying this technique to microstructural characterization,

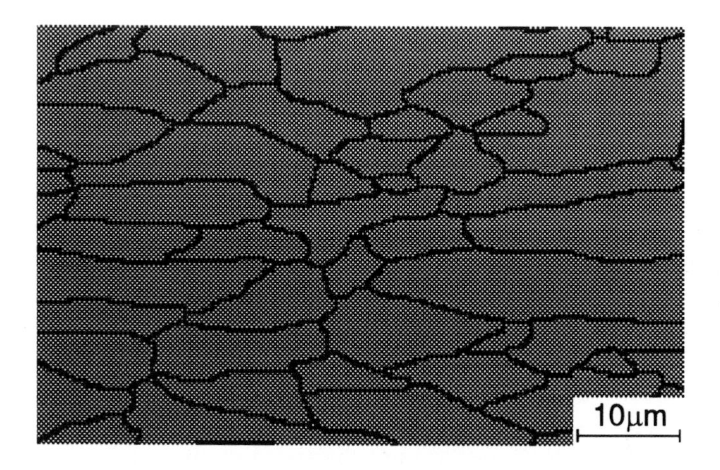

Figure 4.44 Example of the microstructure of Ni-2% Mn alloy after deformation.

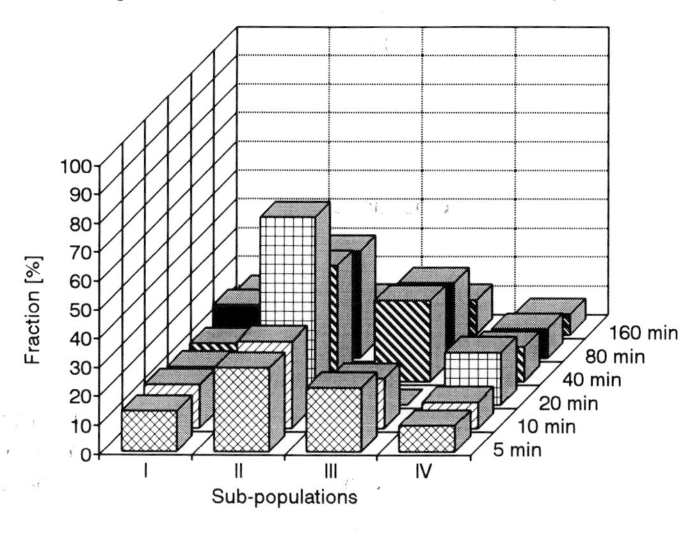

Figure 4.45 Changes in subpopulation fractions with the time of annealing.

the Q_i points can represent:

a) points in real, 1-,2-, or 3-D space describing positions of second-phase particles, grains, etc.;
b) points in an abstract space defined by the quantified properties of microstructural features in a given material.

In each case this technique assumes that the space is divided into polygons by the following procedure:

1. lines are drawn from each point to all the others;
2. the lines are bisected by a system of planes;

3. the smallest polygon formed by these planes is ascribed to the given point as a Dirichlet cell (also called a Voronoi polygon).

This procedure is repeated at each Q_i point, yielding a set of polygons, V_i, which fill the space of interest. This set is studied subsequently from the point of view of homogeneity in the size and shape of individual polygons. An example of different divisions is given in Figure 4.46.

The concept of Dirichlet tessellations has been used to study the distributions of second-phase particles. A recent review of these applications can be found in Reference 16. In two dimensions, the position of a particle can be described by a point in XY coordinate space and the population of particles by a set of points. Dirichlet tessellations result in a division of the plane into a set of polygons exemplified by the one shown in Figure 4.46. These figures have edges numbering six on average, disregarding the details of their shape and arrangement.

The individual polygons forming the tessellation can be characterized by adopting the same parameters as those used for grains. These, among others are

a) area, A_i;
b) perimeter, p_i;
c) shape factor, q_i, defined as the ratio of A over p^2

$$q_i = 4\pi \frac{A}{p^2} \tag{4.68}$$

d) number of edges, n_i.

Accordingly, the population of the polygons is described by a set of distribution functions, f(x), where x is one of the parameters a) to d). It can be characterized by means of the mean values, E(x), the coefficients of variation,

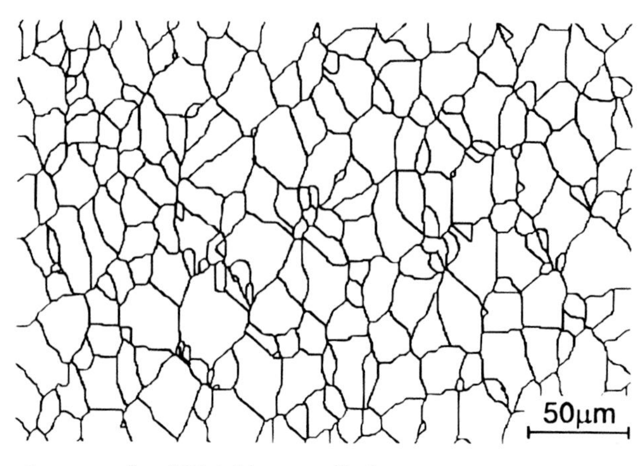

Figure 4.46 An example of Dirichlet tessellations.

CV(x), and possibly higher-order moments. Values of these parameters are characteristic of a given tessellation and, in turn, of a given distribution of particles in the microstructure studied.

Some specific particle distributions are used as reference structures for studies of their arrangements. One of them is the distribution that is obtained in the case of a totally random positions of particles. Such a distribution has been investigated thoroughly in a number of papers (see, for example, Reference 17) and found to be characterized by:

$VAR(n_i) = 1.78$;
$CV(A) = 0.64$.

At the other extreme to a totally random position of particles is the structure consisting of particles forming a regular array, represented by a honeycomb structure. This structure is distinguished by the following parameters:

$VAR(n_i) = 0$;
$CV(A) = 0$.

Between these two extreme cases, there are a range of patterns characteristic of particular microstructures containing particles of different origins, which have been subjected to a variety of treatments. However, using the concept of tessellations and their characteristics, the microstructures can be ranked according to the degree of the randomness in the arrangement of particles.

Tessellation analysis can be modified to include more parameters and variables. For example, much attention has been attracted to the distribution function of the distances, d_{ij}, between a given particle, p_i, and its neighbors, p_j. In particular, one can confine the analysis to the smallest values of d_{ij} or, in other words, the interparticle spacing, d. For a random distribution of particles this parameter is given by the following distribution function:

$$f(d) = 2\pi d N_A \exp\left[-\pi d^2 N_A\right] \tag{4.69}$$

where N_A is the planar density of particles.

On the other hand, for a honeycomb arrangement of particles this distance is constant:

$$f(d) = const \tag{4.70}$$

and its variance $VAR(d) = 0$. Again, the distribution of distances between the nearest neighbor particles is expected to fall between these two cases and measurements of the distance variance can be used to rank the degree of randomness exhibited by a particular microstructure.

Examples of advanced applications of the tessellation method are given in References 18 and 19. The following example is used mainly to illustrate some aspects of the method.

Example 4.20

The systems of images shown in Figures 4.47 and 4.48 are characteristic of two-phase materials after rolling when observed on the sections parallel and perpendicular to the rolling direction. Images of the microstructures were digitized and processed so as to obtain a representation of the particles by their centers. As a result, each system of particles has been represented by a matrix of the form:

$$Z_{ij} = \begin{matrix} x_1 & y_{11} \\ x_2 & y_2 \\ \dots & \dots \\ x_k & y_k \end{matrix} \tag{4.71}$$

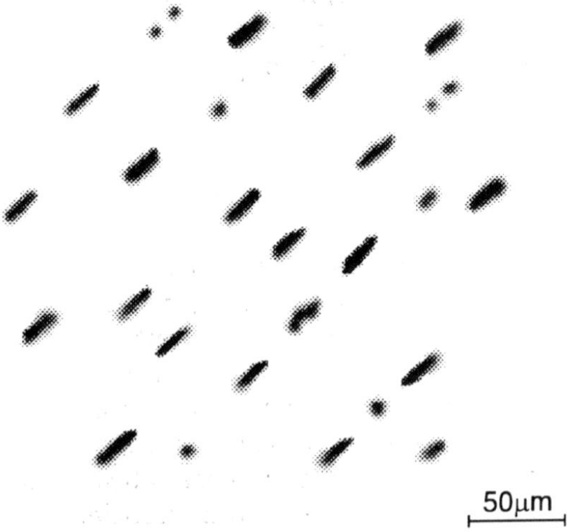

50μm

Figure 4.47 Example of a system of particles on the section parallel to the rolling direction.

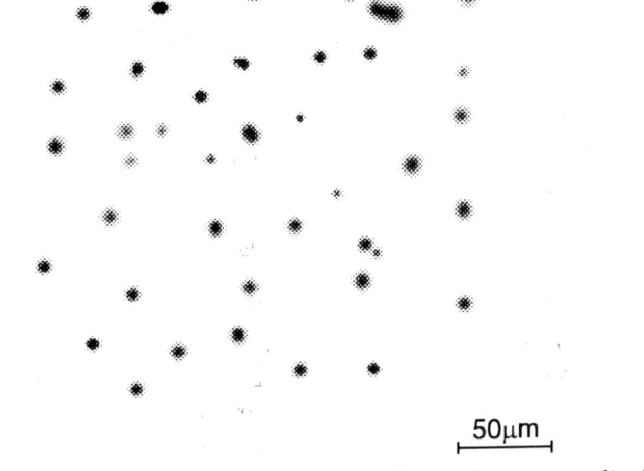

Figure 4.48 Example of a system of particles on the section perpendicular to the rolling direction.

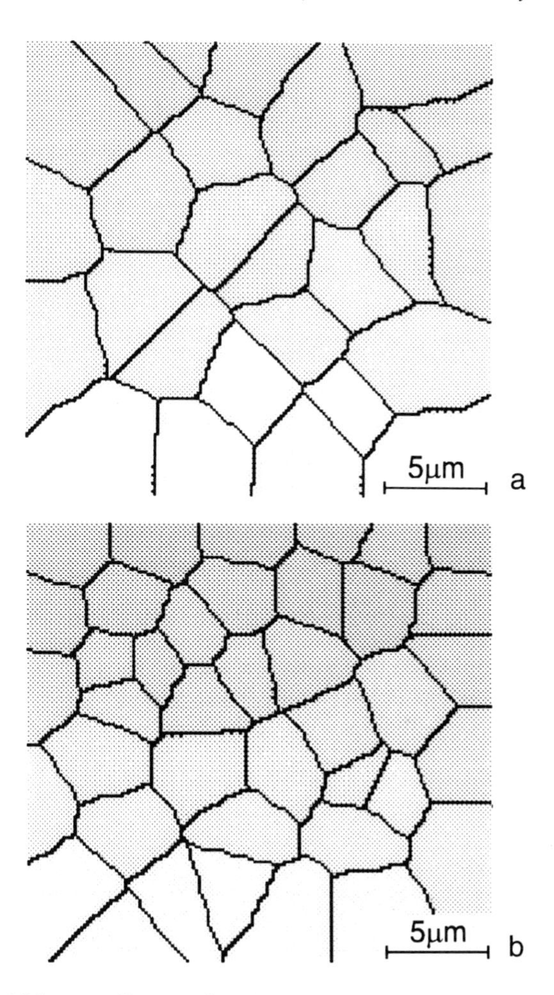

Figure 4.49 Dirichlet tessellations for the systems of particles from Figure 4.47 (a) and Figure 4.48 (b).

These matrices were subsequently used to obtain divisions of the X,Y space into tessellations, which are shown in Figure 4.49. These tessellations have been subjected to extensive measurement of the geometry of the cells. The following parameters were measured:

A_i cell area;
p_i perimeter;
n_i number of sides;
d_{max} maximum dimension (chord);
F_i Feret diameter.

These parameters were studied in terms of their distribution functions, mean values, and coefficient of variations.

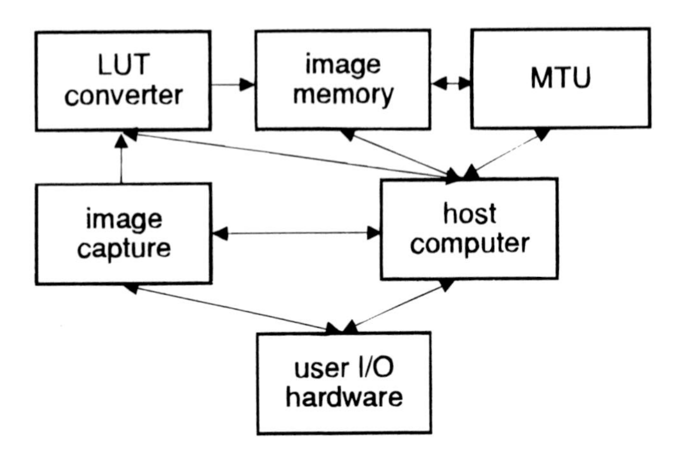

Figure 4.50 Schematic diagram of the main units of an image analysis system.

4.3 *Elements of an image analysis system*

A schematic diagram of an image analysis system is presented in Figure 4.50. By analyzing the functions that can be distinguished in the series of steps performed by the system, the following units can be defined:

—image capture unit;
—look-up table unit;
—image memory;
—matrix transformation unit;
—user input/output hardware;
—host computer.

4.3.1 *Image capture unit*

The image capture unit consists of an imaging device, an input device, and an analog-to-digital converter. The imaging device is, for instance, a light or electron microscope, or a macroviewer. The input device is usually a TV camera. In the scanning electron microscope (SEM), one or more of the signals produced by the detectors may be used directly, so an SEM can be seen as a combination of imaging and input devices. The analog-to-digital (A/D) converter translates an analogue TV signal into a digital grid of pixels, usually by sampling the TV signal at preset time intervals. In modern systems a scanner may combine the functional requirements of imaging, input device and A/D converter.

One of most important features of the imaging and input devices is their ability to provide stable images with a constant background intensity over the whole area of the image. This is very important for a proper thresholding operation described earlier, and critical for densitometric measurements and "gray scale" transformations/operations.

4.3.2 *Look-up table unit*

The look-up table (LUT) converter provides an easy way to convert digitized video signals according to user needs. Typical possibilities of using LUT are negation (inversion) described in mathematical morphology operations earlier in this chapter, thresholding, logarithmic conversion described earlier, etc. The LUT translates the value of a pixel into another value as defined by the LUT registers. It works on any pixel independently of any other. LUTs may be installed on either the input and/or the output sides of the instrument. In some systems there is a so-called "pseudo-color" LUT, which transforms pixel values into several values of red, green, and blue to be displayed on a color monitor as an artificial color. LUT operations, if performed by hardware, are always real-time transformations.

4.3.3 *Image memory*

Image memory is a dedicated area of the RAM. These days, the image memory is usually larger than is necessary to hold just one digital image. Thus, it may hold not only files of images, but also overlays, binary images, multiple images, etc. For instance, at least 256kB (kilobyte = 1024 bytes) of memory is necessary to store one image of 512 × 512 pixels and 256 gray levels, 1.5MB (megabyte = 1024 kilobytes) are necessary to store one full-color image (256 red, 256 green, and 256 blue levels) of 1024 × 512 pixels, etc.

4.3.4 *Matrix transformation unit*

The matrix transformation unit (MTU) is a dedicated electronic device (usually based on a very fast integer mathematics unit), which facilitates real-time (or near real-time) operation on images stored in the memory. The operations of the MTU usually involve looking at the neighborhood of each pixel on the image. Examples of such operations are some of those associated with mathematical morphology, such as erosion, dilation, etc. MTUs are usually based on a specially designed expensive VLSI chip, produced for a particular system. MTUs are currently used only in the more expensive systems, because the same effect (although taking a much longer time) can be obtained by using software and the processor of the host computer.

4.3.5 *User input/output hardware*

On the input side a user can use the keyboard of the host computer. Additional input devices include a mouse (almost standard on every system), a light pen (usually on low resolution systems), a digitizing tablet, etc. On the output side there is usually a dedicated image monitor (aside from that of the host computer monitor), a host computer printer, a video printer (capable of printing the TV signal directly, with high resolution in multiple shades of gray or multiple colors), a video tape recorder for inexpensive storage of images, etc.

4.3.6 Host computer

This can be a simple, inexpensive PC-based system (for those cases where time is of less importance) or a powerful workstation for advanced applications. Images are also analyzed by more powerful computers (supercomputers), but one can hardly say that they are image analyzers, because the processing steps in these cases are divorced from the acquisition and presentation steps.

References

1. Bradury, S., *An Introduction to Optical Microscopy*, Oxford Univ. Press/Royal Microscopical Society, 1989.
2. Briggs, A., *An Introduction to Scanning Acoustic Microscopy*, Oxford Univ. Press/Royal Microscopical Society, 1985.
3. Sharpe, R. S., Projection microradiography, *Journal of Microscopy*, 117, 123, 1979.
4. Żebrowski, G., Bucki, J. J., and Kurzydłowski, K. J., A method for objective threshold selection in area fraction measurements, 6th European Congress for Stereology, Prague, Czech Republic, September 93, 1993.
5. Richardson, L. F., The problem of contiguity: an appendix to statistics of deadly quarrels, *General Systems Yearbook*, 6, 139, 1961.
6. Mandelbrot, B. B., *Fractals—Form, Chance and Dimension*, W. H. Freeman & Co., New York, 1983.
7. Flook, A. G., The use of dilation logic on the Quantimet to achieve fractal dimension characterization of textured and structured profiles, *Powder Technology*, 21, 295, 1978.
8. Kurzydłowski, K. J., On the use of some geometrical measurements to study the plastic deformation of polycrystalline materials, *Acta Stereologica*, 10, 175, 1991.
9. Kurzydłowski, K. J., McTaggart, K. J., and Tangri, K., On the correlation of grain geometry changes during deformation at various temperatures to the operative deformation modes, *Philosophical Magazine A*, 61, 61, 1990.
10. Kurzydłowski, K. J., Sangal, S., and Tangri, K., The effect of small plastic deformation and annealing on the properties of polycrystals, *Metallurgical Transactions*, 20A, 471, 1989.
11. Kurzydłowski, K. J., The changes in the mean intercept length distribution as a function of strain and their implications to the Hall-Petch analysis, *Materials Science and Engineering Letters*, 9, 788, 1990.
12. Slater, J., Quantitative Investigation of the Transformation Kinetics of Low-Alloy Steels, Ph.D. Thesis, Cambridge University, 1976.
13. Slater, J. and Ralph, B., Determination of particle shape and size distributions using image analysis techniques, Fourth International Congress for Stereology, Gaithersburg, Maryland, U.S.A., 177, 1975.
14. Kucharczyk, J., *Algorithms of Cluster Analysis in Algol 60*, PWN, Warszawa, 1982.
15. Kaczmarczyk, E. and Warchoł, J. B., Analysis of similarity/dissimilarity of cell nuclei patterns on the example of thymocytes, *Advance in the Biology of Cells* (in Polish), 14, 1, 33, 1987.

16. Parse, J. B. and Wert, J. A., A geometrical description of particle distributions in materials, *Modelling and Simulation in Materials Science and Engineering*, 1, 275, 1993.
17. Szala, J., Cwajna, J., and Maciejny, A., Stereological criteria and measures of grain size homogeneity, *The Stereological Methods in Materials Science*, Ossolineum, Wrocław, 1988.
18. Lewandowski, J. J., Liu, C., and Hunt, W. H., Effects of matrix microstructure and particle distribution on fracture of aluminum metal matrix composite, *Materials Science and Engineering*, A107, 241, 1989.
19. Spitzig, W. A., Effect of various sulfide stringer populations on the ductility of hot-rolled C-Mn steels, *Acta Metallurgica*, 33, 175, 1985.

chapter five

Characterization of linear and planar microstructural elements

The microstructures of most materials contain linear and planar elements. Examples of linear microstructural elements are dislocations in crystals, fibers in composites, and grain boundary edges in polycrystalline materials. In all these cases the elements listed can be reduced to a line, which means that an individual element can be characterized in terms of a tangent vector. The sizes of linear elements are defined by their length.

The list of planar elements in microstructures contain features such as: (a) interfaces, (b) grain boundaries, and (c) external surfaces. These elements can be characterized by a vector normal at a given point, P, on their surface. The size of an individual planar element is defined by its surface area.

The present chapter considers methods for the quantitative characterization of linear and planar defects. Dislocations and grain boundaries are used as representative cases for these two categories of microstructural elements. However, the methods described here are of a general applicability to any 1- and 2-D elements. The difference between the methodology to be used in studies of dislocations and, for instance, fibers lies basically in the imaging techniques. In the case of dislocations transmission electron microscopy is widely used for imaging, whilst in the case of fibers in composites, good results can be obtained by using light microscopy. Dislocations differ in some geometrical properties in that they do not end inside the material (the same is true for grain boundary edges) and exhibit considerable line tension that in general makes them minimize their length. However, the methods described in this chapter are not based on these special properties of dislocations.

5.1. Dislocations

Dislocations are linear defects encountered in crystalline materials. They are known to take part in the plastic deformation of crystals. Dislocations also modify the electrical and thermal properties of materials. Due to their importance in understanding the properties of materials they have become the subject of extensive research in recent years. A comprehensive treatment of their properties can be found in Reference 1.

An individual dislocation is defined by its Burgers vector, **b**, which describes the displacement associated with a dislocation, and its tangent vector, **t**, which describes its geometry. These two vectors, **b** and **t**, can be used to explain the behavior of a given dislocation under the action of stress and temperature. In the case of fibers, or other linear defects, the parameter **b** can be replaced with any other parameter that could be used to distinguish different types of fibers. For example, fibers can be differentiated in terms of their diameters, d_i, or chemical compositions, c_i.

The parameters **b** and **t** are difficult to obtain for the large number of dislocations that are frequently observed in materials. Moreover, such basic information about individual linear microstructural elements is not required at the present level of modelling of materials' properties. The focus of the present consideration will be placed on the methods of dislocation characterization in terms of the parameters that can be used in the analysis of the bulk properties of materials, such as flow stress, hardness, etc.

Dislocations in crystals are usually observed by means of transmission electron microscopy (TEM) of thin foils where they are seen as linear segments distinguished by an oscillating contrast (Figure 5.1). The dislocation segments on TEM images are seen as projections of the true segments embedded in the thin slice of the material examined. However, it should be stressed that as a result of the complexity of electron image contrast formation, not all dislocations present in the specimen are necessarily recorded. The dislocations revealed are characterized by special orientations of the Burgers vector to the operating reflection. This is one of the important limitations of this technique. An absence of dislocations on the TEM image does not imply their absence in the studied slice of material. Another limitation is related to the possibility of

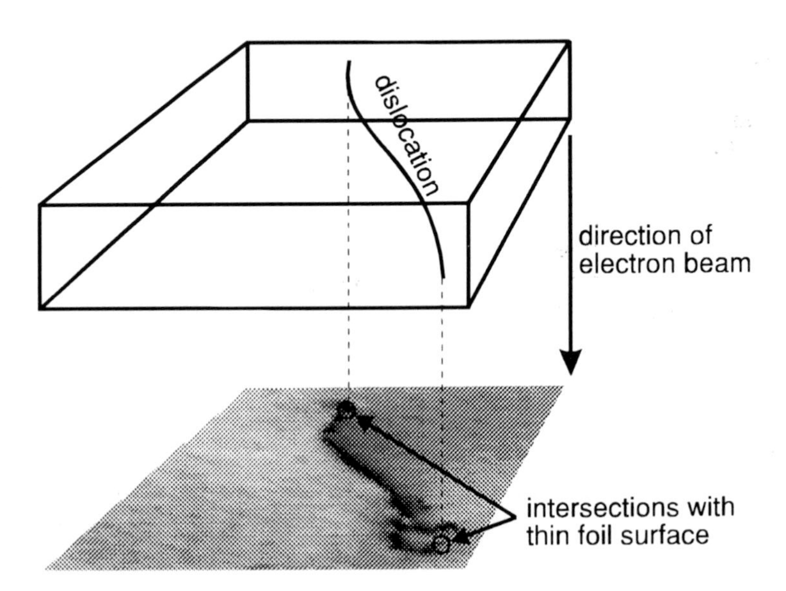

Figure 5.1 An example of the image of a dislocation in a crystal and a schematic explanation of the relationship between the orientation of the image with respect to the orientation of the dislocation.

changing the true density of dislocations in the process of thin foil preparation. It has been found that dislocations tend to escape to free surfaces that act as dislocation sinks and the thin foils required for TEM studies are of thicknesses varying from hundreds to tens of nanometres, depending on electron energy, contain significantly less dislocations than the bulk material. On the other hand, careless handling of foils of such a thickness leads to their deformation and an increased number of dislocations might be observed. In quantitative studies of dislocations by TEM observations these two restrictions need always to be addressed.

Dislocations in crystals can also be studied via surface observations. Sections of the material are polished and etched in a way that reveals the intersections of the dislocation lines with the surface of observation in the form of so-called etch pits (see Figure 5.2). The pits are formed due to the higher energy of atoms in the dislocation core sectioned by the surface of observation. This technique is less powerful than TEM observations but is an easier method of dislocation imaging from an experimental point of view.

The basic parameter used to describe the amount of linear microstructural elements is the density. Dislocation density is an important parameter used to explain the properties of crystalline materials (see Chapter 1). It should be noted that there are two general definitions of dislocation density. Since dislocations are 1-D microstructural elements, they can considered to be distributed either in space (in 3-D reference space) or over surfaces (2-D reference space). From this point of view, the useful parameters are

1. dislocation volume density, L_V, usually designated in the physics of plastic deformation as ρ;
2. dislocation surface density, L_S.

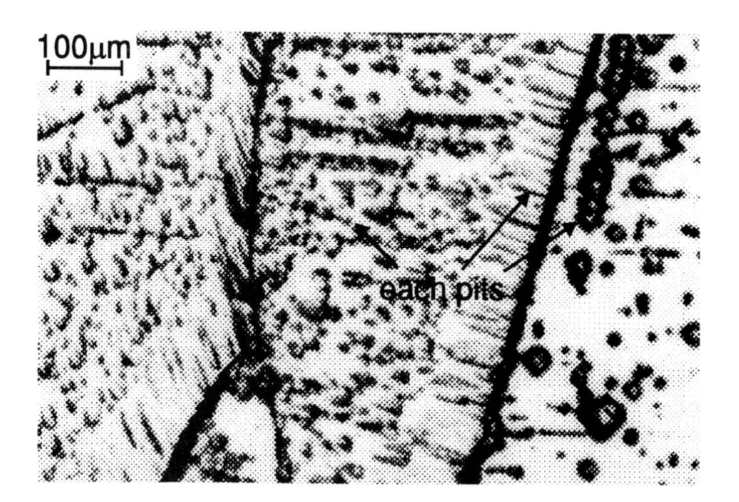

Figure 5.2 Etch pits indicating dislocations emerging on a surface of a tungsten cross-section.

Dislocation volume density, L_V, defines the total length of dislocation per unit volume and its dimension is m/m^3 or m^{-2}. Dislocation surface density, L_S, is given as the length of dislocation per unit surface area and is expressed in m/m^2 or m^{-1}.

5.1.1 *Measurements of dislocation volume density*

Dislocation density, L_V, may be determined by a simple counting method of their intersections with probing surfaces. According to the formula given in Chapter 2, the volume density of linear elements can be estimated from the density of intersection points with randomly positioned and oriented sections using the following equation:

$$L_V = 2P_A \quad (5.1)$$

where P_A stands for the density of intersection points.

In the case of surface observations, where the intersection points are revealed in the form of etch pits, the application of Equation 5.1 is obvious. However, it should be remembered that the fields of observation should be randomly positioned and oriented in space. In order to meet these requirements, one can use the methods described in Chapter 3, such as the orientator.

Example 5.1

Studies of dislocations in an annealed polycrystalline material were carried out on three, mutually perpendicular sets of sections. On each section, two fields have been examined with a total area of 1 × 1 mm. A characteristic image of the microstructure revealed is shown in Figure 5.3. The data obtained for various sets of sections is given in Table 5.1. It was concluded that the orientation of the sections does not affect the dislocation density. The density of dislocations was estimated to be

$L_V = 1.5\ 10^4\ cm^{-2}$.

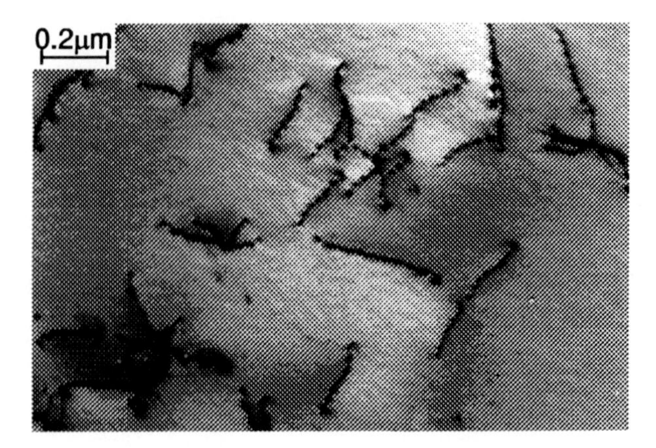

Figure 5.3 An example of the dislocation substructure revealed in a well-annealed nickel sample.

Table 5.1 The Results of Dislocation Density Measurements Described in Example 5.1

Section	Field	P	P_A [10^2 cm^{-2}]	Density [10^4 cm^{-2}]
I.	1	76	76	1.52
	2	72	72	1.44
II.	1	73	73	1.46
	2	78	78	1.56
III.	1	74	74	1.58
	2	76	76	1.52
Total	–	448	74.8	1.5

In TEM studies of dislocation density it is convenient to assume that the probing surfaces are the thin foil surfaces and counting is carried out of dislocation intersections with the thin foil surfaces. These intersections are depicted as ends of dislocation segments (see Figure 5.1). The number of intersections of dislocation lines with the thin foil surface, P_A, is half the counted number of the dislocation segment ends. This is related to the fact that the end points are located on the two surfaces of the thin foil penetrated by the electron beam. Also, the angle θ should be taken into account between the normal to the foil surface and the electron beam. This leads to the following equation:

$$L_V = (P_A)_{TEM} \cos\theta \tag{5.2}$$

where θ is the angle between the direction of the projection (the electron beam in TEM observations) and the normal to the foil.

As in the case of surface observations of etch pits, this equation is true for thin foils that are cut from bulk materials in a way that ensures random orientation of the thin slices with respect to the geometry of the artifact studied; this requirement cannot be met by random orientation of the thin foils in the microscopic stage. The orientation of slices can be selected using the principle of the trisector, described in Chapter 3.

However, it is a common practice that the thin slices are cut perpendicular to some predefined direction. In this case Equation 5.2 can be used only if the dislocation segments are random in their orientations. In studies of microstructures that are characterized by a considerable level of alignment of linear elements, one can use the methods described in Chapter 2.

For dislocation lines oriented preferentially along a certain direction, **r**, in the coordinate system attached to the artifact studied, thin slices should be cut parallel and perpendicular to **r**. Then there should be a difference in the two respective densities of intersections of thin foils with the dislocations:

$$(P_A)_{\parallel} - (P_A)_{\perp} \neq 0 \tag{5.3}$$

This difference is an indication of the texture in the orientation of line segments. The total length of the lines in such a system is the sum of the two classes of segments and this leads to the following equation:

$$(L_V) = (L_V)_{\parallel} + (L_V)_{\perp} \tag{5.4}$$

Example 5.2

Figure 5.3 shows a dislocation substructure in an annealed crystal. A total of 25 slices of the material has been cut and out of this set 5 randomly selected. These five slices have been further thinned and polished to make them suitable for TEM observations. The foils were examined and the density of dislocations obtained from Equation 5.2 for each thin foil. A relatively low scatter in the values for the individual thin foils was found (see Table 5.2). The values of L_V for each thin foil were used to compute the mean for the material studied.

Table 5.2 Some of the Results of Measurements Described in Example 5.2

Foil	P	A [μm²]	P_A [μm⁻²]	cos θ	L_V [μm⁻²]
1	45	37.5	1.2	0.997	1.2
2	26	32.5	0.8	0.997	0.8
3	40	36.4	1.1	0.997	1.1
4	35	35.0	1.0	0.997	1.0
5	37	35.2	1.05	0.997	1.05

The density of dislocations made visible in TEM observations can also be estimated from the measurements of the length of the projected dislocation segments as schematically explained in Figure 5.4. For isotropic orientations of dislocation lines it can be shown (see, for instance, Reference 2) that the true length, L, and projected length, L_p, are related by:

$$L = \frac{4}{\pi} L_p \tag{5.5}$$

Thus, the dislocation density, L_V, can be obtained from the equation:

$$L_V = \frac{L}{At} = \frac{4}{\pi} \frac{\cos\theta}{A_p t} \tag{5.6}$$

where t is the thickness of the foil measured in the direction of the electron beam, A and A_p the area and the projected area of observation, and θ the angle between the normal to the foil and the electron beam. This equation can also be given in the following form:

$$L_V = \frac{4}{\pi} (L_A)_p \frac{\cos\theta}{t} \tag{5.7}$$

where $(L_A)_p$ is the density of the projected lines (length of the projected lines per unit area of the image).

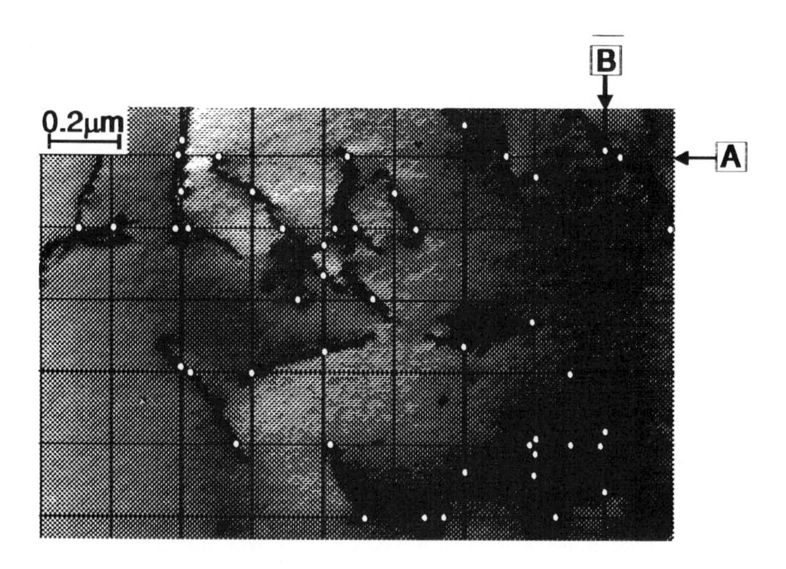

Figure 5.4 TEM image of dislocations in nickel showing the method for estimating the total length of their projected segments.

The density of the projected lines, $(L_A)_p$, can be measured directly using an automatic image analysis system. It can also be determined by the linear probing technique described in Chapter 2. In this case a grid of test lines is used and the density of intersection points of the projected images of the dislocations with the test lines, P_L, is measured. This density is related to $(L_A)_p$ by:

$$\left(L_A\right)_p = \frac{\pi}{2} P_L \tag{5.8}$$

and finally one can obtain the following formula:

$$L_V = 2P_L \frac{\cos\theta}{t} \tag{5.9}$$

where:

P_L is the density of intersections of dislocations lines with the test lines;
θ is the angle between the normal to the foil and the direction of the projection;
t is the local foil thickness measured in the direction of the electron beam.

Example 5.3

Figure 5.4 shows a projected image of dislocations in a thin foil of a nickel crystal. The area of the image shown is 1.8 × 1.2 μm. Using an image analysis system the total projected length of the dislocation lines was found to be

$L_p = 3.81\ \mu m^{-1}$.

Also, grids of test lines, designated as A and B in Figure 5.4, were used to estimate the total length of the projected lines. The following values were obtained:

$(L_A)_P = 4.07\ \mu m^{-1}$;

$(L_B)_P = 3.44\ \mu m^{-1}$.

These two estimates yielded the following average:

$(L_{A-B})_P = 3.77\ \mu m^{-1}$.

It may be noted that the two methods employed give similar estimates of the projected line lengths.

Sampling of the material for dislocation observations, in general, requires a large number of thin foils. Each such foil is characterized by its thickness, t_i. The thin foils are examined at different angles to the electron beam. This results in various projection angles, θ_j. As a result, the application of Equation 5.9 requires averaging over all fields of observations, with each of them characterized by different values of t and θ.

Example 5.4

The density of dislocations in an annealed polycrystal has been studied by TEM. A total of 30 thin slices of the material were cut—10 for each of three mutually perpendicular directions defined in the coordinate system related to the geometry of the polycrystalline specimen. For each ten thin slices, three were randomly selected for TEM observations. The thin foils were examined under different diffraction conditions at various angles θ. Some of the data from these specimens are given in Table 5.3. A total of 30 fields were studied and the mean dislocation density was estimated from the following equation:

$$L_V = \frac{2}{30}\left[\left(P_L\right)_1 \frac{\cos\theta_1}{t_1} + \ldots + \left(P_L\right)_{30} \frac{\cos\theta_{30}}{t_{30}}\right] \quad (5.10)$$

Table 5.3 Some of the Results of Dislocation Density Measurements Described in Example 5.4

Foil	Field	P_L [1/μm]	cos θ	t [nm]
X_1	1	0.3	0.94	1.0
	2	0.4	0.97	1.5
	3	0.7	0.77	2.0
X_2	1	0.3	0.92	1.2
	2	0.8	0.81	1.5
Y_1	2	0.5	0.73	0.8
Z_1	2	0.2	0.88	0.7

The equations derived can be used assuming that the thickness of the material slice is known. The problem of foil thickness measurements is addressed extensively in textbooks on electron microscopy, for example, in Reference 3. Among other methods of thickness measurement, one is based on simple geometrical considerations. This method can be employed if two different angles θ are used, the axis of tilting is known, and two points can be selected on opposite surfaces of the slice, which remain visible for both foil orientations.

5.1.2 *Average and local density of dislocations*

Dislocation density can be used to define the average dislocation content in unit volume of a material. However, in most cases dislocations are distributed in a nonuniform way, either in space or on surfaces. In particular, dislocations are known to form so-called mantle zones in the vicinity of grain boundaries in plastically deformed polycrystalline materials.

This nonuniformity can be described by defining the dislocation density in some specified test volume B (or test surface area A) positioned at point Q in the reference space. The density function thus obtained, $(L_V)_B\,(Q)$, will describe spatial fluctuations of dislocation density.

Example 5.5

Figure 5.5 shows a distribution of dislocations inside grains of an austenitic stainless steel deformed by a 1% tensile strain at room temperature. It may be noted that the dislocation density in the vicinity of the grain boundaries is higher than in the grain interior. The local dislocation densities were studied in a projected area, A_p, of size 0.1×0.1 μm. This density is plotted as a function of the projected area position with respect to the nearest grain boundary in Figure 5.6.

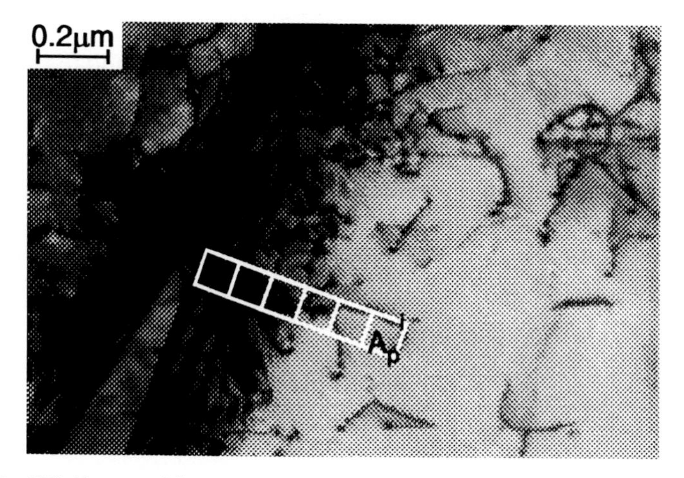

Figure 5.5 TEM image of the arrangement of dislocations in the vicinity of grain boundaries in an austenitic stainless steel specimen. It may be noted that the regions adjacent to the grain boundary plane are characterized by an increased density of dislocation lines.

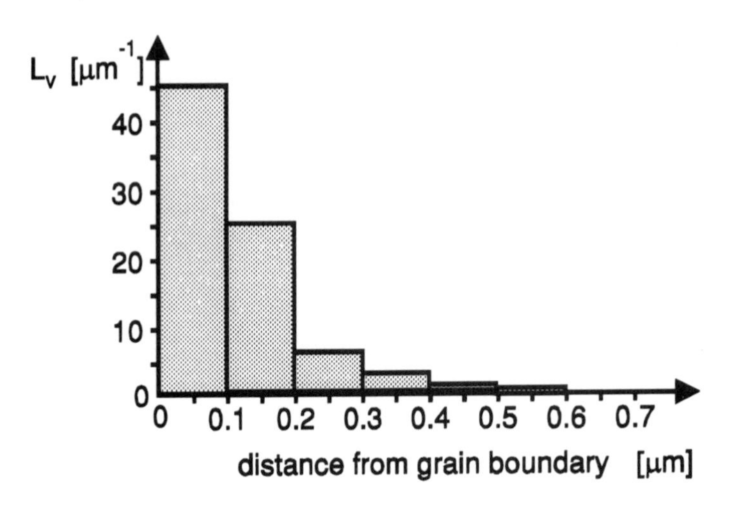

Figure 5.6 Variation in the density of dislocation lines as a function of distance from the grain boundary in the material described in Example 5.5 (see also Figure 5.5).

5.1.3 Special arrangements of dislocations

Dislocations are frequently observed not only to be aggregated in some space but also they are often organized into some specific patterns. There are planar arrangements of dislocations such as pile-ups, subboundaries, cell walls, and low angle grain boundaries. Dislocations also form bundles, walls of a finite thickness, etc. Some characteristic arrangements are shown in Figure 5.7.

For planar and spatial dislocation arrangements one can also introduce parameters that define size, shape, and other properties of specific dislocation patterns. In the case of planar configurations the useful parameter is the planar density of dislocations, L_S.

Figure 5.8 shows an array of dislocations in a grain boundary plane. These are so-called grain boundary dislocations, GBDs. Dislocations in grain boundaries can be characterized by their density, L_{GB}. This quantity has been suggested to be an important parameter in the theory of the plastic flow of polycrystals.[4-6] A relatively simple method of L_{GB} estimation is depicted in Figure 5.8a, b. In this method the projected image of a microstructure is used for both the measurements of grain boundary specific area, $(S_V)_{GB}$, and for estimation of GBD volume density, $(L_V)_{GBD}$. These two parameters are subsequently used to calculate the surface density of the GBDs from the following relationship:

$$\left(L_{GB}\right)_{GBD} = \frac{\left(L_V\right)_{GBD}}{\left(S_V\right)_{GB}}. \tag{5.11}$$

Using this approach, the specific surface area of grain boundaries must be determined initially. This can be achieved by using test lines as shown in Figure 5.8b and the equation given in Chapter 2:

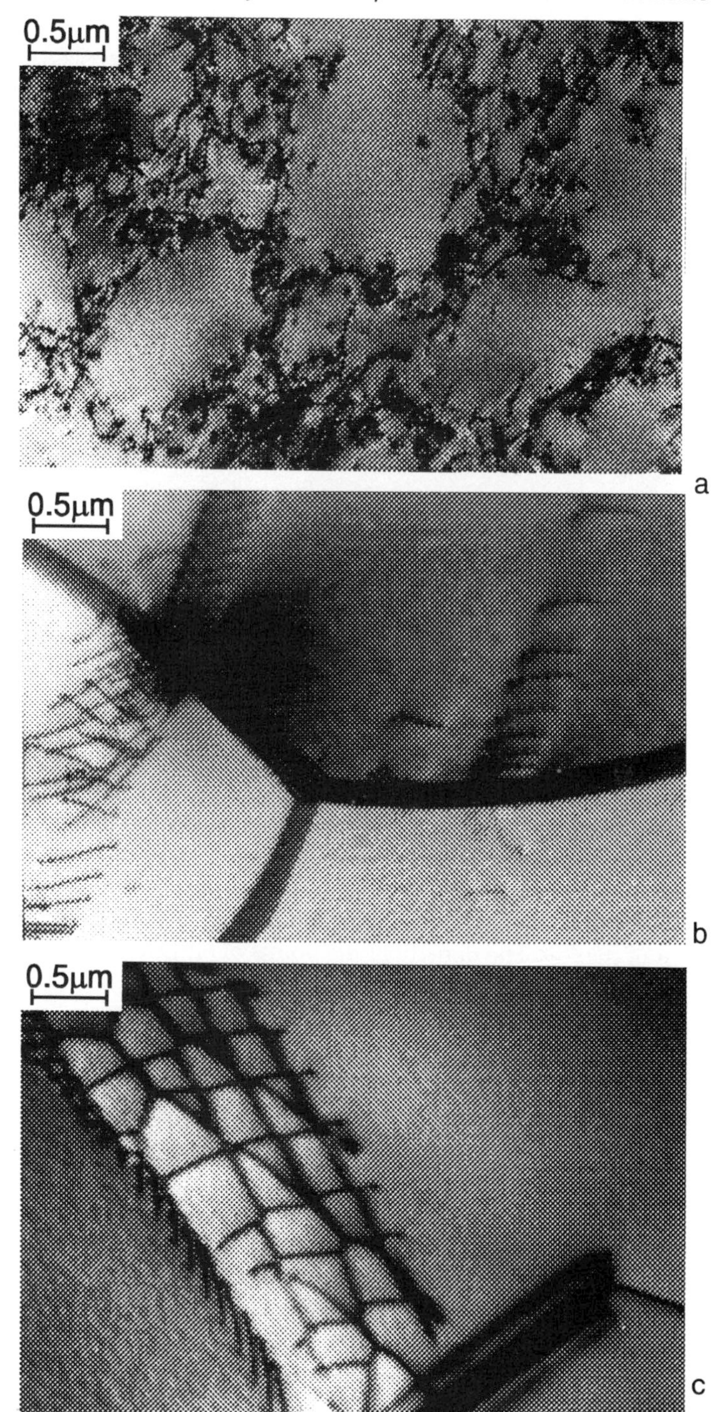

Figure 5.7 Characteristic arrangements of dislocations in crystals: (a) dislocation tangles; (b) dislocation pile-ups/emission profiles; (c) low angle dislocation subboundary; (d) (following page) high-density dislocation walls.

Figure 5.7 (*Continued*)

$$(S_V)_{GBD} = 2(P_L)_{GBD} \tag{5.12}$$

(See also Chapter 6 for other methods for the estimation of the grain boundary surface area per unit volume.) In Equation 5.12, $(P_L)_{GBD}$ is the number of intersections with the boundaries per unit test line length. In order to avoid the difficulties with the estimation of $(P_L)_{GBD}$, related to the possible tilting of the thin foil with respect to the electron beam, a grid of lines parallel to the tilt axis may be used. In the next step, the specific length of GBDs is estimated by counting the number of grain boundary dislocations emerging on the surface of the foil. The density of the points of GBD intersections with the surface, $(N_A)_{GBD}$, can be used to determine $(L_V)_{GBD}$:

$$(L_V)_{GBD} = 2(N_A)_{GBD} = \left[(N_A)_{GBD}\right]_p \cos\theta \tag{5.13}$$

where $[(N_A)_{GBD}]_p$ is the density of the intersection points of the GBDs with the surface of the foils for the projected image and θ is the foil normal angle, or in other words—the angle of projection.

The parameter $(S_V)_{GBD}$ can also be estimated from measurements on the sections of the material studied and methods described in Chapter 6 may be used to this end. However, it should be noted that TEM images of grain boundaries make it possible to distinguish in the population of grain boundaries those boundaries that contain or do not contain GBDs. As a result, the length of dislocations on grain boundaries, depending on the theoretical needs, can be referred to the total area of grain boundaries or to the total area of grain boundaries containing these dislocations. This is not possible if the surface area of the grain boundaries is determined from the surface observations of the grain boundary network.

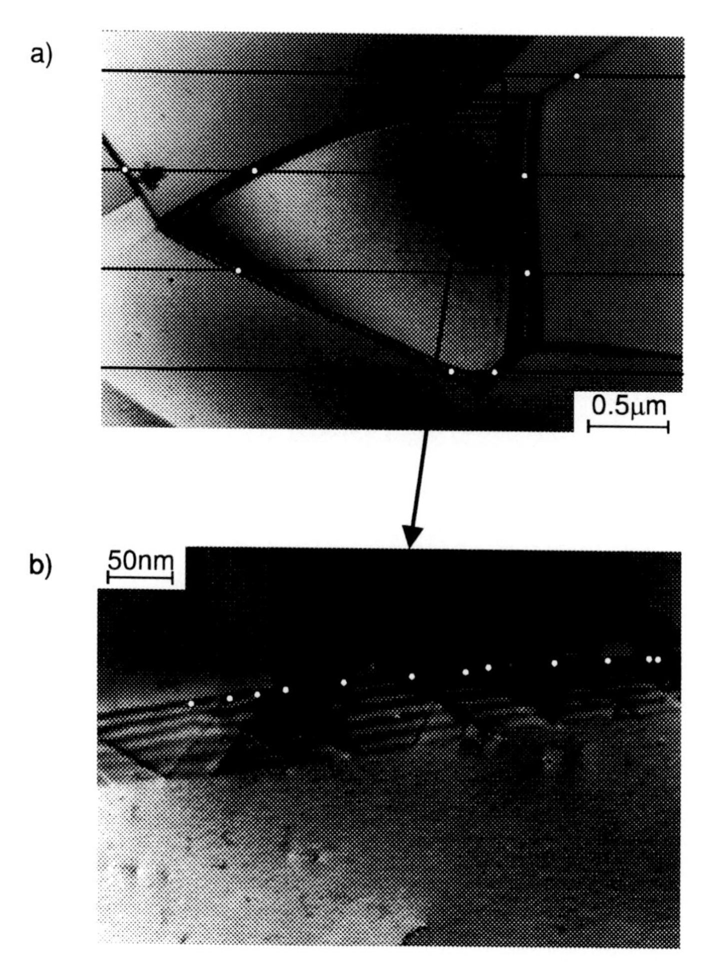

Figure 5.8 Explanation of the method for measurement of the grain boundary dislocation density: (a) a low-magnification TEM image of the grain boundaries where the test lines are used to estimate the area of the grain boundaries per unit volume; (b) a high-magnification image of a grain boundary used for counting the dislocation segments emerging on one surface of the foil.

Example 5.6

Figure 5.8b shows a TEM image of the grain boundaries in an austenitic stainless steel. Figure 5.9 depicts the grain boundary network revealed on a section of the material using a scanning electron microscope. A system of test lines imposed on SEM images of grain boundaries was used to estimate that the total surface area of the grain boundaries per unit volume, $(S_V)_{GB}$ = 55 μm^{-1}. TEM images of the microstructure were used to estimate the total length of the GBDs. It was found that:

$$(L_V)_{GBD} = 0.4\ (\mu m)^{-2}.$$

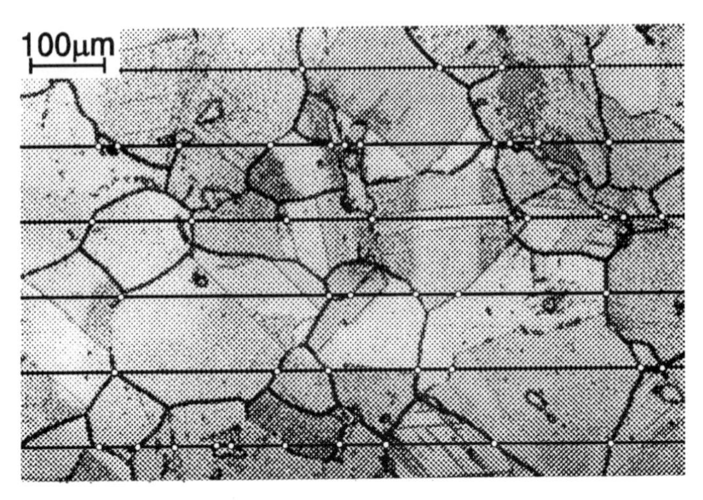

Figure 5.9 Scanning electron microscope image of the grain boundaries in the fine grained austenitic stainless steel used in Example 5.6.

Using these two estimates the surface density of the grain boundary dislocations is given as:

$(L_{GB})_{GBD} = 7.3\ 10^{-3}\ \mu m^{-1}$.

It was found also that a significant fraction of the grain boundaries do not contain dislocations. From the TEM images, the surface area of the grain boundaries without GBDs was estimated to be $(S_V) = 32\ \mu m^{-1}$. Thus, the surface density of the GBDs on these grain boundaries which do contain GBDs can be estimated as being over 70% larger.

The procedures described above are of particular use in the characterization of low-angle grain boundaries (see Figure 5.7).

5.1.4 *Complex arrangements of dislocations*

Crystalline materials subjected to plastic deformation usually acquire large dislocation densities. In fact, the dislocation densities in materials strained at a low temperature is orders of magnitude higher than in annealed crystals and is known to be related to the amount of plastic strain. Frequently, in cold worked metals the dislocation density is so high that it is impossible to observe individual dislocations. On the other hand, dislocations in such materials tend to form thick walls and bundles, which outline regions of a significantly lower dislocation density called dislocation cells.

In principle, dislocation cell walls can be considered to be planar microstructural elements characterized by a nonzero thickness. Accordingly, the size of dislocation cells can be studied using the procedures developed for measurements of the grain size and size of particles. These procedures are discussed in Chapters 6 and 8. It means that the size of dislocation cells can be

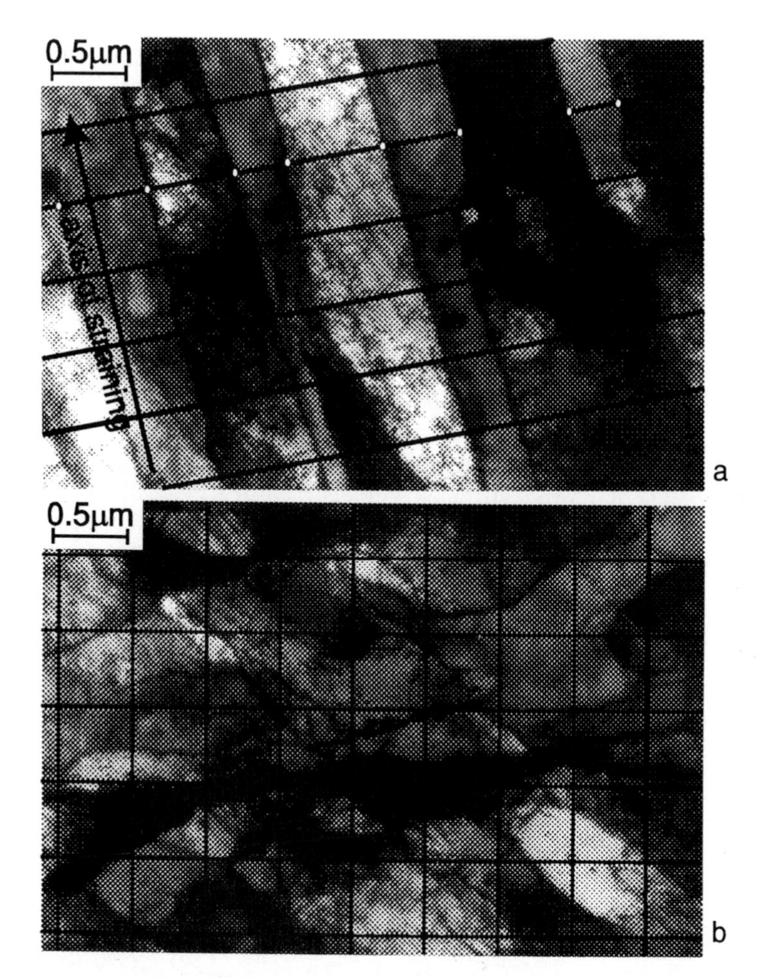

Figure 5.10 Examples of dislocation cell size measurements: (a) measurements of the surface area of cell walls perpendicular to the direction of compression axis, $(S_V)_c$—the system of lines is oriented perpendicular to the walls studied and the density of the intersection points, P_L, determined, which yields $(S_V)_c = 2P_L$; (b) similar measurements for dislocation walls of a less regular arrangement.

described in terms of their equivalent diameter, cell wall surface area in unit volume, etc. Two examples of such measurements are shown in Figure 5.10.

However, it should be noted that the parameters defining the geometry of the dislocation cells are difficult to estimate and require some additional assumptions that may not be true in the general case. The most important assumptions are that: (a) the thickness of the walls is much smaller than the cell diameter and (b) the cell diameter is larger than the thickness of the foil. These assumptions help to resolve the problem of an unknown orientation of the wall with respect to the foil normal. The validity of these assumptions may be tested by examination of the cells under different tilting angles.

5.2 Internal surfaces

Internal surfaces are planar elements of the microstructure of materials. They are defined as the surfaces between 3-D regions in the material volume. Examples of such surfaces are

a) grain boundaries (GBs);
b) interphase boundaries (IBs);
c) surfaces of pores (PS);
d) cracks.

It has been established experimentally that internal surfaces are important elements that control the properties of a large number materials (see, for example, References 7 and 8). Their size and shape controls such properties as yield and flow stress, fatigue strength, and toughness of materials.[9] An example of image of internal surfaces is given in Figure 5.11. More examples can be found in Chapters 6 and 8, which discuss the geometry of grain boundaries and interphase boundaries, respectively.

These internal surfaces in the microstructure can be revealed due to their physical properties. In most cases, these internal surfaces show significantly different strengths, diffusivities, corrosion resistances, etc. This is related to the fact that the atoms at internal surfaces have, on average, much higher energy than other atoms in the material. These distinct physical properties of the internal surfaces make it possible to obtain images from them and study their geometry.

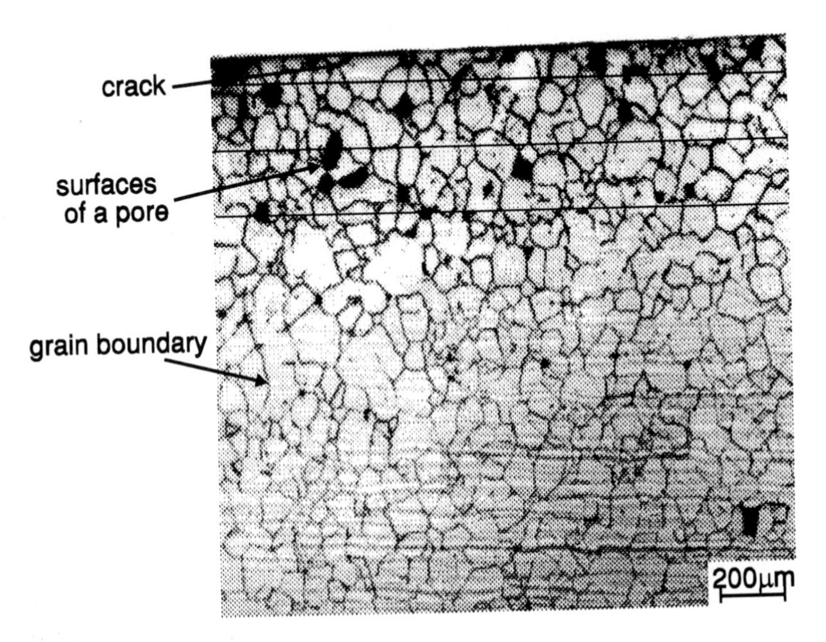

Figure 5.11 Microcracks revealed in an austenitic stainless steel sample and a system of test lines used in Example 5.7.

In studies of the geometry of internal surfaces one may focus attention on the whole population, without addressing possible differences between the individual surface elements. In this case a useful parameter is internal surface area per unit volume, $(S_V)_{IS}$. This parameter can be determined by the methods described in Chapters 2, 3, and **6**.

Example 5.7

Microcracks in a metal specimen were studied via observations on randomly cut sections of the specimen. An image showing the typical pattern of cracks revealed in the material is shown in Figure 5.11. A grid of lines was superimposed on the images of the cracks. The total length of the test lines was 5400 μm. These lines intersected 72 cracks. This gives the density of intersections, P_L, equal to 13.34 mm^{-1}. From the basic stereological relationship, the surface area of cracks per unit volume, $(S_V)_{crack}$, is given as:

$(S_V)_{crack} = 2\,(P_L) = 26.68\ mm^{-1}$.

In more detailed studies, the population of internal surfaces can be seen as consisting of elements having different properties. An example of such a situation is the population of grain boundaries in a polycrystal. As a result of differences in the spatial orientation of individual grains in a polycrystal, grain boundaries differ in their crystallography, microstructure and in turn, in their properties.

A grain boundary in a polycrystal can be characterized by a set of at least five parameters, s_i, defining its crystallography and defect content. This large number of parameters makes the analysis difficult and boundaries are often grouped into broader classes. These classifications frequently are based on their properties to a larger extent than on their microstructure. It is assumed, in principle, that the microstructure and properties of grain boundaries are in a functional relationship. For example, it has been shown that random grain boundaries have higher energy and diffusivity and that these depend weakly on their orientation. Further, segregation of alloying elements strongly changes the properties of boundaries. From the point of view of grain boundary characterization these two sets of parameters, describing the microstructure and properties, to a certain extent can be used interchangeably.

For simplicity of further analysis we may assume that the boundaries, (and other internal surfaces), in a material can be categorized into a finite number of property/microstructure classes, C_i, $i = 1,2,\ldots, n$.

The boundaries in a material form a population. Each element of the population is distinguished by its class, C_i, and its size, expressed as its surface area, which again can be divided into a system of size classes, S_j, $j = 1,2,\ldots, k$. The population of boundaries is described by a joint distribution function of two random variables:

$$f(X,Y) = f(C_i, S_j) \tag{5.14}$$

This distribution function defines the number of boundaries (internal surfaces) with a given size and property range. This function is related to two marginal distributions, $f(C_i)$ and $f(S_j)$, defining the number of boundaries with a given property and of a given size, respectively. It should be pointed out that these two distributions are expected not to be independent and as a result the variables cannot be separated:

$$f(C_i, S_j) \neq f(C_i) f(S_j) \tag{5.15}$$

However, separation of the marginal distributions is possible if the boundaries are of constant size $S_j = S_o$. In the general case, the distribution function characterizing grain boundaries should be based on measurements that take into account both their size and their properties.

5.2.1 *Populations of grain boundaries*

Grain boundaries in polycrystals are an example of internal surfaces which form populations characterized by distributions of geometrical and physical properties. A single grain boundary is distinguished by the area of its surface, S_i, its curvature, H_i and a set of parameters, P_{ij}, defining its structure. These parameters include:

a) misorientation angle, θ, (one parameter);
b) misorientation axis, **l**, (two independent parameters);
c) normal to the grain boundary plane, **n**, (two independent parameters);
d) rigid body translations (two parameters).

The population of grain boundaries in a polycrystal can be described by means of distribution functions of the respective parameters. The distribution function of the misorientation angle, $f(\theta)$, has attracted a great deal of interest since it has been postulated that this parameter controls the energy, diffusivity, and other properties of grain boundaries.

The distribution function $f(\theta)$ can be defined as the number distribution, $f_N(\theta)$, or surface weighted distribution, $f_S(\theta)$. The number distribution function specifies the fraction of grain boundaries distinguished by a given value of θ in the total population. This is a correct description from the point of view of mathematical statistics, but has some important disadvantages when applied to experimental studies. It should be noted that this function cannot be estimated from counting the grain boundaries with a given misorientation on the cross-sections of polycrystals. This is due to the fact that the boundaries are cut by the plane of sectioning with a probability that depends on the size of the boundary. As a result, the experimentally established fraction of grain boundaries counted with a given misorientation, $f_{exp}(\theta)$, depends both on their number, $f_N(\theta)$, and their area, $S(\theta)$. The number distribution function, $f_N(\theta)$, is therefore justified mainly in the modelling of polycrystals where all the boundaries are assumed to have the same area.

In experimental studies, a population of grain boundaries can be described effectively in terms of the surface-weighted distribution function of misorientation angle, $f_S(\theta)$. This distribution function specifies the fraction of the grain boundary area made up by boundaries with a misorientation θ. This function, $f_S(\theta)$, may be estimated by counting the number of intersections per unit length, P_L, of random lines with the boundaries of a given misorientation. From the basic stereological relationships the surface area $S(\theta)$ of such boundaries is given by:

$$\frac{S(\theta)}{S} = \frac{P_L(\theta)}{(P_L)_{total}} \tag{5.16}$$

where S is the total area of all grain boundaries.

As said before, in some applications it is more convenient to arrange the set of possible grain boundary characteristic parameters, for example, possible misorientation angles, into some classes (misorientation intervals) represented by a single value (characteristic misorientation or its function). In this way a continuous space of grain boundary parameters is reduced to a discrete set of some characteristic parameters. Such a situation is exemplified by the procedure based on the coincidence site lattice (CSL) model.[10] In this case the space of angles θ is divided into intervals represented by parameters Σ, which are computed for each misorientation angle.[7] These Σ values for grain boundaries in cubic materials have odd integer values from three upwards. Those grain boundaries with $\Sigma \leq 29$ usually are assumed to have special properties[10] and these special properties of grain boundaries are retained despite small deviations from the exact misorientation characteristic of an exact CSL Σ relationship. The maximum value of this deviation, $\Delta\theta_{max}$, or in other words the size of the θ intervals represented by given Σ values, is commonly calculated from the Brandon criterion:[11]

$$\Delta\theta_{max} = 15° \, \Sigma^{-\frac{1}{2}} \tag{5.17}$$

Further, it has been shown that, with some additional assumptions regarding the spatial arrangement of the grains of a given orientation, the distribution function, $f(\theta)$, and in turn $f(\Sigma)$ are related to the texture of the polycrystal.[12,13]

The populations of grain boundaries can also be described by grain boundary energy distribution functions, $f_N(E)$ and $f_S(E)$, which describe, respectively, the relative number and relative surface area of the grain boundaries of a given energy, E, in the population.

To a first approximation, the population can be divided into two broad categories:

1. random boundaries (R) that show generally higher energy and lower sensitivity to the orientation of the grain boundary plane;
2. special boundaries (S) that have lower energy, which is usually strongly dependent on the orientation of the boundary plane.

In this case, the distribution functions, $f_N(E)$ and $f_S(E)$, are defined by two ratios:

$$R_N = \frac{N_{spec}}{N_{ran}} \tag{5.18}$$

$$R_S = \frac{S_{spec}}{S_{ran}} \tag{5.19}$$

The ratio R_S can be determined experimentally by employing the same general rules as for $f_S(\theta)$ and, for example, by thermal etching of the grain boundaries. Both ratios, R_N and R_S, can be obtained from the distribution function of the misorientation angle, $f(\theta)$, using relationships between the energy and the misorientation of the boundaries.

5.2.2 *The system of joined planar elements of a microstructure*

Planar elements within a microstructure frequently form complicated spatial arrangements. This is exemplified by the geometry of grain boundaries in polycrystals. Grain boundaries in polycrystals form a system of surfaces connected along threefold (grain) edges, TEs, and fourfold (grain corner) points, FPs. On a cross-section, TEs are revealed as so-called triple points, TPs. Grain corners, FPs, would only very rarely be seen and are not identified on a section from a polycrystal, but may be observed on projected images obtained in the TEM very occasionally unless the grain size is extremely small.

These threefold edges and fourfold points play an important role in a number of processes taking place at grain boundaries, especially in micro- and nanograin-sized polycrystals. This is related to:

a) topological restrictions that their presence imposes on the mobility of grain boundaries;
b) interactions of the grain boundaries joined along a common edge.

This is the justification for the need to describe populations of grain boundaries in terms of the distribution functions of TE properties in a way similar to that adopted for the individual boundaries.

The properties of the population of common edges can be described by distribution functions of the misorientation angles of the three boundaries meeting along a common edge. Again number, $f_N(\theta_1,\theta_2,\theta_3)$, and length weighted, $f_L(\theta_1,\theta_2,\theta_3)$, distribution functions may be used. The first distribution function may be determined easily by computer modelling. On the other hand, the length-weighted distribution function, $f_L(\theta_1,\theta_2,\theta_3)$, can be obtained by counting the density of triple points on the sections of the polycrystal studied. The density, $N_A(\theta_1,\theta_2,\theta_3)$, of the triple points connecting the boundaries of orientations $\theta_1,\theta_2,\theta_3$ (with the θs taken in any order) is related to $f_L(\theta_1,\theta_2,\theta_3)$ by the formula:

$$f_L(\theta_1,\theta_2,\theta_3) = N_A(\theta_1,\theta_2,\theta_3) \tag{5.20}$$

In a simplified approach four specific types of TEs can be distinguished:

a) RRR or random-random-random, ($q = 0$);
b) RRS or random-random-special, ($q = 1$);
c) RSS or random-special-special, ($q = 2$);
d) SSS or special-special-special, ($q = 3$);

where q is the number of special grain boundaries at a given TE.

This division takes into account the facts that:

a) in general the misorientation angles of the boundaries joined along a common axis, ($\theta_1,\theta_2,\theta_3$), need not sum up due to the presence of disclinations;
b) deviations from the exact misorientation for a given Σ might sum up.

Consequently, to a first approximation, a system of grain boundaries in a polycrystal can be described additionally by a distribution function, $f_{TE}(q)$, with q values equal to 0,1,2, or 3 and the length-weighted distribution, $f_{TE}(q)_L$, may be estimated via counting the different types of triple points.

5.2.3 Physical and geometrical properties of grain boundaries

Under quasi-equilibrium conditions, for example in annealed polycrystals, the geometry of grains can be expected to be related to their physical properties. Simple considerations show that in a polycrystal with all grain boundaries with the same properties, grains assume regular geometry, which can be approximated by a set of equal-volume tetrakaidekahedra (see Chapter 6). In this case the grain boundaries are joined along the common edges at an angle close to 120°.

The angles between grain boundaries, frequently termed dihedral angles, under near equilibrium conditions tend to depend on the differences in the energy of the grain boundaries (see Chapter 6). The larger are these differences in this energy, the larger is the departure of the angles from 120°. It may be further shown, on purely geometrical grounds, that a departure from 120° in equiaxial structures implies a larger variation in the grain size (Figure 5.12).

For polycrystals characterized by a variation in grain boundary energy, it should be expected that under near equilibrium conditions grains will differ in their size; the effect is reflected by larger values of CV(V). This can be rationalized in terms of the relation between grain boundary energy and misorientation parameters. Therefore, a relationship can be expected to exist between the distribution function of the misorientation parameter and the degree of uniformity of grain size. It has been suggested that the linking element between the distribution of the misorientation parameter and the distribution of the grain size is the distribution function describing the

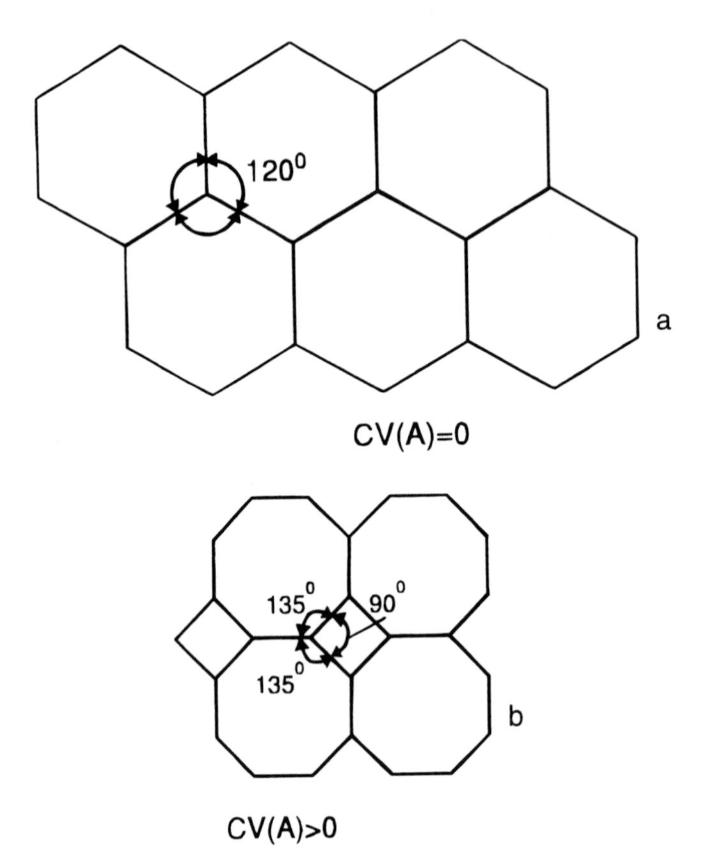

Figure 5.12 Schematic explanation of the dependence of the diversity in the grain size on the angles between the grain boundaries: (a) a grid of hexagons of a constant size A—all angles are equal to 120°; (b) a grid of squares and octagons—the angles between the grain boundaries differ from 120°, and as a consequence the figures differ in their size.

character of the grain edges. This idea can be schematically described in the following way:

$$Texture \rightarrow f(\theta,l) \rightarrow f(\Sigma) \rightarrow f(TE) \rightarrow D\text{-}angles \rightarrow GG \tag{5.21}$$

where D-angles stands for dihedral angles and GG for grain geometry.

Example 5.8

The relation between the texture and geometry of grain and grain boundaries was studied for high-purity aluminum samples (99.99%). Polycrystalline specimens of this material were extruded and then recrystallized for 1 h at various temperatures. The microstructures obtained differed in the grain size distributions,[12] annealing textures, and the distribution of their grain boundary diffusivity properties.[13]

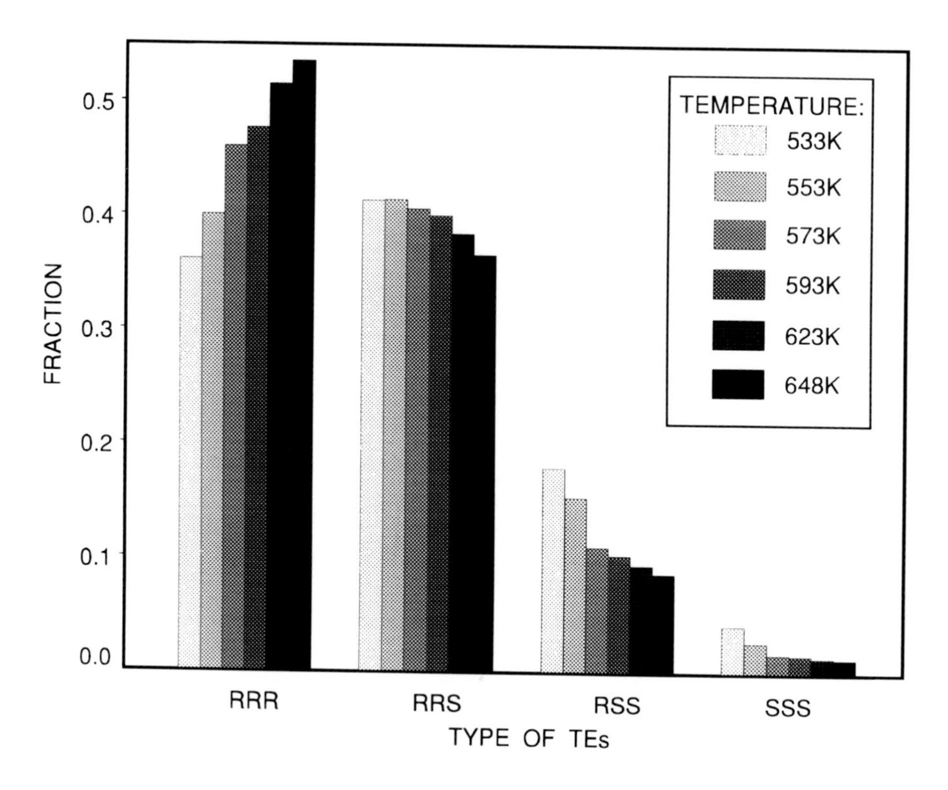

Figure 5.13 Histograms of triple-edge characteristics in terms of the number of random to special grain boundaries for aluminum annealed at temperatures in the range from 533 to 648 K.

The texture of the material was determined by X-ray diffraction.

The method developed by Garbacz and Grabski[14] was used to model the distribution function of grain boundary misorientation parameters and Σ values. The characteristics of the triple edges, TEs, of the grain boundary system for each texture was obtained. Each CSL boundary with coincidence Σ ranging from 1 to 29 was counted into the "S" category. The common edges in the system of grain boundaries were examined in terms of the number of special/random boundaries which they join. The results obtained from modelling the TE distribution function for different annealing textures/temperatures are presented in Figure 5.13. This figure presents histograms of TEs of a specified number of random/special boundaries. It may be noted that there is a clear dependence of the TE character distribution function with the texture and grain size of the polycrystals. Generally, the fraction of TEs without special grain boundaries, ($q = 0$), increases with increasing mean grain size, while the fractions of all other TEs, ($q = 1,2,3$), decrease.

The grain geometry in this series of aluminum polycrystals was examined with an automatic image analyzer and described as a function of the annealing temperature. The degree of grain size uniformity was measured for

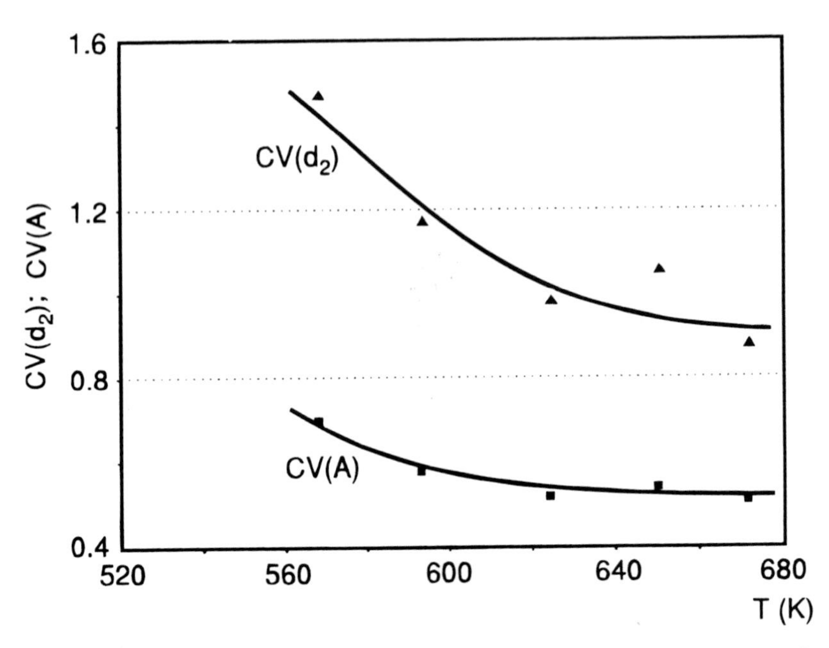

Figure 5.14 Alterations in the coefficients of variations, CV(A) and CV(d_2), as functions of the annealing temperature for specimens of aluminum annealed in the temperature range from 533 to 648 K.

each annealing temperature from the grain section areas, A, and equivalent circle diameters, d_2. The coefficients of variation of both of these parameters are plotted against the annealing temperature in Figure 5.14.

It may be noted that extensive changes in the grain shape, the degree of grain size uniformity, and in the characteristics of the threefold edges take place as a function of the annealing temperature in the range 533 to 648 K. These changes are accompanied by a systematic decrease in the sharpness of the material's texture.

The choice of parameters, p_i, used to characterize individual boundaries depends on the needs of theory and on the possibility of making experimental measurements of given grain boundary properties. The same applies to the decision as to whether to analyze functions $f_N(p_i)$ or $f_S(p_i)$. It should be noted that in a large number of applications the information on the total surface area of a given type of boundary is more important than information on their number, and hence the second function is more appropriate.

An example of grain boundary characterization is the application of TEM observations to the spreading kinetics of grain boundary dislocations, GBDs. This phenomenon, illustrated by the sequence of micrographs shown in Figure 5.15, takes place during annealing of thin foils from polycrystalline materials. The kinetics of the changes in the GBD contrast depend on the diffusivity of the grain boundary studied. The results of such studies are usually presented as a fraction of grain boundaries exhibiting this effect, f_S, against the annealing

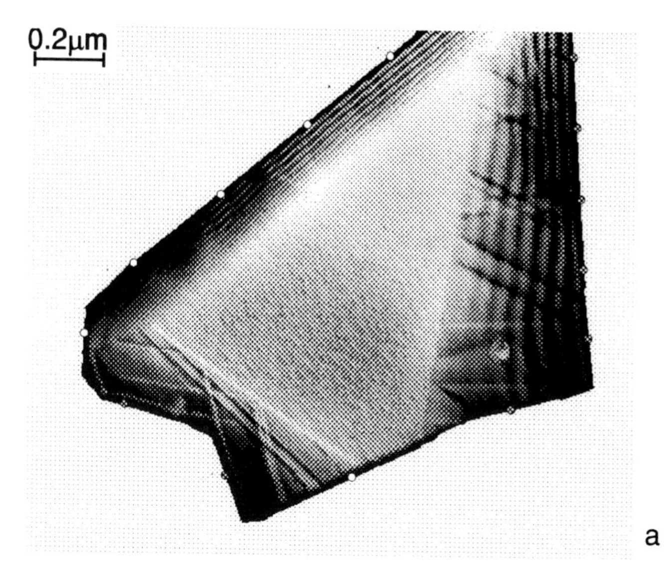

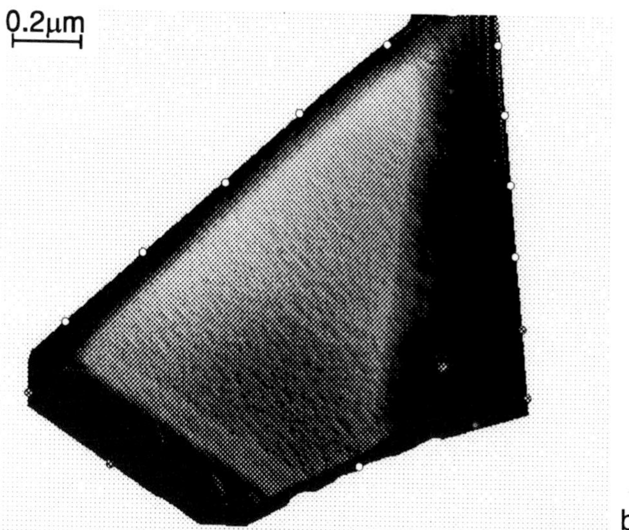

Figure 5.15 An example of the disappearance of grain boundary dislocations during *in situ* annealing of a thin foil: (a) the initial image of the dislocations; (b) the same area after annealing for 30 s at 700°C.

temperature, T_S. In most cases f_S is understood as the fraction of the number of grain boundaries studied, $(f_s)_n$, exhibiting this effect. However, this is not a proper stereological parameter and can be used only if all the grain boundaries have the same size. A better description can be obtained if the fraction f_s is defined as the surface fraction, $(f_s)_S$.

Example 5.9

In studies of the grain boundaries in an austenitic stainless steel after 5% straining at room temperature, it was found that initially all the grain boundaries contained GBDs. A thin foil was annealed at 800 K for 1 h. Out of a total of 74 grain boundaries in the foil, 38 were found to be dislocation-free after the annealing process. This makes the fraction $(f_s)_n$, for this particular condition, equal to 51%.

Test lines were used to measure the surface area of the grain boundaries that were free of dislocations. By measuring the density of the intersection points of these lines with the dislocation-free grain boundaries, it was found that:

$(f_s)_S = 0.68$ or 68%.

5.2.4 *Application to the study of twin boundaries in metals*

Twin boundaries constitute a distinct group of the system of boundaries in metallic materials. They are distinguished by:

1. properties: low surface energy, extreme mobilities (low for coherent and high for noncoherent twin boundaries);
2. microstructure: can be described in terms of a special misorientation or a crystallographic reflection;
3. morphology: flat, step-like surfaces and low dihedral angles with other boundaries.

Twin grains, TGs, and twin boundaries, TBs, in a given material can be described quantitatively using the same procedures as those developed for general grains and grain boundaries. For instance, the parameter defining the specific surface area of twin boundaries, S_V, can be determined from 2-D intercept measurements. The value of $(S_V)_{TB}$ is given as:

$$(SV)_{TB} = 2(P_L)_{TB} \tag{5.22}$$

where $(P_L)_{TB}$ is the density of TB intersection with the test lines.

In attempts to measure the content of twin boundaries, some authors have used the so-called twin frequency, f_{TB}, defined as:

$$f_{TB} = \frac{(P_L)_{TB}}{(P_L)_{GB}} \tag{5.23}$$

where $(P_L)_{GB}$ is the density of intersections of test lines with general boundaries. However, it should be noted that this parameter is not a measure of frequency but simply relates the surface area of twin and grain boundaries per unit volume:

$$f_{TB} = \frac{(S_V)_{TB}}{(S_V)_{GB}} \tag{5.24}$$

Another parameter frequently measured in studies of twin boundaries is their number per cross-sectional area, $(N_A)_{TB}$. However, this parameter neither measures the size nor the number of twin boundaries. Stereologically, it is related to the length of twin boundary intersections with general boundaries $(L_V)_{T/GB}$.

The density of intersection edges, $(L_V)_{T/GB}$, between TBs and grain boundaries can be estimated from 2-D counts of twin boundary trace intersections with the traces of grain boundaries:

$$\left(L_V\right)_{T/GB} = \left(N_A\right)_{T/GB} \tag{5.25}$$

Instead of counting triple points one may also count twin boundary traces and use the relationship between the number of such traces and the triple points:

$$\left(N_A\right)_{T/GB} = 2\left(N_A\right)_{TB} \tag{5.26}$$

This last equation can be used to rationalize the counting of twin boundary traces on 2-D sections of materials.

Example 5.10

Figure 5.16 shows traces of twin boundaries revealed on a polycrystalline high-purity copper sample. Grids of lines were used to determine the surface area of the twin and general grain boundaries in unit volume, $(S_V)_{TB}$ and $(S_V)_{GB}$. Also, the density of the triple points with at least one twin boundary was determined. These measurements yielded the following values:

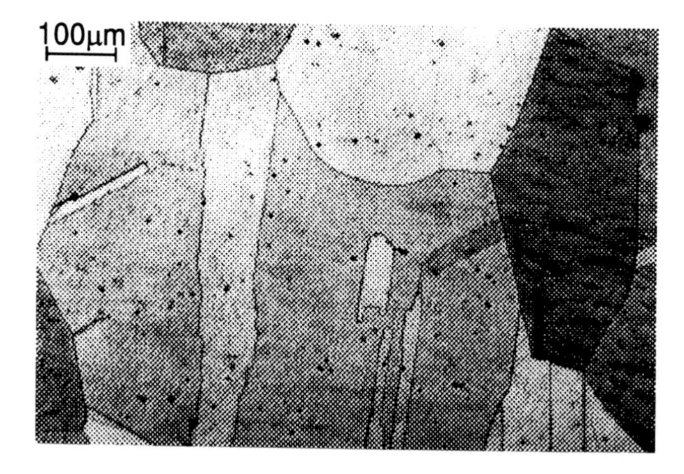

Figure 5.16 Twin boundaries revealed in a polycrystalline copper sample. Test lines were used to determine the surface area of the twins and general grain boundaries per unit volume, $(S_V)_{TB}$ and $(S_V)_{GB}$.

$(S_V)_{TB} = 24.4\ mm^{-1}$;

$f_{TB} = 0.26$;

$(L_V)_{T/GB} = 3.5\ mm^{-2}$.

5.3 External surfaces

A special case of 2-D interfaces are external surfaces. The microstructures of external surfaces are either formed during processing or as a result of the material wear and, ultimately, failure. External surfaces are also produced by surface engineering technologies, which include coatings and heat-treated layers. An example of an external surface created as a result of material failure is a fracture surface. Studying the microstructures of external surfaces over recent years has become an important field of materials science as well as of mechanical and chemical engineering (see, for example, Reference 15).

The geometry of an external surface is macroscopically described by a function that defines the shape of the artifact studied. In the case of a flat specimen, see Figure 5.17a, this is a combination of linear functions of spatial coordinates x,y,z. By an appropriate choice of the coordinate system, the surface of interest, say G, can be described by the following formula:

$$G_{macro}: \quad z = 0 \tag{5.27}$$

When examined at higher magnification, the surface of the artifact usually reveals characteristic microscopic geometrical features. These microscopic features can be uniquely described by a function of two spatial variables, (in the case of the surface G, these variables are x and y), which defines the topography of the surface with respect to the macroscopic geometry, G_{macro}. A natural choice for a function describing the microscopic geometry is the height function, H(x,y), which defines the distance of a given point on the surface of the artifact from the reference, macroscopic surface:

$$G_{micro}: \quad H = H(x, y) \tag{5.28}$$

In an experimental approach, the height function, H(x,y), can be estimated with a precision dependent on the resolution limit. An example of the resolution effect is given in Figure 5.17b. This problem of accounting for the resolution effect can be solved using the concept of fractal dimensions (see Chapter 4 and further text in the present chapter). A general recommendation is that studies of this geometry should be carried out under magnifications varying over a relatively wide range, for instance, from 100 to 10,000, and the trend in the data obtained should be analyzed as a function of magnification/resolution limit.

Modern experimental techniques now make it possible to obtain an estimate of the height function, H(x,y), with the required precision. However, a large number of the experimental methods still used are based on studies of

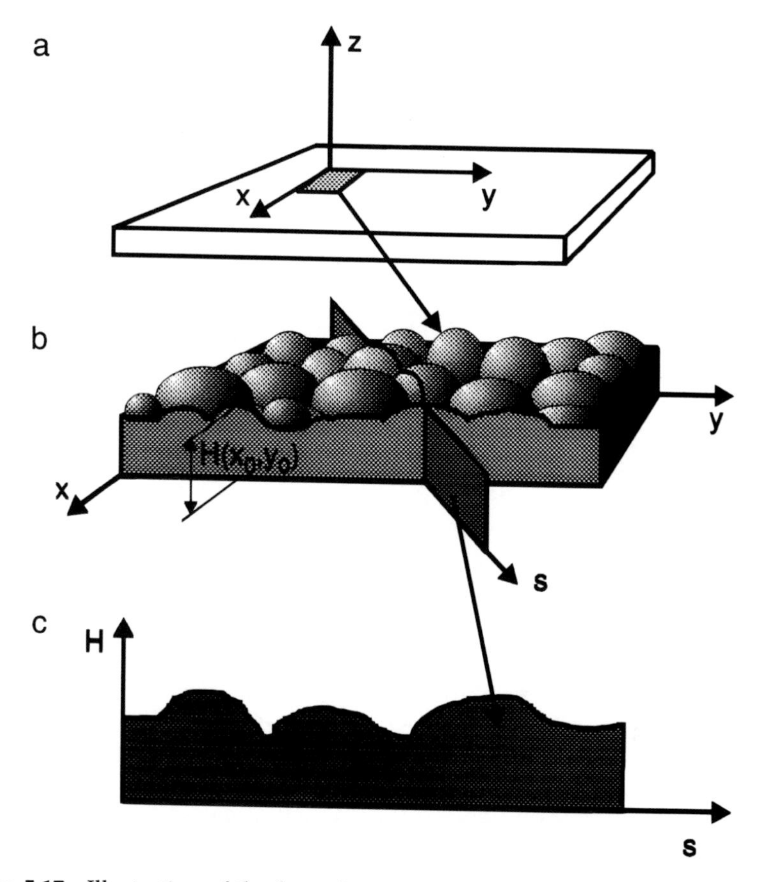

Figure 5.17 Illustration of the formalism used in the description of surface topography: (a) system of coordinates used in the analysis of the macroscopic geometry of the surface; (b) schematic representation of the microscopic aspects of the surface geometry as a function H(x,y); (c) surface profile for the orientation of the section indicated in (b).

surface profiles, H(s). The surface profile function, H(s), defines the height function, H(x,y), along a line placed on the reference plane—see Figure 5.17c.

This profile function, H(s), provides a less complete description of the surface studied. The height function, H(x,y), determines properties of all its profiles, $H_i(s)$. On the other hand reconstruction of the function H(x,y) from the profile measurements is possible only if the profiles are systematically taken along parallel lines placed close one to another. However, in many cases the description H(s) is found to be satisfactory and, what is even more important, it can be obtained with less sophisticated equipment. In fact, the profile function, H(s), can be obtained from direct microscopic observations on the perpendicular sections of the surface studied.

The surface height functions, H(x,y), as well as the surface profile functions, H(s), in most cases show a considerable level of complexity and elements of randomness. This complexity and randomness make it virtually impossible to list the height or profile functions for a given type of surface. Moreover, such a description is not required in most materials science applications (it is extensively used in military applications—e.g., driving a missile over the surface of Poland!). In materials science applications we are usually concerned with some selected properties of the surface such as:

a) surface roughness;
b) density of microscopic features (such as cracks, pits, etc.);
c) degree of randomness/order in the geometry;
d) size and shape of grains and particles emerging on the surface.

These and other properties are functionally related to the height function, H(x,y). However, these properties can be described quantitatively without the need for obtaining the height function, H(x,y). In most cases a satisfactory description can be obtained from the surface profiles, H(s).

The methods of external surface characterization are discussed in the present text for two special examples:

a) a fracture surface;
b) a thin film.

The first example, a fracture surface, exemplifies surfaces created by a process that is of a stochastic nature. These surfaces are frequently characterized by significant variations in the values of height function, H(x,y), overlaps, etc. Some examples of fracture surfaces are shown in Figure 5.18.

The second case, a thin film, shows an external surface produced in the fully controlled process of surface engineering. This process leads to much smoother and more regular geometry of the surface and reveals the microstructure of the material deposited on the surface of an artifact (see Figure 5.19).

5.3.1 *Roughness of fracture surfaces*

Fracture surfaces are formed as a result of a material decohesion under the action of applied load. Depending on the material's microstructure, and in turn its resistance to fracture, these surfaces differ in their geometrical features. Fracture surfaces in brittle materials are relatively flat while in the case of high-toughness materials the decohesion surfaces are usually characterized by complicated geometry. Differences in the properties of fracture surfaces are frequently described in terms of their roughness.

It has been shown in a number of papers[16–18] that a natural surface roughness parameter of great importance is R_S. It is defined as the true surface area, S, divided by the apparent projected area, A, according to:

$$R_S = \frac{S}{A} \tag{5.29}$$

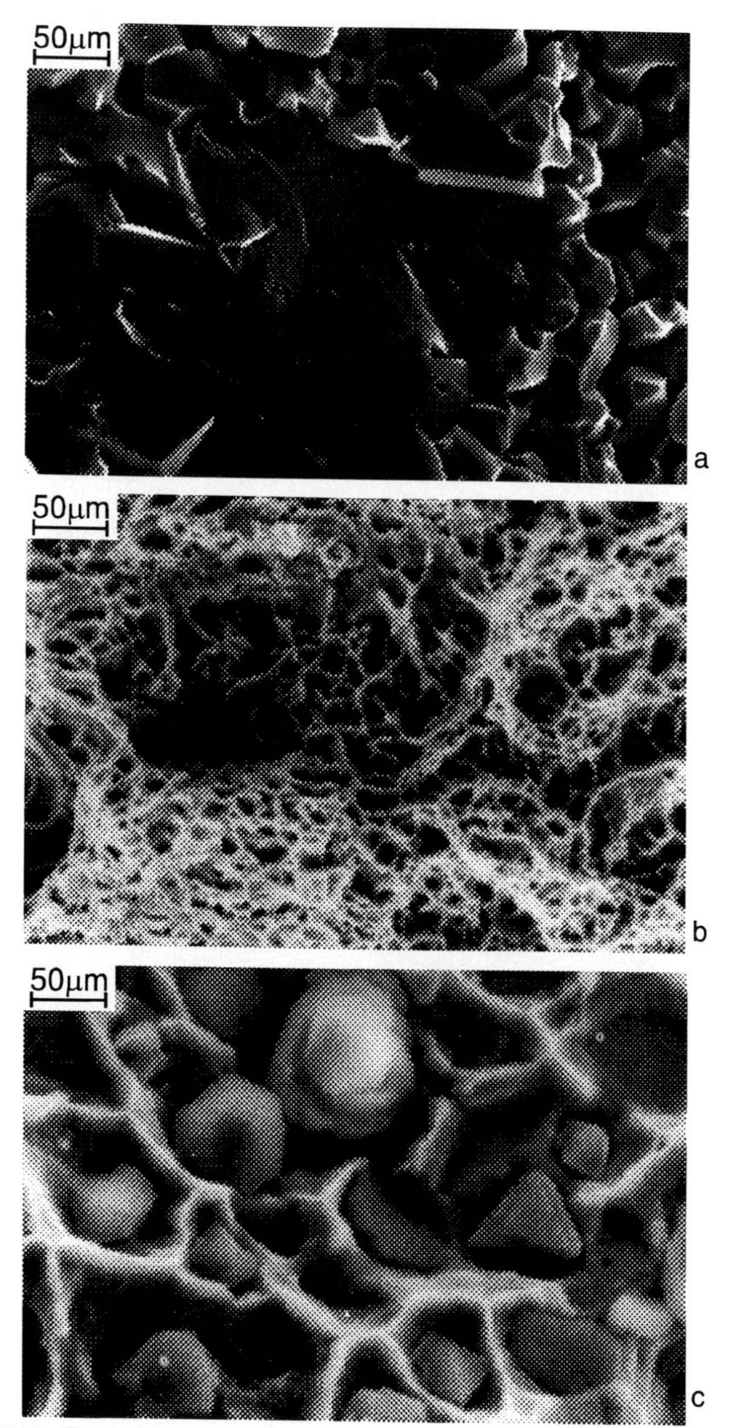

Figure 5.18 Examples of fracture surfaces: (a) a brittle fracture surface; (b) a ductile fracture surface; (c) inclusions on a fracture surface.

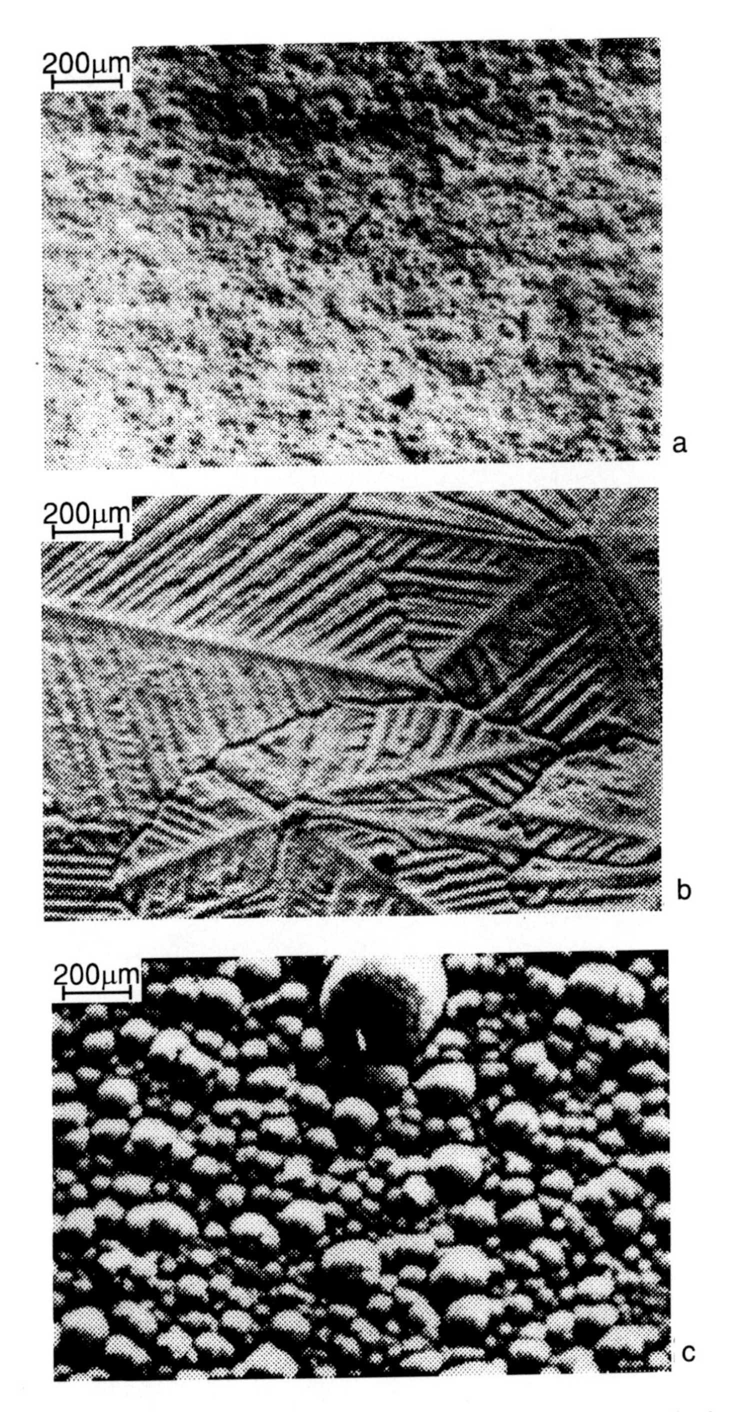

Figure 5.19 Examples of surfaces produced by surface engineering methods: (a) a layer of a paint; (b) a surface produced by a conventional galvanizing; (c) a TiN film produced by plasma-assisted chemical vapor deposition (PACVD).

This parameter can be used directly to characterize the fracture surface and it has been shown (for example, Reference 19) that for some materials a correlation exists between the roughness parameter, R_S, and the toughness. The roughness parameter can also be used as a scaling factor which relates results of the measurements performed on 2-D images to the results of measurements on the fracture surfaces.

Usually, the fracture surface will intersect many types of other microstructural features, such as grain boundaries, particles, voids, etc. In many cases one will wish to know, from a mechanistic point of view, whether there is a lower or higher than average association of these features with the fracture surface. For instance, it is common for the fracture surface to follow grain boundaries in embrittled polycrystals and in materials that have undergone creep failure. Thus, it is important to be able to describe this association between these microstructural features (elements) and a fracture surface.

The number of elements related to unit fracture area yields the number of elements per unit fracture surface area, N_S:

$$N_S = \frac{N}{S} \tag{5.30}$$

and, when related to the surface area of the projection, it expresses the number of elements per unit of projected fracture surface area, N'_A:

$$N'_A = \frac{N}{A'} \tag{5.31}$$

In an analogous manner, one can define the average length of linear elements per unit area of fracture surface (Equation 5.32) or to the projection of the fracture surface area (Equation 5.33):

$$L_S = \sum_i \frac{L_i}{S} \tag{5.32}$$

$$L'_A = \sum_i \frac{L'_i}{A'} \tag{5.33}$$

Similarly, in the case of the surface area of arbitrary elements located on the fracture surface, one can define the surface share of these elements on the fracture surface:

$$S_S = \sum_i \frac{S_i}{S} \tag{5.34}$$

and on the projection of the fracture surface:

$$A'_A = \sum_i \frac{A_i}{A'} \tag{5.35}$$

Based on definitions expressed by Formulae 5.29 to 5.31 one can easily derive the equation which allows N_S to be computed:

$$N_S = \frac{N'_A}{R_S} \tag{5.36}$$

Equation 5.36 is valid without assuming any simplifications. Similar equations relating to L_S and S_S remain valid only in the case of elements that exhibit a completely random distribution on the fracture surface:[20,21]

$$L_S = \frac{L'_A}{R_S} \tag{5.37}$$

$$S_S = A'_A \tag{5.38}$$

In addition to this, on the projection of the fracture surface one can determine the values of various parameters known from the classical geometry of solids (for example, References 22 and 23), such as degree of orientation of linear elements, mean free path, mean chord, etc. However, many of these values are rather difficult to determine because of the necessity to identify and delineate individual components on the geometrical structure of the fracture surface.

5.3.2 *Profile roughness*

The basic measure of a surface profile is its length, L, which, when related to the length of the projection L′, yields the coefficient of development of the profile line, R_L:

$$R_L = \frac{L}{L'} \tag{5.39}$$

The fracture profile can also be treated as a string of peaks and valleys (Figure 5.20). In this case the degree of development of the fracture surface can be characterized by specifying the ratio R_V of mean relative peak height, h, to the mean horizontal distance, I, between the peak and the bottom of the valley:

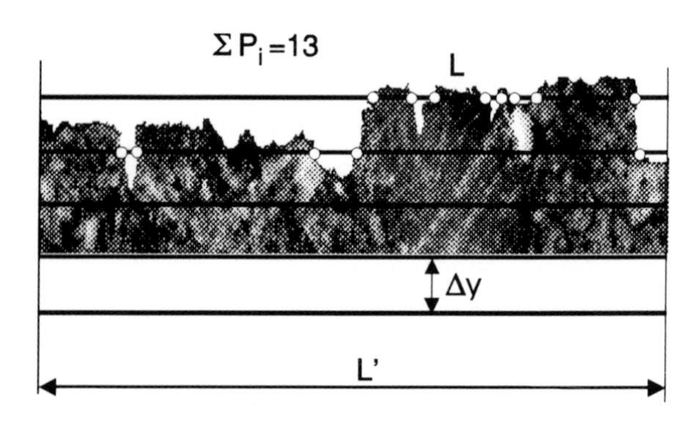

Figure 5.20 Basic description of surface profile geometry.

$$R_V = \frac{\sum_i^n h_i}{\sum_j^n I_j} \tag{5.40}$$

where n is the sum total of the number of valleys and peaks. The equation for determining the value of this coefficient R_V assumes the following form (see Figure 5.20):

$$R_V = \left(\frac{\Delta y}{L'}\right)\sum_i P_i \tag{5.41}$$

The coefficient R_V is a dimensionless parameter and does not depend on the size of the peaks, but only on the proportion of them. The individual peaks can be described by specifying their average height or deviation from the mean value, similar to the practice employed in measuring roughness.

5.3.3 *Relationship between surface and profile roughness*

From the moment when the coefficient R_S was defined it attracted the attention of researchers dealing with quantitative fractography. The surface area of the fracture should be related positively with the fracture mechanism, as well as with the amount of energy consumed during destruction of the material. Scientists, such as Underwood,[23] have even regarded R_S to be the most important and fundamental parameter of quantitative fractography. However, attempts to use the value of R_S for analyzing fracture processes did not yield the expected results[24] and this problem is still of considerable theoretical importance.

El Soudani[16] had already demonstrated that the coefficient R_S is proportional to the degree of development of the profile R_L. This concept has been developed by Coster and Chermant[19] as well as by Wright and Karlsson,[18] and has led to the well-known relationships:

$$R_S = \left(\frac{2}{\pi - 2}\right)(R_L - 1) + 1 \tag{5.42}$$

$$R_S = 1 + \frac{\pi(R_L - 1)}{2} \tag{5.43}$$

Both of the relationships have the form of a one variable function of the type:

$$R_S = f(R_L) \tag{5.44}$$

The above considerations lead to the following conclusions:

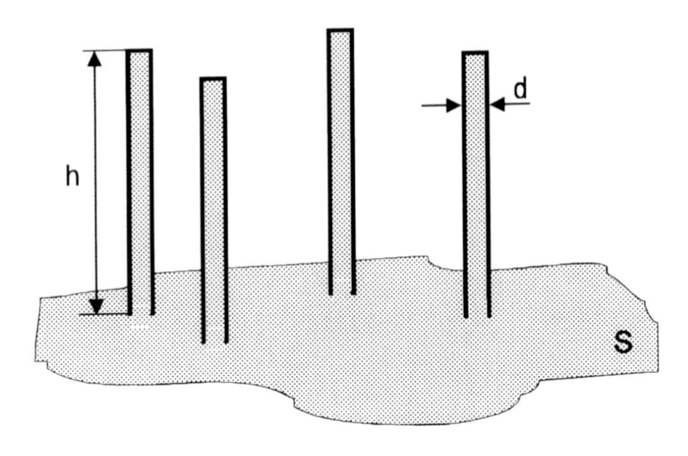

Figure 5.21 An example of the surface topography used in the Wojnar model.

1. Depending on the configuration of the fracture surface, different R_S values can correspond to the same R_L values.
2. Theoretically, there may exist the largest and smallest R_S values for each R_L—in other words, there may exist such functions (Equation 5.44) that constitute the lower boundary and the upper boundary of all theoretically possible pairs R_L–R_S.

The lower boundary is obvious, and for an ideally flat surface it is described by the following equation:

$$R_S = R_L \tag{5.45}$$

The existence of the upper boundary was a matter of controversy for some time. Different formulae have been suggested (see, for example, References 24 to 26). In order to determine the upper boundary, Wojnar[24] analyzed a surface shown schematically in Figure 5.21. The following simplifying assumptions, which will facilitate calculations without affecting the form of the final result, are made:

1. all chimneys have cylindrical envelopes of diameter, d, and height, h;
2. the height of the chimneys is very large in comparison to the other dimensions of the model surface;
3. the fracture surface between chimneys is of a random curvature.

Assumption 3 is important to the extent that it significantly simplifies calculations, since the surfaces mentioned have constant R_L and R_S values, equal to $\pi/2$ and 2, respectively. However, this does not have any effect on the final result, because the area of the model fracture surface is determined by the lateral surface area of the cylinders (compare assumption 2). For N'_A chimneys per unit area of projected fracture surface, R_S can be determined as follows:

$$R_S = 2 + N'_A\, P\, d\, h \tag{5.46}$$

In an analogous manner, assuming that N'_L chimneys are cut by a vertical secant plane, R_L can be expressed as:

$$R_L = \frac{\pi}{2} + 2h\, N'_A \tag{5.47}$$

The surface share of the chimneys in the projection of the fracture plane, A'_A, is defined by the dependence of Equation 5.35, and it can be computed based on the model from Figure 5.21 by using Equations 5.48 and 5.49 (see, for example, Reference 27):

$$A'_A = N'_A \frac{\pi d^2}{4} \tag{5.48}$$

$$A'_A = L'_L = N'_L\, I = N'_L \frac{\pi d}{4} \tag{5.49}$$

where I is the mean length of the secant through the chimney. This yields the following set of equations:

$$R_S = 2 + \frac{4hA'_A}{d} = 2 + 4\left[\frac{hA'_A}{d}\right] \tag{5.50}$$

$$R_L = \frac{\pi}{2} + 8h\frac{A'_A}{\pi d} = \frac{\pi}{2} + \frac{8}{\pi}\left[\frac{hA'_A}{d}\right] \tag{5.51}$$

These, in turn, lead to the equation:

$$\frac{1}{4}\left(R_S - 2\right) = \frac{\pi}{8}\left(R_L - \frac{\pi}{2}\right) \tag{5.52}$$

which, after some simple transformations, assumes its final form:

$$R_S = \frac{\pi}{2} R_L - \frac{\left(\pi^2 - 8\right)}{4} = 1.57\, R_L - 0.467 \tag{5.53}$$

The method for deriving Equation 5.53 as presented above has been published by Wojnar.[21]

5.3.4 Measurements of the coefficient R_S by the vertical section method

The surface roughness parameter, R_S, can be effectively estimated using the method of vertical sections described in Chapter 3. Vertical sections are a set of secant planes that are oriented parallel to an arbitrarily chosen direction designated as the vertical (Figure 5.22). In particular, the vertical axis can be perpendicular to the surface studied. It is then equivalent to the preparation of fracture profiles.

According to the stereological relationship (see Chapter 2), the area of an arbitrary surface can be estimated from the following equation:

$$S_V = 2P_L \tag{5.54}$$

where S_V is the surface area per unit volume and P_L is the number of intersections per unit length of the test lines. This equation requires that the test lines be randomly and evenly distributed in space. It should be noted that a random distribution in space is not equivalent to a random distribution of test lines on vertical sections. In fact, the random orientation of test lines in space requires placing on the vertical sections a system of cycloids as shown in Figure 5.23. (The method of vertical sectioning is described in Chapter 3.)

The application of the vertical sectioning method to the estimation of R_S has been described by Wojnar.[21]

This method requires the following steps:

1. Prepare a series of vertical cross-sections, of which there should be not less than six, and these should be uniformly distributed around the vertical axis. In the case where we are sure of the absence of linear orientation and the individual vertical cross-sections are statistically equivalent, it suffices to examine one vertical cross-section.
2. Prepare a suitable measurement grid with cycloids. The cycloids should be uniformly distributed across the entire surface of the measurement grid. The measurement grid should be supplemented by a system of horizontal lines spaced by the length of the individual cycloids. An example of such a grid is shown in Figure 5.23.
3. Apply the measurement grid to the vertical section image, such that the horizontal aligns exactly with the fracture macrosurface.
4. Select a measurement rectangle, which includes the whole profile to be analyzed and whose height is equal to h, which is a multiple of the cycloid length (in Figure 5.24, h = 4). Because of this, the actual length of the cycloid does not affect the result.
5. Determine the number of intersections, N, of the system of cycloids with the line of the fracture profile (in Figure 5.24, N = 13) and calculate R_S from the formula:

$$R_S = 2\frac{Nh}{m} \tag{5.55}$$

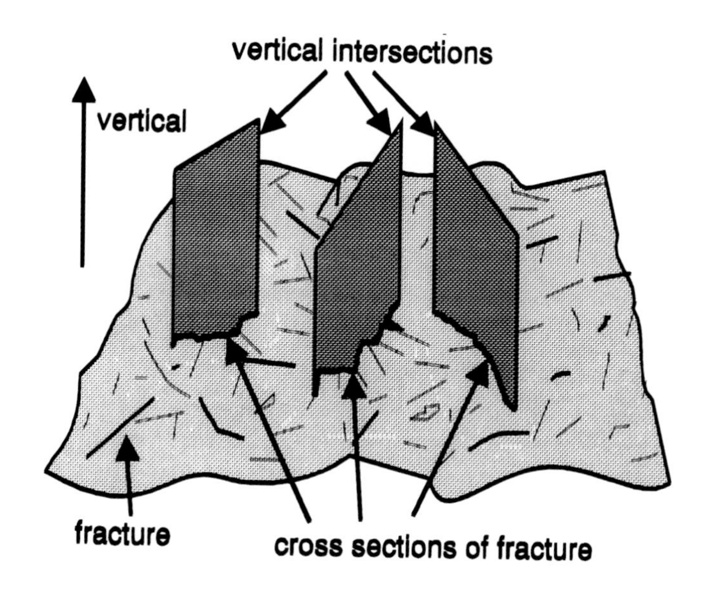

Figure 5.22 Schematic explanation of the vertical sectioning approach.

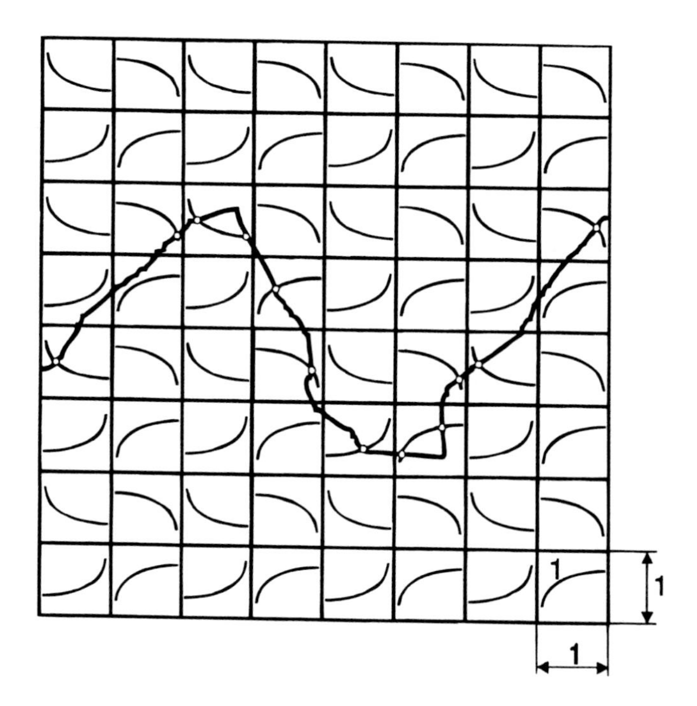

Figure 5.23 Explanation of the measurement technique based on a grid of cycloids.

where:

N—total number of intersections of the profile with the cycloids;
h—height of the measurement rectangle expressed as a multiple of the cycloid length;
m—total number of cycloids contained in the measurement rectangle.

It should be noted that the grid needs to be applied many times in order to ensure that the measurements have a random character, so as to eliminate the possible effect of local inhomogeneities of the fracture surface.

The application of vertical sectioning method measurements of R_S is illustrated by the following examples.

Example 5.11

Figure 5.24 shows the profile of a fracture surface observed in a low-alloy steel. A total number of six vertical sections of the fractured specimen were studied. For each fracture profile obtained by vertical sectioning, the roughness parameter, R_S, was estimated using the system of cycloids shown in Figure 5.23. After averaging over all the profiles the following value of R_S was obtained: $R_S = 2 \times 13 \times 4/48 = 2.13$. The profile lines were also digitized (see Chapter 4) and the profile roughness parameters, R_L, computed using an image analysis system. The following estimate was found: $R_L = 1.67$.

Example 5.12

Figure 5.25 shows two profiles of the fracture surfaces of α-Fe strained to fracture at different temperatures. Measurements based on the vertical sectioning method yielded the following values of the roughness parameters for the fracture surfaces studied: $(R_S)_a = 2.61$, $(R_S)_b = 2.58$.

The examples presented illustrate the advantages and shortcomings of the characterization of surfaces via the roughness parameter, R_S. The important

Figure 5.24 The profile of a fracture surface with a superimposed grid of cycloids used to estimate the surface roughness of a fracture surface in a low-alloy steel sample.

Figure 5.25 Fracture surface profiles for specimens of α-Fe strained to fracture in a tensile test carried out at: (a) 293 K and (b) 873 K.

benefit is the possibility of obtaining numbers that can be used to rank materials in terms of the complexity of their surface development. On the other hand, as shown in Example 5.12, two significantly different profiles can yield similar values of R_S.

Gokhale and Underwood[26] have independently developed another method for estimating the value of the roughness coefficient, R_S. In this method R_S is estimated based on the measurement of R_L and an additional parameter Y, which is a function of the distribution of angles along the profile:

$$R_S = \overline{R_L \times Y} \tag{5.56}$$

On the right side of this equation there is the arithmetic mean of the products R_L and Y for the individual vertical sections. Parameter Y is calculated based on the knowledge of the distribution of angles, α, between the normal to the profile and the vertical. Function f(α) represents the distribution of the probability density of these angles:

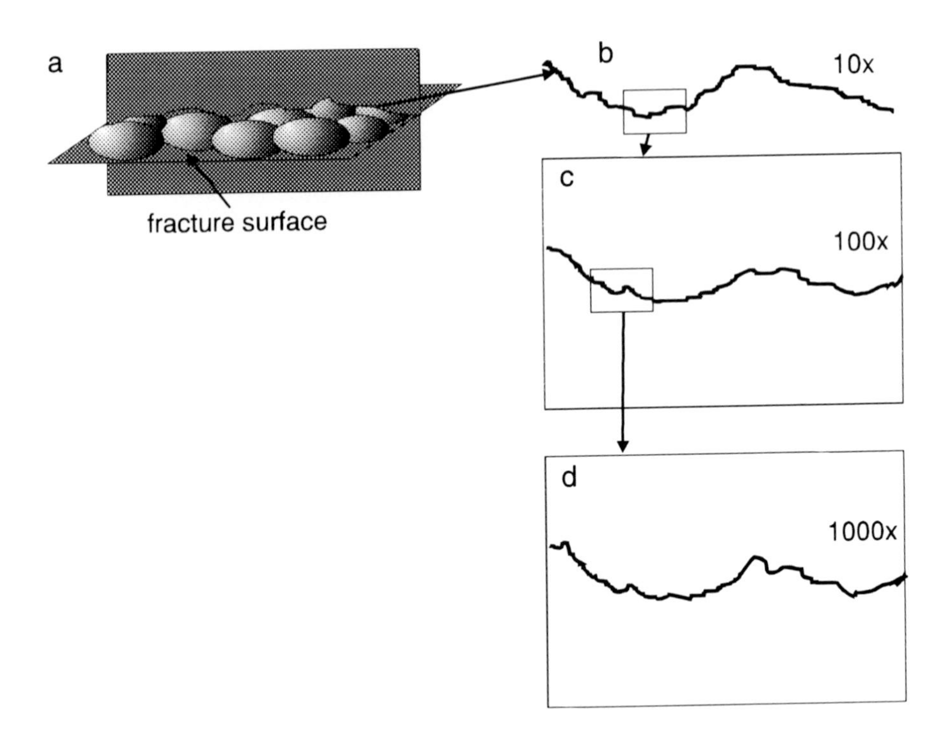

Figure 5.26 Schematic explanation of the fractal character of a fracture surface: (a) the fracture surface; (b) fracture surface profile observed at a magnification of 10×; (c) fracture surface profile observed at a magnification of 100×; (d) fracture surface profile observed at a magnification of 1000×.

$$Y = \left[\sin\alpha + (\pi/2 - \alpha)\cos\alpha\right] \times f(\alpha)\, d\alpha \tag{5.57}$$

It may be noted that the method based on the principle of vertical sections seems to be simpler and requires fewer calculations of intermediate values, which probably decreases the errors in estimating the value of R_S.

5.3.5 *Fracture surface and fractals*

It has been well known for some time that many irregular lines look similar at various magnifications. This characteristic is called self-similarity, and the objects exhibiting it are called fractals (see Chapter 1). Fractals are frequently encountered in nature; fracture surfaces have shown themselves to be fractals as well.[28–37]

The topological dimension of linear segments is 1 and of surface elements is 2. In the case of flat fractals (such as shown in Figure 5.26) their topological dimension is larger than 1 and smaller than 2. This is because, although the fractal does occupy a certain surface, it itself does not have a surface since it is made up exclusively of 1-D elements: segments and curves. On the other hand, the length of these 1-D elements is infinite, despite the fact that the entire fractal occupies a limited surface, in the case of so-called mathematical fractals. Mandelbrot[37] defined the dimension, D, of a fractal as:

$$D = \frac{\log(N)}{\log(1/r)} \tag{5.58}$$

where N is the number of segments constituting the initial structure of the fractal, and r—the degree of reduction of the elements constituting the fractal during subsequent stages of its construction.

The fractal dimension of a curve can thus be interpreted as a measure of the degree of its complexity. For an extremely uncomplicated curve, such as a straight-line segment, D = 1. In fractographical studies one uses profile analysis at various measuring intervals, h, to determine the fractal dimension. Fractal profiles exhibit an exponential dependence between the measuring interval and the geometrical parameter under study, e.g., length.[36] In the case of fracture surfaces, it is most common to use the relation between the measuring intervals and coefficient R_L:

$$R_L(h) = R_L(0) \times h^{-(D-1)} \tag{5.59}$$

After taking the logarithm of both sides of Equation 5.59 one obtains the relationship:

$$\log R_L(h) = \log R_L(0) - (D-1)\log h \tag{5.60}$$

It follows from this equation that all results from the measurements of R_L obtained with various values of the measuring unit, h, should lie on a straight line in a double logarithmic system. These types of graphs are called fractal diagrams.

Underwood and Banerji[35] carried out such an analysis taking advantage of the fact that every fractal diagram for a fracture surface has two asymptotes: $R_L = 1$ for very large values of measuring intervals, and $R_L = (R_L)_0$ for an infinitely small measuring interval (in practice this interval is comparable to the dimensions of a single atom). To interpret the fractal diagrams Underwood and Banerji[35] proposed the use of the transition curve whose equation is

$$\log\log\left[\left((R_L)_0 - 1\right)/\left(R_L(h) - 1\right)\right] = c - (D_b - 1)\log h \tag{5.61}$$

Equation 5.61 represents a linearized form of the function describing the dependence between h and R_L. This form is convenient for calculating the value of the modified fractal dimension, D_b.

5.3.6 *Advanced analysis of fracture profiles*

The description of the fracture profile by a single number like R_V or R_L gives, in some applications, insufficient information about the fracture surface topography. For example, these parameters are not sensitive to the relative positions of the surface irregularities. A more comprehensive description of the surface topography may require information about the spatial distribution of peaks and valleys across the fracture surface. Two functions—the autocorrelation function, C(β), and the power spectral density, P(ω)—may be used to extract this information.

The autocorrelation function, C(β), is defined by:

$$C(\beta) = \frac{1}{L}\int_0^L h(s)\,h(s+\beta)\,ds \tag{5.62}$$

where h is the height of the surface, s is the distance along the profile line, L is the overall length of the profile under examination, and β is the displacement distance along the surface profile.

The value of the autocorrelation function for some displacement distance, β, measured along the surface is therefore derived by the following operations:

a) shifting two trial points P_1 and P_2, a distance β apart, along the fracture surface profile;
b) multiplying the height at the shifted points, P_1 and P_2;
c) calculating the area beneath the resultant product curve.

It can be shown that when the displacement β is zero, the value of the function C(β) has its maximum.

The autocorrelation function provides a measure of the correlation between the positions of the heights on the fracture surface profile. Large values of C(β) for given values of the distance β indicate that there is a tendency for the heights to be a particular distance apart. In other words, the function provides information on the characteristic spacings of the surface features. Regular undulations on the surface show up as a series of maximum points in the autocorrelation function.

The second function used for advanced descriptions of surface profiles is based on the concept of the power spectral density function, P(ω). This function conveys direct information about the spatial frequencies present in the surface profile. It can be shown that the power spectral density is the Fourier transform of the autocorrelation function:

$$P(\omega) = \frac{2}{\pi}\int_0^\infty C(\beta)\cos(\omega\beta)\,d\beta \tag{5.63}$$

Some of the properties of the autocorrelation function, C(β), and the power spectral density function, P(ω), are shown in Figure 5.27. The figure gives

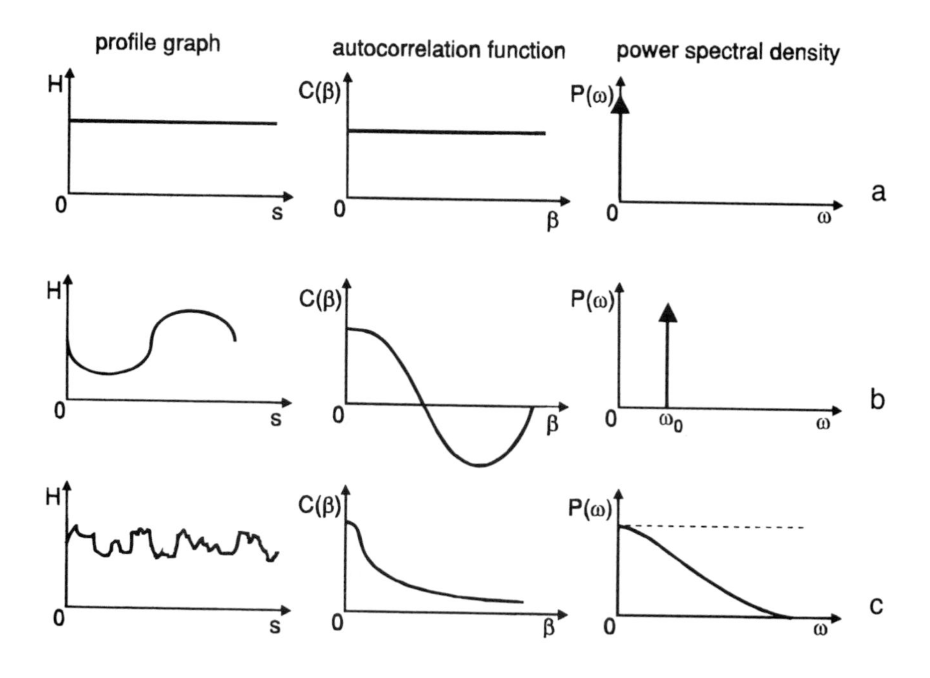

Figure 5.27 Characteristic examples of profile graphs and the related autocorrelation functions and power spectral densities: (a) totally flat surface; (b) a surface of strictly regular geometry; (c) a surface geometry with elements of randomness in the geometry of the profile/surface.

examples of specific types of the surface profiles and the corresponding functions.

Example 5.13

Figure 5.28a, b shows two fracture profiles observed for a single-phase metal strained at two different temperatures. Depending on the temperature of straining, the material showed different degrees of tendency for brittle fracture reflected also in the differences in the fracture profiles. However, these two distinct profiles yield almost the same values of the profile roughness parameter, R_L. In this situation, in order to quantify the differences in the geometry of the profiles, the spectral power function, $P(\omega)$, was studied. The results are presented in Figure 5.28c, d. It may be noted that there is a clear difference in the functions obtained for different profiles.

5.4 Quantitative characterization of thin film structures

Properties of thin films, such as corrosion resistance, hardness, and resistance to wear are functions of their structure. The structure of thin films typically

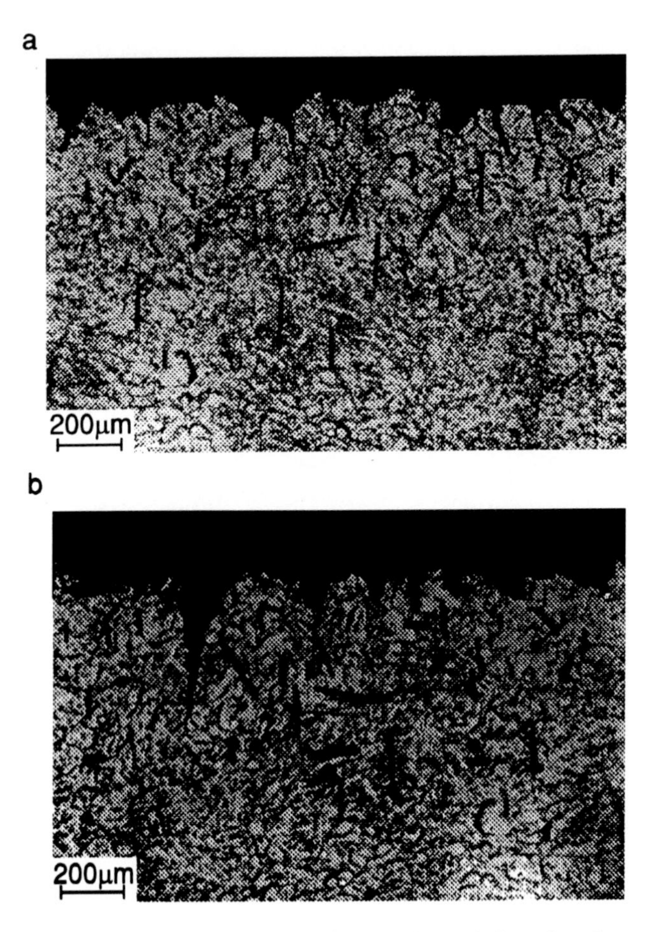

Figure 5.28 Two profiles of fracture surfaces (a,b) and the related power spectral densities (c,d).

consists of a variety of macro- and microscale elements. Examples of macroscale elements are

a) linear cracks and steps on the surface;
b) pits.

Among the important microscale elements are

a) grains and particles;
b) grain boundaries.

A quantitative description of thin film structures defines the structural elements present, their number, size, shape, and spatial distribution. A number of parameters can be used to describe the size of the structural elements of thin films. The choice of the parameters to be used in a given study depends

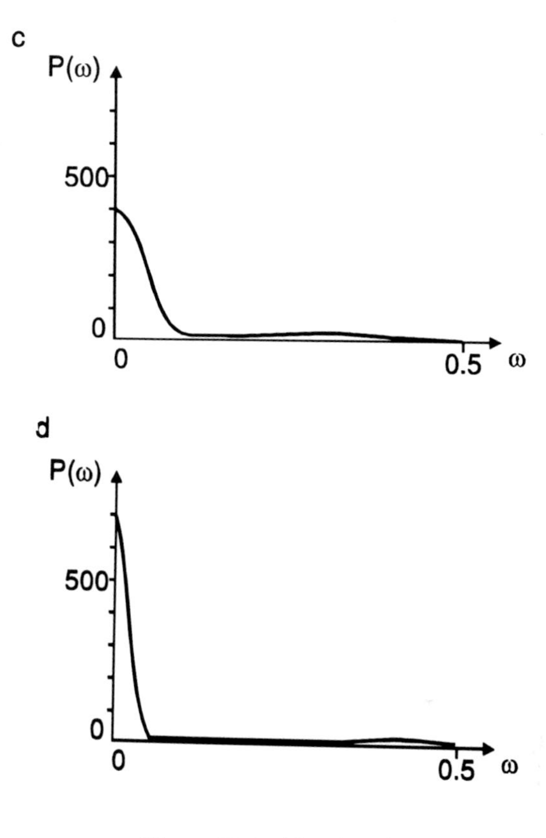

Figure 5.28 (*Continued*)

on the dimensionality of the elements (1-D and 2-D elements) and the needs of the theoretical models. The following two parameters can be used to describe the average size of the elements in thin film structures:

N_A—number of structural elements per unit area;
L_A—length of structural elements per unit area.

The size of individual 1-D elements (steps, cracks, etc.), can be described by their length, L. The size of 2-D elements (e.g., grains) is defined by their area, A, or equivalent circle diameter, d_2.

The elements making up thin film structures differ in size. As a result parameters L, d_2, and A, are random variables. The populations of these elements can be described by distribution functions such as f(A), f(L), and $f(d_2)$. The distribution of these sizes can be presented in the form of histograms.

Populations of structural elements of a given type, for example, grains, can be characterized in terms of the mean values of the size parameters: E(L), $E(d_2)$, and E(A). The diversity in the grain size can be described by standard deviations: SD(L), $SD(d_2)$, and SD(A) or by the dimensionless coefficients of variation, CV(L), CV(d_2), and CV(A) defined as the ratio of the appropriate standard deviation to the mean value. For example, the coefficient of variation of the area, A, is given as CV(A) = SD(A)/E(A).

The size of individual elements revealed in the structure of thin films frequently varies over a wide range. This justifies the application of the logarithmic scale for element size measurements. In this case the distribution functions f(A), f(L), and f(d_2) are transformed into f(lnA), f(lnL), and f(lnd_2). Accordingly, the mean size of the structural elements is characterized by E(lnA), E(lnL), and E(lnd_2). The diversity in the size is well represented by SD(lnA), etc.

Example 5.14

Figure 5.29 exemplifies a scanning electron microscope (SEM) image of the surface of a TiN film obtained by a plasma-assisted chemical vapor deposition method. The characteristic topography reveals grains emerging on the surface of the thin film studied.

Simple counting procedures can be used to estimate the number of grains per unit area, N_A. For the observation field shown in Figure 5.29 it can be shown that:

$$N_A = 2.5 \times 10^5 \text{ mm}^{-2}$$

The sizes of individual grains depicted in Figure 5.29 were measured using a special program developed for an image analyzer. This program determines the length of maximum chord, L_{max}, and computes the area of a given grain assuming its shape is circular. The histogram of grain size for the observation field from Figure 5.29 is given in Figure 5.30. It may be noted that the distribution function of the equivalent grain diameter, f(d_2), has nonzero values in the range from 0.5μm to 16 μm. The following parameters can be estimated for the grains shown in Figure 5.29:

1. E(A) = 4 μm^2
2. SD(A) = 6.52μm^2
3. CV(A) = 1.63

The observation field shown in Figure 5.29 can be divided into two subfields, A and B. If the parameters E(A) and SD(A) are determined for these subfields, they are likely to differ in value. The differences in the values of these parameters are due to the randomness in the geometry of grains. The unbiased description of a thin film requires a random selection of the observation fields. Such a selection of the observation fields can be realized by a number of procedures. One method is based on random selection of the coordinates of the centers of the observation fields. If the first observation field has random position, the subsequent fields can form a regular pattern on the surface of observations.

Studies of the thin film microstructures provide information important to the understanding of their properties and their relation to the technological process. This is illustrated by the following example.

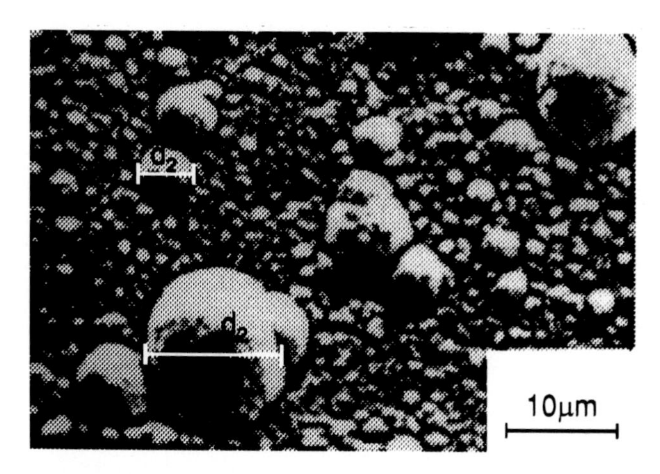

Figure 5.29 Explanation of the measurement technique for estimating the size of grains on a surface of TiN.

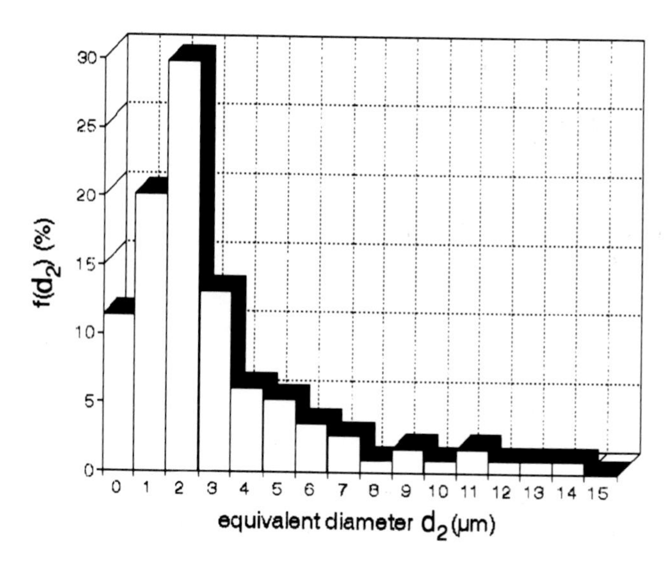

Figure 5.30 Distribution function of the particle diameters for a typical layer of TiN produced by the PACVD method.

Example 5.15

The size of grains was studied for a series of specimens of TiN obtained by PACVD processes. The mean values of grain area logarithm, E(lnA), as well as SD(lnA) were measured as functions of process conditions. The results of the measurements are summarized in Table 5.4.

Table 5.4 Mean Values of Grain Area Logarithm (E[lnA]) and Standard Deviations of Grain Area Logarithm (SD[lnA]), for 5 and 10 h Treatment at Different Flow Rates (Example 5.15)

	Time of the treatment			
	5 h	10 h	5 h	10 h
flow rate [10^{-3} m/s]	E[lnA]		SD[lnA]	
0.063	0.73	0.79	0.56	0.68
0.154	1.07	1.15	0.82	1.15
0.252	1.17	1.75	0.95	1.19

It may be noted that the mean values, E(lnA), and standard deviations, SD(lnA), systematically depend on the process parameters. Lower flow rates and shorter times result in smaller grain sizes and a more uniform distribution of grain size.

References

1. Hirth, J. P. and Lothe, J., *Theory of Dislocations*, Johns Wiley and Sons, New York, second ed., 1982.
2. Underwood, E. E., *Quantitative Stereology*, Addison Wesley, Massachusetts, 1970.
3. Hawkes, P. W., *Electron Optics and Electron Microscopy*, Taylor & Francis Ltd., London, 1972.
4. Varin, R. A., Kurzydłowski, K. J., and Tangri, K., Analytical treatment of grain boundary sources for dislocations, *Materials Science and Engineering*, 85, 115, 1987.
5. Wyrzykowski, J. W. and Grabski, M. W., The Hall-Petch relation in aluminum and its dependence on the grain-boundary structure, *Philosophical Magazine*, A53, 505, 1986.
6. Sangal, S., Kurzydłowski, K. J., and Tangri, K., The effect of extrinsic grain boundary dislocations on the grain size strengthening in polycrystals, *Acta Metallurgica et Materialia*, 39, 1281, 1991.
7. Ralph, B., Interfaces in engineering materials, *University of Wales Review*, 4, 29, 1988.
8. Peckner, D. ed., *The Strengthening of Metals*, Reinhold Publishing Corporation, New York, 1964, 163.
9. Armstrong, R. W., The yield and flow stress dependence on polycrystal grain size, *Yield, Flow and Fracture of Polycrystals*, Baker, T. N., Applied Science Publishers, London, 1983, 1.

10. Ranganathan, S., On the geometry of coincidence—site lattices, *Acta Crystallographica*, 21, 197, 1966.
11. Brandon, D. G., The structure of high-angle grain boundaries, *Acta Metallurgica*, 14, 1479, 1966.
12. Kurzydłowski, K. J. and Garbacz, A., A model of the interaction between a dislocation and a sliding grain boundary, *Philosophical Magazine*, A52, 689, 1985.
13. Kwiecinski, J. and Wyrzykowski, J. W., Kinetics of recovery on grain-boundaries in polycrystalline aluminum, *Acta Metallurgica*, 37, 1503, 1989.
14. Garbacz, A. and Grabski, M. W., The relationship between texture and CSL boundaries distribution in polycrystalline materials, *Acta Metallurgica et Materialia*, 41, 469, 1993.
15. Hutchings, I. M., *Tribology: Friction and Wear of Engineering Materials*, Edward Arnold, London, (Chap.2) 1992.
16. El-Soudani, S. M., Profilometric analysis of fracture, *Metallography*, 11, 247, 1978.
17. Underwood, E. E. and Chakrabortty, S. B., Quantitative fractography of fatigued Ti-28%V alloy, *Fractography and Materials Science*, American Society for Testing and Materials, STP 733, 337, 1981.
18. Wright, K. and Karlsson, B., Topographic quantification of nonplanar localized surfaces, *Journal of Microscopy*, 130, 37, 1983.
19. Coster, M. and Chermant, J. L., Recent developments in quantitative fractography, *International Metals Review*, 28, No.4, 228, 1984.
20. Cwajna, J., Szala, J., and Maciejny, A., Contemporary state and development trends of quantitative fractography, *Inżynieria Materiałowa*, 6, 161, 1984.
21. Wojnar, L., *Quantitative Fractography. Basic Principles and Computer-aided Research*, Cracow Techn. Univ., Cracow, 1990 (in Polish).
22. Saltykov, S. A., *Stereometriches metallographie*, VEB Deutscher Verlag für Grundstoffindustrie, Leipzig, 1974.
23. Underwood, E. E., Directed measurements and heterogenous structures in quantitative fractography, in *Proceedings of 3rd Conference on Stereology in Materials Science*, Cracow-Szczyrk, Poland, 100, 1990.
24. Wojnar, L., On some misunderstanding of fracture roughness parameters, in *Proceedings of 3rd Conference on Stereology in Materials Science*, Cracow-Szczyrk, Poland, 94, 1990.
25. Baddeley, A. J., Gundersen, H. J. G., and Cruz-Orive, L. M., Estimation of surface area from vertical sections, *Journal of Microscopy*, 142, 259, 1986.
26. Gokhale, A. M. and Underwood, E. E., A new parametric roughness equation for quantitative fractography, *Acta Stereologica*, 8, 43, 1989.
27. Ryś, J., *Metalografia ilościowa*, AGH, Cracow, 1982.
28. Dauskardt, R. H., Haubensak, F., and Ritchie, R. O., Overview No.88. On the interpretation of the fractal character of fracture surfaces, *Acta Metallurgica et Materialia*, 2, 143, 1990.
29. Hornbogen, E., Fine, coarse and fractal microstructures, *Practical Metallography*, 91, 258, 1986.
30. Jost, N. and Hornbogen, E., On fractal aspects of metallic microstructures, *Practical Metallography*, 157, 1988.
31. Mandelbrot, B. B., *The Fractal Geometry of Nature*, W. H. Freeman, San Francisco, 1982.
32. Mecholsky, J. J., Passoja, D. E., and Feinberg-Ringel, K. S., Quantitative analysis of brittle fracture surfaces using fractal geometry, *Journal of the American Ceramic Society*, 1, 60, 1989.

33. Pande, C. S., Richards, L. R., and Smith, S., Fractal characteristics of fractured surfaces, *Journal of Materials Science Letters*, 295, 1987.
34. Przerada, I. and Bochenek, A., Microfractographical aspects of fracture toughness microalloyed steel, in *Proceedings of 3rd Conference on Stereology in Materials Science*, Cracow-Szczyrk, Poland, 88, 1990.
35. Underwood, E. E. and Banerji, K., Fractals in fractography, *Materials Science and Engineering*, 80, 1, 1986.
36. Wright, K. and Karlsson, B., Fractal analysis and stereological evaluation of microstructures, *Journal of Microscopy*, 2, 185, 1983.
37. Mandelbrot, B. B., *Les Objects Fractals. Forme, Hasard et Dimension*, Flammarion, Paris, 1975.

chapter six

The geometry of grains and its effect on the properties of polycrystals

6.1 *Introduction*

Polygrained materials are aggregates of a large number of smaller volumetric elements called grains. The grains fill almost entirely the volume of polygrained materials. An example of a polygrained material is a polycrystal. Polycrystals are aggregates of a large number of space-filling crystals, all of the same phase. Grains are distinguished by the orientation of their crystal lattices with respect to some reference system XYZ. They are joined along common surfaces termed grain boundaries that have the property of transferring normal and shear stresses. Polycrystalline materials constitute a significant fraction of functional and structural materials of technical importance. Examples of such materials include metals, ceramics, and intermetallics. Examples of images of grains in polycrystals are given in Figure 6.1. These show characteristic features of grains seen on a fracture surface of a ceramic material, on a cross-section of an intermetallic and in a transmission electron micrograph of a fine-grained metal.

The concept of a polycrystal has been successfully used for the description of polymers and liquid crystals also. From a practical point of view this concept can also be extended to 2- and 1-D structures. Examples of 2-D polycrystals are layers produced by surface engineering (an example of such a structure is shown in Figure 6.2). An example of a 1-D polycrystal is schematically shown in Figure 6.3c. Such a "bamboo" structure is known to exist in well-annealed thin wires in tungsten lamp filaments.

Grains, and the grain boundaries between them are the main microstructural elements in polycrystals. The properties of grain boundaries have been discussed in Chapter 5 and the present discussion is focused on the properties of the grains. The physical properties of grains can be described in terms of:

a) the type and orientation of their crystal lattices;
b) the density and spatial distribution of the crystal lattice defects which they contain such as dislocations and points defects.

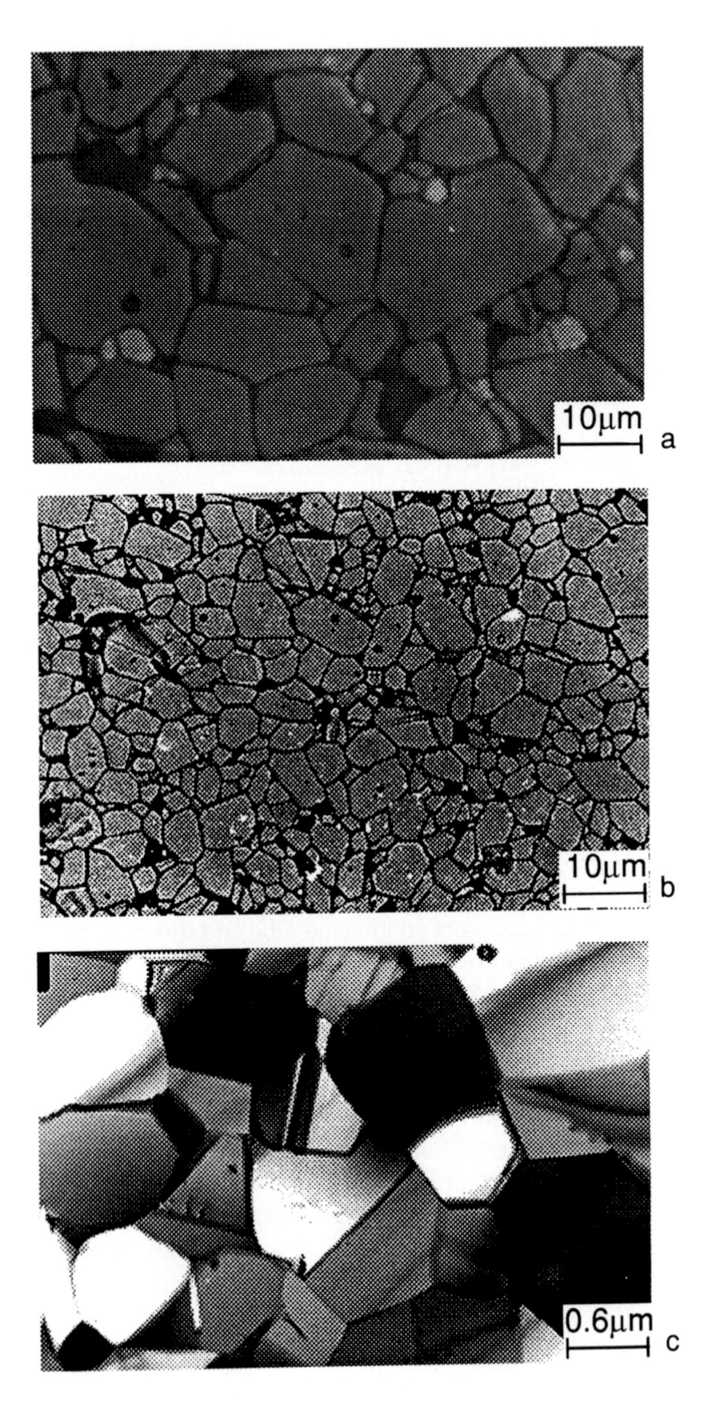

Figure 6.1 Characteristic images of polycrystals: (a) fracture surface of a ceramic material; (b) grain boundary network revealed on a section of an intermetallic; (c) transmission electron micrograph of a fine-grained metal.

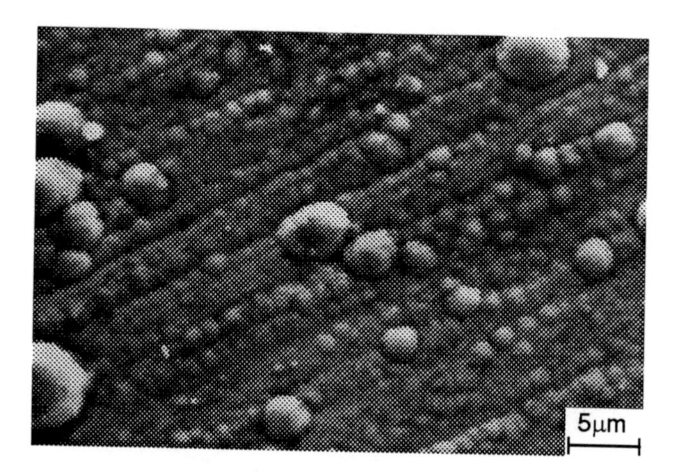

Figure 6.2 An example of a 2-D polycrystal—surface layer of TiN obtained by a chemical vapor deposition method.

The problem of describing the orientation of grains has led to the development of the concept of texture and has become over the years a scientific discipline by itself. A comprehensive treatment of textures can be found in Reference 1. The same can be said about the studies of crystal dislocations and point defects (see References 2 and 3). The question of the geometrical properties of grains has received relatively less attention so far. On the other hand, a proper description of grain geometry gives a basis for a quantitative characterization of the physical properties of grains and the geometry of grains, and by itself has been found to influence the properties of polycrystals. In view of this it was decided to make grain geometry a major topic in the present text.

6.2 *Grains as space-filling volumetric elements*

Grains forming a 3-D polycrystal are volumetric elements. They can be described primarily in terms of their volume, V_i, and surface, S_i. As a result of the space-filling requirement, the surfaces of the grains are divided into faces, S_{ij}, and each face is shared by two neighboring grains with indices i and j. The term *grain boundary* in most cases is used to refer to any of the faces of a grain. Faces of grains meet along common edges (grain edges), E_{jkl}, at which at least three grains (j,k,l) are joined. The edges E_{jkl} are connected at points, C_{pqrs}, termed vertices (grain corners). If observed on sections, 3-D polycrystals give images that have characteristics of a 2-D polycrystalline appearance.

2-D polycrystals contain planar grains characterized by grain area, A_i (or equivalent circle diameter, d_2, such that $4A = \pi d_2^2$), and grain perimeter, p_i. Grain perimeter is divided into grain boundary segments, l_{ij}, which are joined at triple points, T_{ijk}. A 1-D polycrystal is described by the length of its grains, L. All these parameters are further explained in Figure 6.3 and listed in Table 6.1.

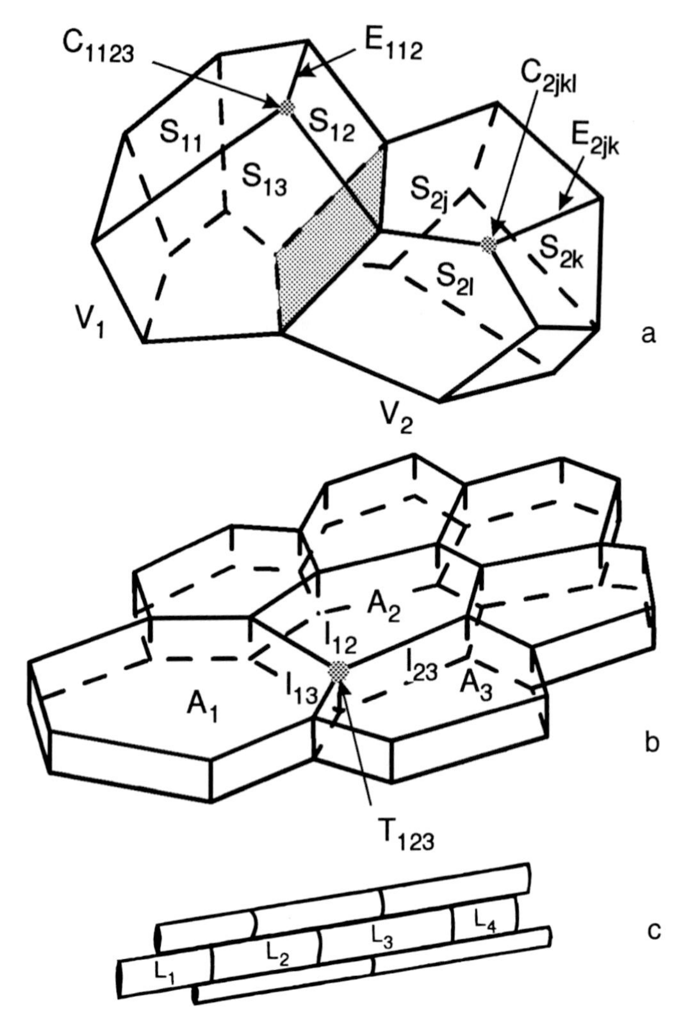

Figure 6.3 Schematic representation of the geometry of grains in polycrystals: (a) 3-D polycrystal (V_1 and V_2 are volumes of the grains, S_{ij} are grain boundaries, E_{ijk}—grain edges, C_{ijkl}—grain corners); (b) 2-D polycrystal (A_i stand for grain area, l_{ij} designates grain edges, T_{ijk}—triple points); (c) 1-D polycrystal (L_i designates the length of the i-th grain).

Table 6.1 List of Parameters Used to Describe Geometry of Grains in Polycrystals

Type of polycrystal	3-D parameter	2-D parameter	1-D parameter	0-D parameter
3-D	Volume V	Surface S	Common edge E	Vertex C
2-D		Area A	Perimeter p	Common point T
1-D			Length L	Intersection points I

The system of grains in polycrystals is characterized by the existence of:

a) 2-fold faces, grain boundaries (shared by 2 grains);
b) 3-fold edges (shared by 3 grains);
c) 4-fold vertexes (shared by 4 grains).

The existence of 2-fold faces is a common feature of any space-filling elements. On the other hand, the existence of 3-fold edges and 4-fold vertexes, are unique properties of the grain boundary networks in polycrystals and are related to the surface tension (energy) of the grain boundaries. This property of the grains imposes significant restrictions on their geometry and its possible changes. For example, a single vertex, or an edge cannot disappear from the system of vertices and edges in a polycrystal. These restrictions are formalized by equations which link the number of grains, N, number of their faces, F, edges, E, and vertexes, C, in a system of grains in polycrystals characterized by 3-fold edges and 4-fold vertices. For 3-D polycrystals the following formula holds:

$$N + F - E + C = 1 \tag{6.1}$$

As a result, a disappearance of a grain from the population of grains in a polycrystal, i.e., a change from N to N – 1 in Equation 6.1, must be accompanied by a rearrangement of the neighboring grains in a way that brings down the number of faces, F, edges, E, and vertices C. (In addition, E cannot increase without increasing F and C.)

For a 2-D polycrystal the following formula is valid:

$$N - E + C = 2 \tag{6.2}$$

In the literature a number of 3-D polyhedra have been described that show the property of space filling and form in a spatial system of 3-fold edges and 4-fold vertices. Two of them are based on the polyhedra shown in Figure 6.4 and these and some additional polyhedra are described in Table 6.2. More can be found in Reference 4.

A simple 3-D model of a polycrystal can be obtained by arranging in space a system of polyhedra having a constant volume. Properties of such grains have been studied, for example, in References 4 to 6. Such model polycrystals are reference structures used in theoretical studies of polycrystals.

6.3 *Stochastic character of grain geometry*

Due to the 3-D character of grains, direct studies of their geometry in nontransparent materials would require separation of a polycrystal into individual grains. This has been achieved for some metal systems[4–6] by reducing the strength of grain boundaries.[7] An alternative to grain separation is the method of serial sectioning discussed in Chapter 4 coupled with procedures that make it possible to reconstruct 3-D geometry from 2-D sections.

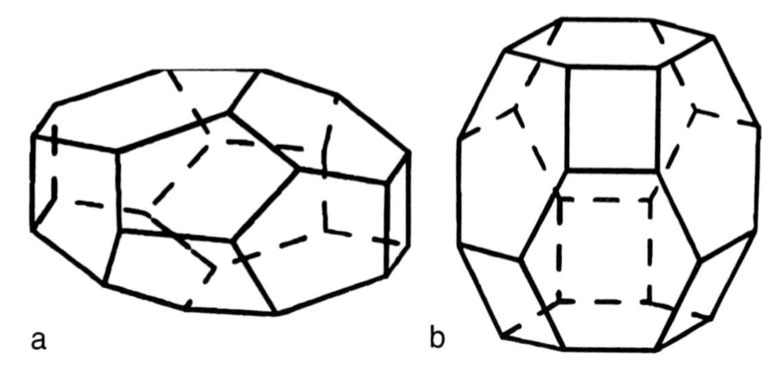

Figure 6.4 3-D polyhedra showing space-filling properties: (a) William's 14-hedron; (b) truncated, hexagonal, double skew pyramid.

Table 6.2 Properties of Shapes Used to Model Geometry of Grains in Polycrystals

Object shape	S	V	q_1
Sphere	$4\pi r^2$	$4\pi r^3/3$	$36\pi \approx 113.10$
Pentagonal dodecahedron	$20.646a^2$	$7.663a^3$	149.87
Truncated octahedron	$26.785a^2$	$11.314a^3$	150.12
Rhombic dodecahedron	$8.486a^2$	$2a^3$	152.76
Octahedron	$3.464a^2$	$0.4714a^3$	187.05
Cube	$6a^2$	a^3	216
Tetrahedron	$1.732a^2$	$0.1179a^3$	373.78

Note: a is a characteristic dimension of the grain, q_1 is a shape factor defined by Equation 6.8.

The geometry of grains in annealed polycrystals has been studied in a series of papers by Smith[8] and later by Rhines.[4] Experimental observations of the geometry of grains have shown that when studied under higher magnification the grains reveal a complicated geometry as illustrated by examples shown in Figure 6.5. The grain faces have been observed to show a measurable curvature and their edges do not join at angles typical of a theoretical structure consisting of simple polyhedra. More importantly, it has also been found that grains differ in their geometry. An indication of such differences is the observation that the number of faces, F_i, edges, E_i, and vertices, C_i, related to a grain varies over a significant range. In the case of F_i, values in the range from 4 to more than 20 have be observed.[4]

Grains in polycrystals differ also in their size. The volume of the smallest grains found in polycrystals in most cases is limited by the resolution limits of the observations (in fact, in the case of polycrystals subjected to grain growth, there should also exist grains with a volume approaching zero).

In short, grains in polycrystals form populations of elements with geometrical features appearing with certain probabilities, and the grains in a polycrystal should be viewed as forming a population. The population of grains

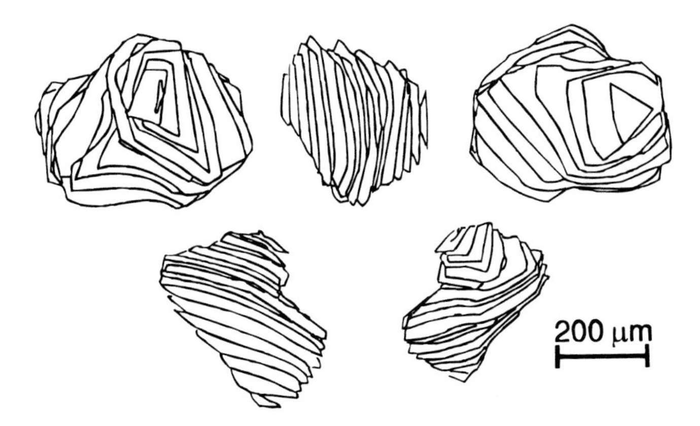

Figure 6.5 Examples of geometry of grains in a polycrystalline metal. The geometry of these grains has been reconstructed from the images on serial sections of polycrystals.

in a polycrystal can be characterized by a set of distribution functions, $f(x_j)$, where x_j are quantities that describe specified properties of their geometry.

The geometrical features of grains are usually discussed in terms of their size and shape. Such a division is based on the assumption that the shape of grains is independent of their size and further, in most cases of practical importance, the grain shape is "statistically constant" and characteristic of a material. The term *statistically constant* means here that the frequency of appearance of different grain shapes in a material remains constant. The separation of the grain size and shape is also related to the differences in the mathematical description of these two aspects of their geometry. It should be noted also that the currently used models for polycrystalline properties pay, in general, more attention to the grain size than to grain shape.

The separation of shape and size in the studies of grain geometry results in two sets of parameters being introduced: size parameters, p_i ($i = 1,2,\ldots,n$) and shape parameters, q_j ($j = 1,2,\ldots,m$).

The fact that grains forming a polycrystal differ in size and shape results in these parameters being random variables and the geometry of grains in a polycrystal can be described in terms of a probability distribution function, $f(p_i,q_j)$. Such a function defines the chances of finding a grain of a given shape and size (p_i,q_j), which is equal to the relative number of such grains in the polycrystal studied. If the population of N grains is divided into a finite number of classes with the geometry of grains in a given class described by certain representative values (p_i,q_j), then the distribution function $f(p_i,q_j)$ is discrete and the number of grains in this class, $N(p_i,q_j)$, is given by:

$$N(p_i, q_j) = f(p_i, q_j)\, N \tag{6.3}$$

For a continuous distribution of the parameters defining the geometry of grains, $N(p_i,q_j)$ can be obtained by integration of $f(p_i,q_j)$:

$$N\left(p_i \in P_i, q_j \in Q_i\right) = \int_{P_i}\int_{Q_i} f\left(p_i, q_j\right) dp_i \, dq_j \tag{6.4}$$

where P_i, Q_i indicate certain specified subsets of size and shape parameters p_i and q_i, respectively; the symbol ∈ means "belongs to".

If the shape and size of grains are studied independently, attention is focused on the marginal probability functions $f(p_i)$ and $f(q_j)$. These functions determine the relative numbers of grains of a given size or grains of a given shape. As in the case of the function $f(p_i,q_j)$, these two can be of discrete or continuous character. A formula analogous to Equations 6.3 and 6.4 can be obtained easily. In particular, the number of grains of a size specified by the parameters $p_1, p_2, \ldots, p_n$ in the discrete case is given as:

$$N\left(p_1, p_2, \ldots, p_n\right) = f\left(p_1, p_2, \ldots, p_n\right) N \tag{6.5}$$

where, for example, $p_1 = V$, $p_2 = S$. In the case of a single parameter approach, say grain volume, V, this reduces to:

$$N(V_i) = f(V_i)\, N \tag{6.6}$$

where $f(V_i)$ is the volume distribution function.

The discussion presented here underlines the fact that the geometry of grains is of a stochastic nature. As a result, its description can be made within the framework of mathematical statistics and is based on the principles of the theory of probability. As a consequence, questions related to the shape and size of grains cannot be simply answered by giving specific unconditional numbers for size description, or by listing figures defining grain shape. Such a description always needs to be supplemented by information on the probability of finding grains with the properties specified in the population of grains. This leads, in a natural way, to a description of the grain geometry in terms of the average values, $E(p_i)$ and $E(q_i)$, standard deviations, $SD(p_i)$ and $SD(q_i)$, and coefficients of variations, $CV(p_i)$ and $CV(q_i)$, of the parameters used to define the grain geometry.

6.4 The relationship between 3-D grains and their 2-D sections

In practical cases the geometry of grains (shape and size) is investigated through the examination of their sections. The sections can be cut systematically as is the case in serial sectioning and then the 3-D geometry of grains can be reconstructed (see, for example, Figure 6.5). However, in most cases sections of the observations are positioned and oriented randomly. In this situation little can be said about the 3-D geometry of a particular grain. On the other hand statistical and stereological methods may be used to characterize the geometry of the grain aggregate.

On sections of polycrystals the grains are seen as 2-D figures. The 2-D sections of grains differ in their size and shape. For a given geometry of grains, the shape and size of sections depends on the distance from the grain center.

If the geometry of grains is examined via observation of their sections, for each specified size and shape of a grain, $f(p_i,q_j)$, there are distribution functions of sizes and shapes of its 2-D intersections:

$$f(p_i,q_j) \rightarrow f^2_{p_i,q_j}(p^*_k,q^*_m) \tag{6.7}$$

where f^2 is a distribution function of the shape (q^*_i) and size (p^*_j) parameters determined for the grain sections. This effect is illustrated by the example shown in Figure 6.6.

If the size and shape of grains are studied separately, grains of the same shape and different sizes can be normalized to one figure of a unit size (size equal to 1). Then, a specified shape of a 3-D normalized grain generates a distribution function of 2-D shapes:

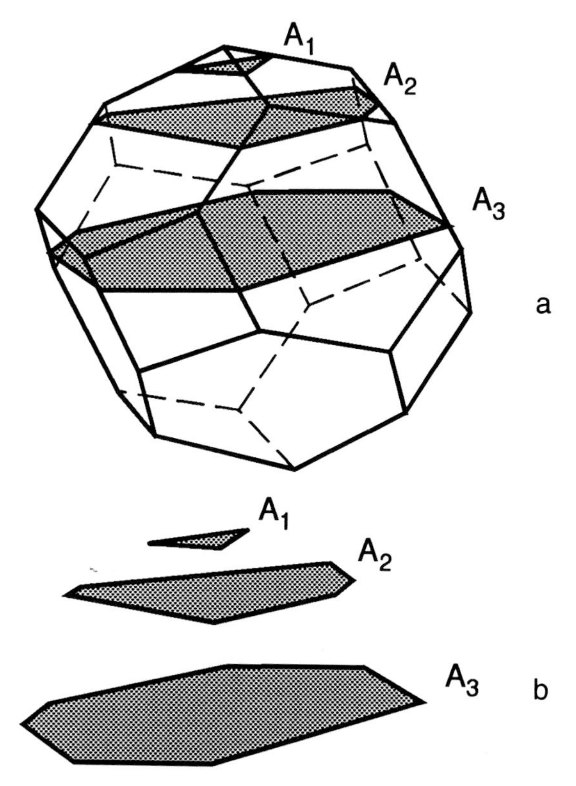

Figure 6.6 Illustration of the relationship between the shape and size of a grain and the shape and size of its sections: (a) model geometry of a grain (shadowed areas are examples of the grain sections; (b) geometry of the sections A_1, A_2, and A_3. The sections differ in their size and the shape, as indicated by the varying number of edges.

3-D grain Specified shape	2-D sections Distribution of shapes
q_i	$f^2(q_j^*)$

An example of the shape distribution for normalized grains of a simple shape is shown in Figure 6.7. The quantitative data for such distribution functions can be obtained either experimentally or by computer simulations.

6.4.1 *Shape of grains*

The data obtained for metallic materials indicate that, to a first approximation, grains in polycrystals can be modelled using polyhedra with a number of faces, F_i, varying from around 6 to more than 20. The following mean values of the number of faces per grain, E(F), edges per grain, E(E), and vertices, E(C), have been found in aluminum by Rhines and co-workers[6] and recently in iron by Liu and co-workers:[9]

$$E(F) = 14, \quad E(E) = 36, \quad E(C) = 24.$$

These values are characteristic of two known 3-D polyhedra: tetrakaidecahedra and truncated octahedra. A number of other geometrical objects have been studied in this context and an extensive review can be found in Reference 10.

The list of 3-D shapes studied include:

a) spheres (S);
b) truncated octahedra (TO);
c) Kelvin's α-tetrakaidecahedra (KT);
d) Williams' β-tetrakaidecahedra (β-W);
e) Williams' simplified tetrakaidecahedra (s-W);
f) rhombic dodecahedra (RD);
g) pentagonal dodecahedra (PD);
h) cubes (C).

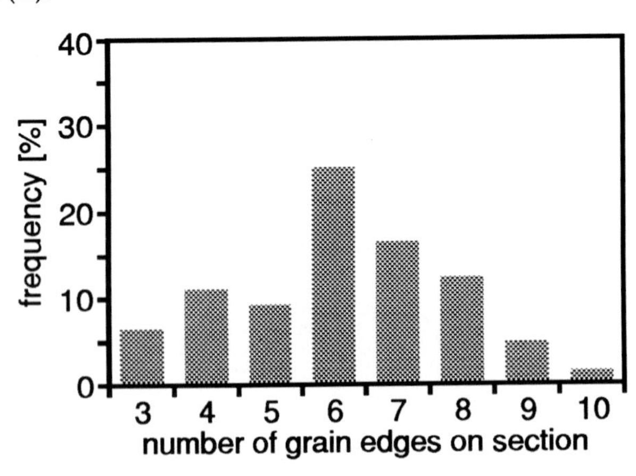

Figure 6.7 Number of edges distribution function for sections of a model 3-D shape of a grain.

Among these shapes, (b) to (f) and (h) have the property of space filling. However, aggregates of cubes are characterized by the existence of 4-fold edges and 6-fold vertices, which are not observed in polycrystals. Cubes and spheres are included in the list for their geometrical simplicity. The other shapes are not so geometrically simple but they have more realistic numbers of faces, edges, and corners.

A parametric approach to grain shape assumes that the properties of a given shape can be described by a series of dimensionless parameters, termed shape factors. Each of these parameters is designed to provide information on some specified aspects of the grain geometry, for example, about grain elongation, grain boundary undulations, etc. These parameters may be obtained in the form of ratios of selected sizing parameters raised to the appropriate powers required for canceling out dimensions. Of special interest are couples of parameters that determine interrelated properties of the geometrical objects such as: (volume–surface), (perimeter–area), (height–width).

An example of a shape parameter is the ratio, q_1, of the surface to volume raised to the third and second powers, respectively:

$$q_1 = \frac{S^3}{V^2} \tag{6.8}$$

This parameter reaches the lowest value for spheres and increases with the elongation of the grains or an increase in complexity of their shape. Table 6.2 lists the values of q_1 for selected simple geometrical objects. It can be shown that the value of q_1 is sensitive to variations in the curvature of the grain surface and to a lesser degree to the grain elongation. This parameter can be normalized in such a way that the normalized value for the sphere, q_1^*, is equal to 1.0 and then for a cube $q_1^* = 1.91$.

Another parameter, q_2, can be defined as the ratio of the maximum chord (maximum intercept) to the cube root of the grain volume:

$$q_2 = \frac{l_{max}}{\sqrt[3]{V}} \tag{6.9}$$

This parameter is a much more sensitive measure of grain elongation.[11] More parameters can be found in Reference 10.

6.4.2 Shape of grain sections

There are a number of parameters that can be measured from individual grain sections revealed on a cross-section of a polycrystal. Among the most frequently used are the following:

a) grain area, A;
b) equivalent circle diameter, d_2:

$$d_2 = \sqrt{\frac{A}{4\pi}} \tag{6.10}$$

c) maximum chord length, d_{max};
d) perimeter, p;
e) Cauchy perimeter, p_c;
f) number of grain boundary segments, n;
g) orientation of the maximum chord, ω;
h) dihedral angles, α_j;
i) vertical and horizontal Feret diameters, FD_v, FD_h.

These are schematically illustrated in Figure 6.8. The combination of these parameters define various shape factors for 2-D sections that are sensitive to different aspects of the grain geometry.

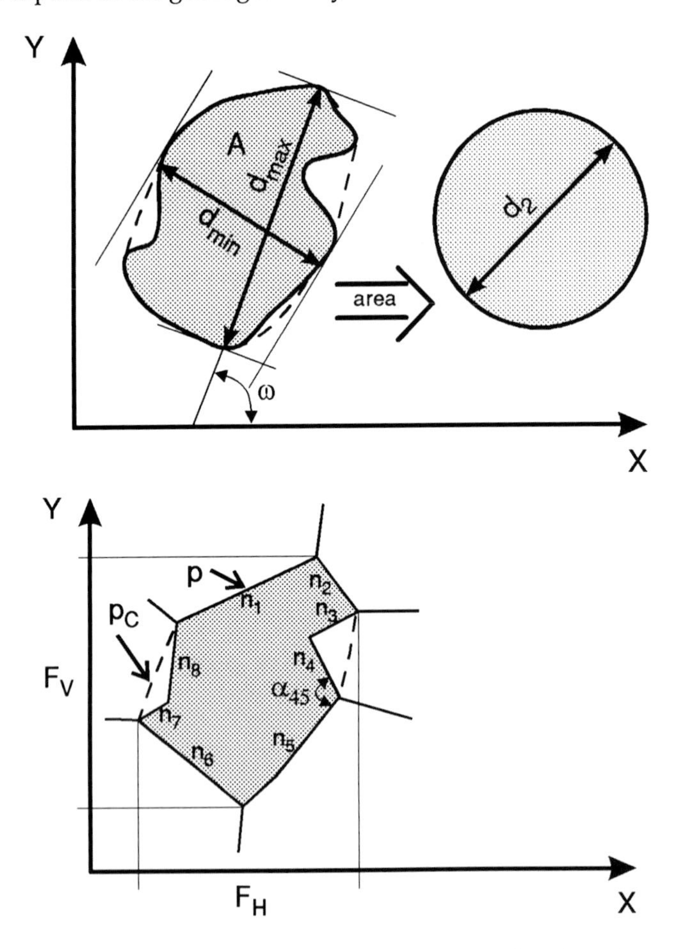

Figure 6.8 Schematic explanation of the parameters used to describe geometry of grain sections.

Table 6.3 Basic Parameters Used for Characterization of the Shape of Grain Sections

Definition	Interpretation	Characteristic values
$\frac{d_{max}}{d_2}$	Measure of grain elongation	1.20–1.30 for annealed polycrystals
$\frac{p}{d_2}$	Measures the variation in curvature	3.30–3.90 for annealed polycrystals
$\frac{p}{p_c}$	Measure of convexity	1.0 for convex figures; > 1 for nonconvex

Three of these factors are listed in Table 6.3.
Other typically used shape factors for 2-D sections include:

a) the ratio of area, A, to squared perimeter, p^2:

$$q_3 = \frac{4\pi A}{p^2} \tag{6.11}$$

b) aspect ratio, AR, which is defined as the ratio of the maximum chord to the minimum chord (the shortest chord passing through the center of the transect):

$$AR = \frac{d_{max}}{d_{min}} \tag{6.12}$$

c) the ratio of the horizontal to vertical Feret diameters, which are projections of the figure onto two vertical directions:

$$q_m = \frac{FD_h}{FD_v} \tag{6.13}$$

These and the shape factors given earlier are used in the following examples.

Example 6.1

One of the parameters listed above is the dihedral angle. The 3-D dihedral angles, α_i, are between the surfaces of boundaries meeting along a common edge while the planar angles, β_i, are between the traces of the grain sections seen on a cross-section meeting at the triple points. The relationship between these two sets of parameters for the same three boundaries is explained in Figure 6.9.

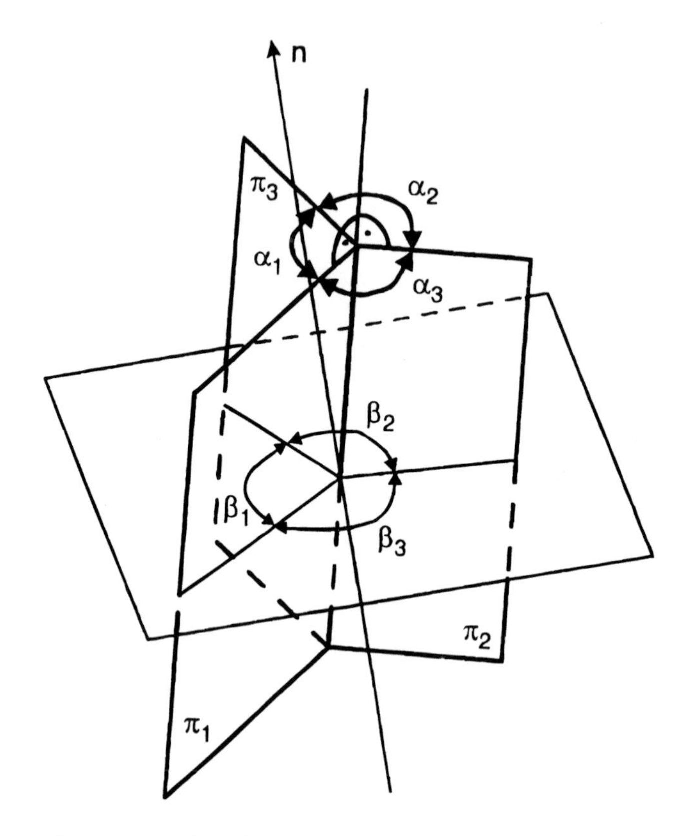

Figure 6.9 Illustration of the relationship between 3-D dihedral angles, α_i, and planar dihedral angles, β_i. These two sets of angles are equal only if the common edge of the grain boundaries is normal to the surface of observations.

The standard deviation of 3-D angles, SD(α), influences the standard deviation of the planar angles, SD(β). The minimum value for the standard deviation of the planar dihedral angles, SD(β), is observed for SD(α) = 0. Computer simulations have shown that for SD(α) = 0, SD(β) = 22°.

Example 6.2

Figure 6.10 shows a digitized network of grain boundaries revealed in a well-annealed polycrystal of aluminum. Four grains, shaded, were randomly selected, from which the required measurements of geometrical properties were made. The values of the computed shape factors are listed as well as some other geometrical properties.

Example 6.2 illustrates that shape factors, defined as the ratio of d_{max}/d_2 and p/d_2, for sections of grains revealed in a cross-section of a polycrystal achieve significantly different values. In fact these shape factors are stochastic

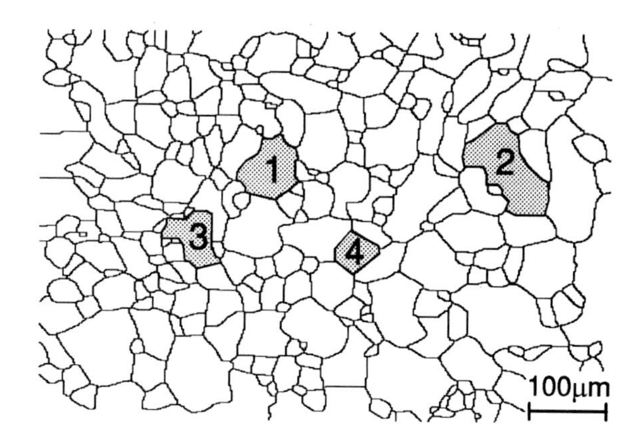

number of the grain	A [μm^2]	d_{max} [μm]	d_2 [μm]	p [μm]	$\frac{d_{max}}{d_2}$	$\frac{p}{d_2}$
1	2850.5	86.7	60.2	213.3	1.440	3.543
2	4550.0	127.4	76.1	304.7	1.647	4.003
3	1971.1	77.5	50.1	197.5	1.574	3.942
4	1170.6	55.5	38.6	132.3	1.438	3.425

Figure 6.10 Digitized image of the network of grain boundaries revealed in a well-annealed polycrystal of aluminum.

variables. The population of grain sections is characterized by a set of distribution functions of the shape factors. Figure 6.11 shows the distribution function of d_{max}/d_2, $f(d_{max}/d_2)$, for the grain structure from Example 6.2.

Since the parameters used to define the geometry of grain sections are random variables, the average shape of grain sections is expressed by the mean values of the shape factors, $E(q_j)$. In some cases one may also use the standard deviations and coefficients of variation of shape factors.

The mean values of some shape factors for grain sections in metallic and ceramic polycrystals are listed in Table 6.4. These mean values can be computed as the number mean, $E_n(q_j)$, or weighted means, $E_w(q_j)$, with section area being one of the weights suggested. The choice between these two approaches should depend on the purpose of the study. It may be noted that the weighted means give better agreement with the perceived views of humans.

The distribution function of the geometrical parameters for grain sections, and thus also the mean values, vary as a function of material characteristics and treatment. In general, it is rather difficult to predict the mean values of shape factors and they are used in most cases mainly to differentiate between some microstructures and to rank them in terms of such properties as grain elongation, convexness, etc. It should be stressed that for equiaxed microstructures the mean values of grain shape factors does not need to be equal

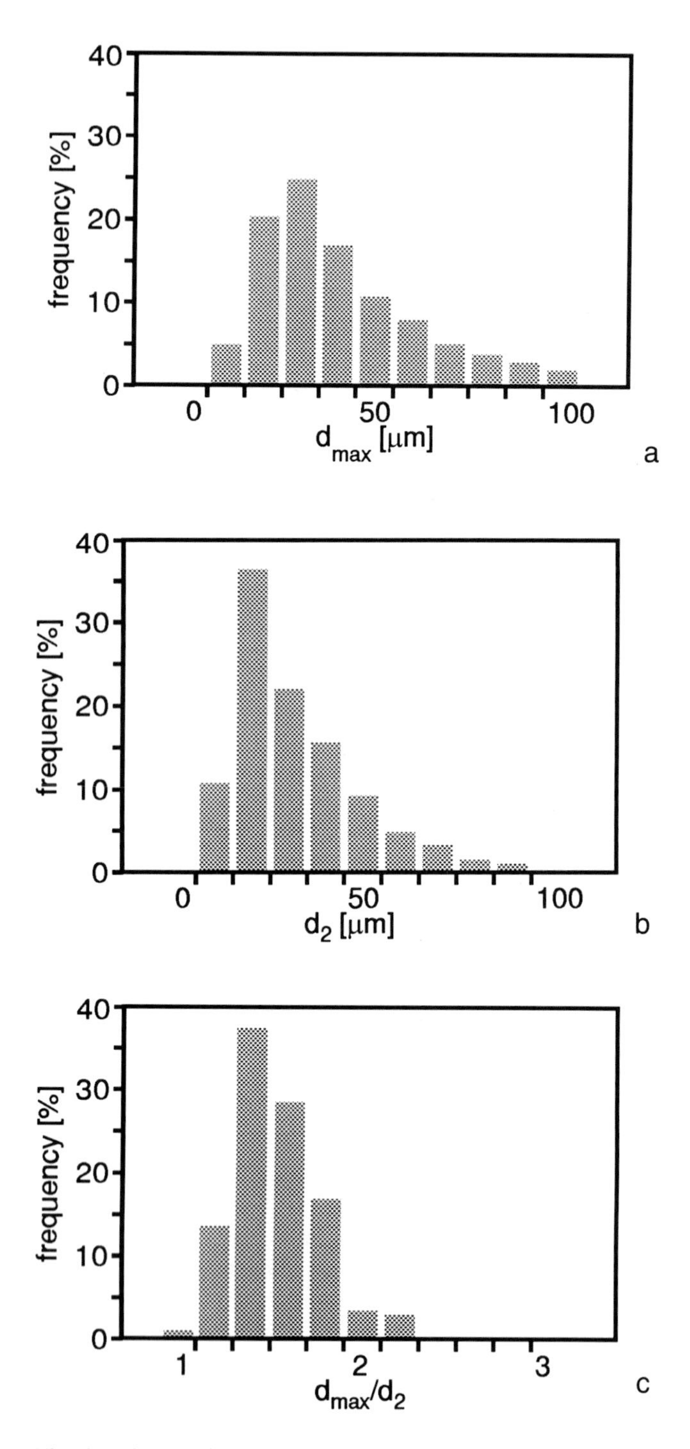

Figure 6.11 The distribution functions for the grains from Figure 6.10: (a) $f(d_{max})$; (b) $f(d_2)$; (c) $f(d_{max}/d_2)$.

Table 6.4 Examples of the Mean Values, in Numbers and Area Weighted, of Shape Factors for Well-Annealed Metal and Ceramic Polycrystals

Material	$E\left(\frac{d_{max}}{d_2}\right)$	$E_W\left(\frac{d_{max}}{d_2}\right)$	$E\left(\frac{p}{d_2}\right)$
Aluminum	1.25	1.30	3.65
γ-Fe	1.25	1.25	3.70
α-Fe	1.30	1.25	3.70
Ni-2%Mn	1.28	1.32	3.80
Al_2O_3	1.20	1.25	3.55

to 1.0. This is related to the effect of grain corners, as shown in Example 6.3 and the data collected in Table 6.5.

Example 6.3

The distribution function $f(d_{max}/d_2)$ for a metallic polycrystal is given in Figure 6.11. The mean value, $E(d_{max}/d_2)$, was found to be equal 1.17. These grains sections were found to have the number of their sides varying from 3 to 15. The values of d_{max}/d_2 for regular polygons are listed in Table 6.5. It may be concluded that the mean value of the shape factor used in this example is close to the shape factor for a hexagon.

Table 6.5 The Values of d_{max}/d_2 for Simple Regular Polygons

Number of edges	3	4	5	6	7	8	∞
$E\left(\frac{d_{max}}{d_2}\right)$	1.35	1.25	1.09	1.10	1.04	1.05	1.0

Among the parameters listed, there are two which have a constant mean value disregarding the grain shape. These are the number of edges, n_i, and dihedral angles, β_i. For 2-D cells on a section, the average number of sides is equal to six and the average dihedral angle is equal to 120°:

$$E(n) = 6.0 \tag{6.14}$$

$$E(\alpha_i) = E(\beta_i) = 120° \tag{6.15}$$

On the other hand, 2-D grain structures differ in their values of the standard deviations SD(n) and SD(β). The standard deviation of the number of edges, SD(n), is a measure of the departure of a given 2-D structure from the all-cells-six-neighbor structure of the honeycomb, which makes a reference structure for systems of cells covering a flat surface. Rhines and Patterson[12]

Table 6.6 The Percentage of Grains with a Given Number of Edges in Micrograined 316L Austenitic Stainless Steel

Number of edges	2	3	4	5	6	7	8	More
%	1	8	23	24	19	11	7	7

proposed that the distribution of the number of edges per grain for sections of polycrystals, f(n), is of a log-normal character. The data for an austenitic stainless steel are presented in Table 6.6. These data generally confirm this assumption. Further, the standard deviation of dihedral angles, SD(β), is a measure of the departure from the reference honeycomb structure, which is characterized by all dihedral angles being equal 120°. Experimental values of SD(n) and SD(β) are plotted for aluminum in Figure 6.12 as a function of the annealing temperature. Some data on the values of SD(β) can be found in Reference 13.

6.5 Grain size

In metallographic practice the term *grain size* usually refers to a parameter that describes the average grain size. In this context the three most commonly used parameters to describe grain size in polycrystals are

a) grain volume, V;
b) grain area, A;
c) intercept length, l.

The ASTM grain-size number, G, is also used, which is a function of either the mean intercept length, E(l), or mean grain area, E(A) (see, for example, Reference 14). For polycrystals G is given as:

$$G \approx -3.29 - 6.644 \log_{10} E(l) \tag{6.16}$$

or:

$$G \approx -2.95 + 3.32 \log_{10} E(A) \tag{6.17}$$

The choice of parameters describing grain size can also be dictated by the needs of a theoretical model of the material and its properties. For example, grain size can be described by:

a) intercepts, if the size of grains is analyzed in the context of dislocation pile-ups;
b) section area, if the size of grains is analyzed in terms of the size of dislocation loops;
c) grain surface, if one is interested in the total amount of grain boundaries per unit volume of the material.

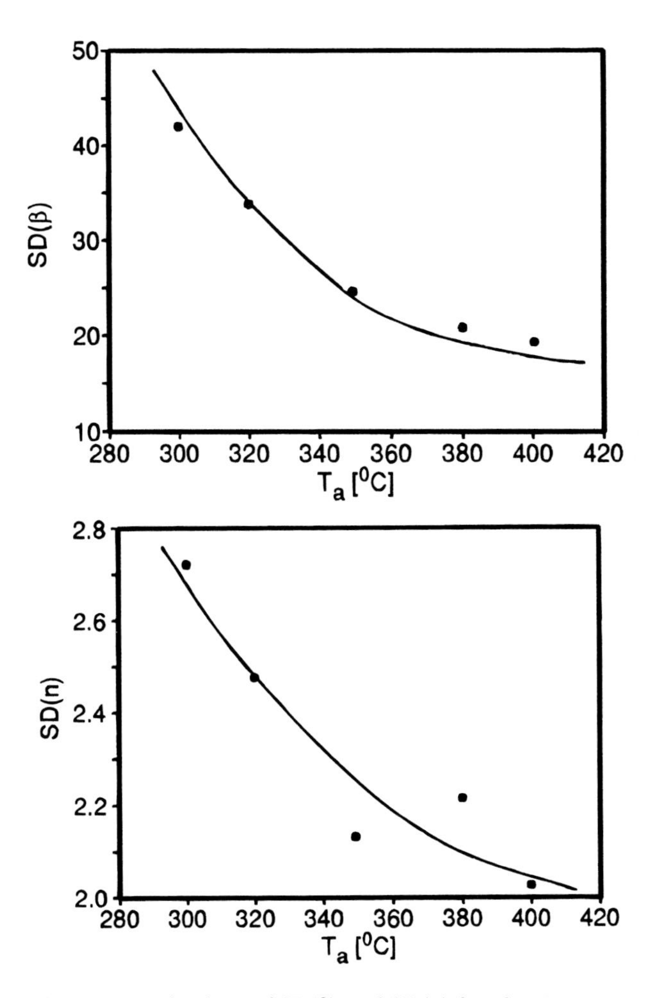

Figure 6.12 Experimental values of SD(β) and SD(n) for aluminum against annealing temperature.

The decision as to the best choice for a grain sizing parameter also must take into account the complexity in measuring a given parameter. Measurements of grain volume and surface are known to impose more technical difficulties and, frequently, grain size is described by other parameters such as grain area and intercept length.

In the general case, if no theoretical guidance to the choice of size parameter is available, grain size in 3-D polycrystals is described by volume, while 2- and 1-D grains are described by their area and length respectively:

Dimensionality	Size of grains
3	$V = 1/6\,\pi d_3^3$
2	$A = 1/4\,\pi d_2^2$
1	l

In order to simplify the interpretation of the measurements, the grain volume and grain area are replaced by equivalent diameters d_3 and d_2, respectively. These diameters are defined as diameters of a sphere (d_3) and circle (d_2) of the same volume or area as that of a given grain.

It should be noted that although grain area and intercept length are also used to characterize grain size in 3-D polycrystals, in contrast to grain volume, they do not describe uniquely the size of individual grains in 3-D polycrystals as illustrated in Figure 6.6. A grain of a specified volume, V, can have ascribed to it a distribution of grain areas, A, measured on its intersections and a distribution of grain intercepts, l.

6.6 Grain area distribution function and its relationship to grain volume distribution

A polycrystalline aggregate of 3-D grains is characterized by a distribution function, f(V). The grain size is studied on sections of polycrystals by grain area measurements leading to a distribution function, f(A). In order to estimate f(V) one has to deal with the convolution of two distribution functions f(V) and f(A).

In the case of grain area measurements on a section of a 3-D polycrystal a given area, A, cannot be directly linked to any specific grain volume, V. As a result the distribution function of grain volume f(V) can be obtained from the grain area distribution function, f(A), measured on a section through a 3-D polycrystal only through special procedures.

Two general types of approach to the relationship between f(V) and f(A) can be identified: "forward" and "backward" procedures. They are schematically shown in Figure 6.13. In the backward procedure, which is exemplified by the Soltykov method (see, for example, Reference 10 and Chapters 2 and

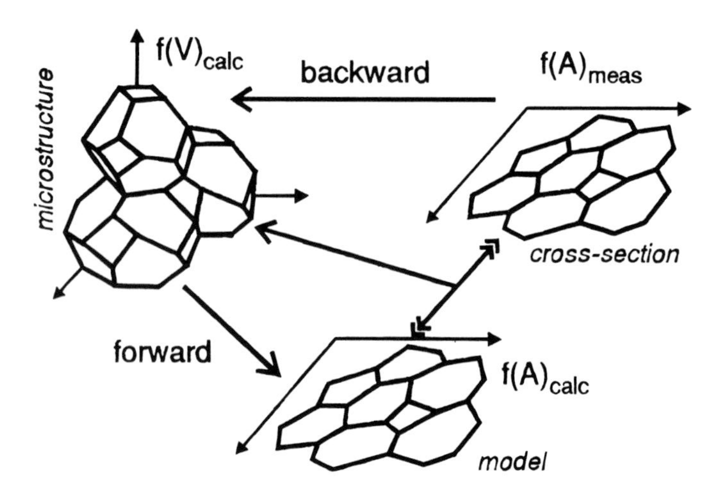

Figure 6.13 Schematic representation of the "backward" and "forward" procedures of the approach to the relationship between the f(V) and f(A) distribution functions.

8), grain area measurements, $f(A)_{meas}$, are adopted as a starting point and on their basis, for a given grain shape (often assumed, rather than known), the distribution function $f(V)_{calc}$ is calculated.

In the forward procedure, one assumes a certain type of distribution function, $f(V)_{theor}$, which has some adjustable parameters. Then, the distribution function $f(A)_{calc}$ is calculated and the values of the f(V) parameters are determined in a procedure based on examination of the fit between $f(A)_{calc}$ and $f(A)_{meas}$.

Apart from numerical details these two methods differ in the fact that the first approach emphasizes the role of the data collected, $f(A)_{meas}$, while in the second, attention is focused on the $f(V)_{theor}$ function. In view of this, the second approach is more convenient in those cases when a model of a polycrystal is the center of attention.

As a starting point in this analysis, an assumption is made that there is a functional dependence, H, which, for given experimental conditions described by a parameter, δ, relates the grain area distribution function, f(A), to the grain volume distribution, f(V), and grain shape distribution function, $f(q_i)$:

$$H[f(V)] \xrightarrow{f(qi)} f(A) \qquad f(V) \xrightarrow{f(q_i)} H^{-1}[f(q_i)] \tag{6.18}$$

General formulae arising from this equation can be simplified if some specific assumptions concerning the measurement technique and the grain shape are made.

The experimental conditions of grain area measurement are defined by the sampling procedure and by the restrictions in the precision of the measurements. In the case of metallographic observations, the population of grains revealed on a section of a polycrystal, Ω_c, differs from the population of grains forming the aggregate, Ω. The grains hit by the section tend to be those that are the larger grains in the population Ω. The probability of a grain being hit by a section plane is proportional to the height of the grain in the direction normal to the section. The grain volume distribution function of the grains cut by a section, $f_c(V)$, is then proportional to $V^{1/3}f(V)$:

$$f_c(V) = V^{\frac{1}{3}} f(V) \tag{6.19}$$

The experimental error in grain area measurements is mainly related to the limitations of the experimental technique used to reveal the grain boundaries and to measure the grain area. This is related to the fact that the trace of a grain boundary on sections of polycrystals has a finite width much larger than its physically rationalized thickness. Another error is introduced in the process of image digitization if the measurements are carried out with an automatic image analyzer (the effects are schematically explained in Figure 6.14).

The limitations imposed by the resolution limit characteristic of the observation technique can be described by a critical value, A_{min}, such that the areas smaller than A_{min} are not revealed and/or are skipped during the

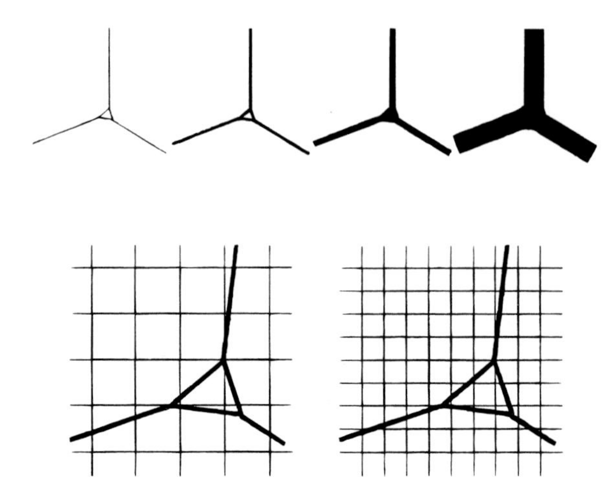

Figure 6.14 Schematic explanation of the resolution limit effect in the studies of grain size on sections of polycrystals.

measurements. For practical reasons, a parameter δ can be defined as the ratio of A_{min} to the mean grain area E(A). A value of $\delta = 0$ indicates an extreme case of unlimited resolution, which provides the reference level in the analysis of the effect of experimental error on grain area measurements.

Computer studies of the relationships between f(A) and f(V) require an assumption with regard to f(V). It has been assumed that the volume distribution function, f(V), is described by log-normal and multimodal distributions. The choice of a log-normal distribution function is based on the data reported in a number of experimental studies on equilibrium structures formed in the process of normal grain growth (see, for example, References 15,16). On the other hand, a bimodal distribution can be used to approximate the grain size distribution in nonhomogeneous structures resulting from abnormal grain growth.

In view of the assumption of a log-normal character of the volume distribution function, it is convenient to transform the variable by changing V into ln(V/[units of V]), or in short, lnV and accordingly A into ln(A/[units of A]) or in short, lnA.

The ln(V) distribution function is uniquely described by the mean value of lnV, E(lnV), and by the standard deviation of lnV, SD(lnV). Thus, for a given grain shape and given value of δ, the H-function relates the following f(A) function to E(lnV) and SD(lnV):

$$\left[E(\ln V), SD(\ln V)\right] = H_0^{-1}\left[f(A)\right] \tag{6.20}$$

The bimodal distribution is defined by a set of two mean values, E_1 and E_2, and two standard deviations, SD_1 and SD_2, and by the ratio θ, which determines the relative number of grains in the two grain populations. Thus, the H-function in this case relates:

$$H_0^{-1}\left[f(A)\right] = H\left[E_1, E_2, SD_1, SD_2, \theta\right] \tag{6.21}$$

From a practical point of view the problem of the relationship between grain volume and grain area distributions can be reduced to the problem of the relationship between the mean values and the standard deviations (or coefficients of variation) of both functions. The problem is then reduced to the functional relationships such as:

$$H_0^{-1}\left[E(\ln A), SD(\ln A)\right] = \left[E_1(\ln V_1), E_2(\ln V_2), SD_1(\ln V_1), SD_2(\ln V_2), \theta\right] \tag{6.22}$$

Other combinations of parameters can also be taken and one can replace the standard deviations SD(lnA) or SD(lnV) by CV(A) or CV(V). The choice should be based on the theoretical needs and numerical precision of the calculations.

Computer programs implemented in Reference 17 to calculate H-functions for a given grain shape, given volume distribution, and experimental conditions executed the following steps:

1. definition of the mean value of grain volume, (E(V), E(lnV)), definition of the distribution function f(V) (log-normal, bimodal, other), definition of the standard deviation SD(lnV);
2. for a given distribution function f(V), division of the grain volume space into discrete classes;
3. calculation of the relative number of grains in all classes;
4. definition of the grain shape (one of those listed earlier);
5. calculation of the number of grains cut by a section in each class of the grain volume;
6. for a given grain shape and for the number of grains determined in step 3, calculation of the grain area distribution function, $f(A)_{I\text{-class}}$, for all classes of the grain volume ($I = 1 \rightarrow N$);
7. definition of the resolution limit, δ, and corrections of the functions $f(A)_{I\text{-class}}$ (deleting those areas smaller than A_{min});
8. for a given resolution limit, δ, stacking and sieving of the functions $f(A)_{I\text{-class}}$ in order to obtain the distribution function $f(A)_\delta$;
9. computation of the statistical parameters of the function $f(A)_\delta$;
10. mapping of the dependence of the statistical parameters of the function f(A) with respect to the statistical parameters of the function f(V).

The results of the computations described in Reference 13 are presented first for the case of an unlimited resolution in Figure 6.15, which shows the relationship between SD(lnV) and CV(A), and SD(lnV) and SD(lnA) for various assumed grain shapes. It can be concluded that the relationship SD(lnV)-CV(A) is much less grain-shape sensitive than the other one. Therefore this relationship is recommended for practical applications.

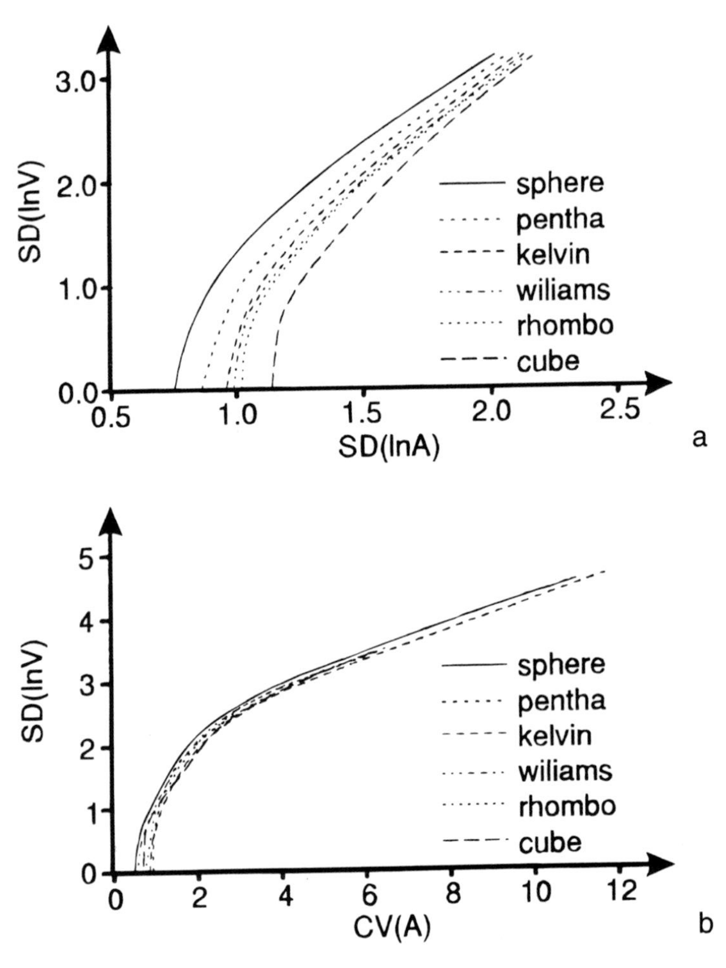

Figure 6.15 The computed relationships between (a) SD(lnV) and SD(lnA) and (b) SD(lnV) and CV(A) for various assumed grain shapes for the case of unlimited resolution.

The parameters of distribution functions f(A) for different grain shapes computed for the extreme case of constant-size-grains (CSG polycrystals) are given in Table 6.7. CSG polycrystals are a reference aggregate, which is characterized by SD(lnV) = 0.

These results indicate that the values obtained for different grain shapes are quite similar, with the results for spheres slightly deviating from the rest. The values of CV(A) and SD(lnA) for spheres are significantly lower than the values for the rest of the shapes. This can be explained in terms of the effect of the number of edges and corners characteristic of a given grain shape. On the basis of these results, the following formulae can be proposed for a polycrystal containing grains of the same shape and size:

Table 6.7 Parameters of the Distribution Function f(A) for Some Reference Structures Consisting of Constant Size and Shape Grains

Parameter/shape	Spheres	PD	C	TO	β-W	RD
E(A)	0.663	0.566	0.498	0.539	0.469	0.545
CV(A)	0.444	0.532	0.588	0.531	0.546	0.561
SK(A)	0.240	0.140	0.070	0.140	0.070	0.120
R(A)	2.140	1.830	1.590	1.850	2.010	1.760
E(lnA)	–0.610	–0.853	–1.038	–0.932	–1.072	–0.974
SD(lnA)	0.794	0.926	0.992	1.020	1.009	1.108
SK(lnA)	2.070	1.350	1.020	1.620	1.550	1.470
R(lnA)	7.430	3.580	2.630	4.640	4.370	3.940
R^3 space		Nonfilling			Filling	

Abbreviations used: PD—pentagonal dodecahedra, C—cubes, TO—truncated octahedra, β-W—Williams' β-tetrakaidecahedra, RD—rhombic dodecahedra.

$$CV(A)_{CSG,\delta=0} \approx 0.55 \tag{6.23}$$

$$SD(\ln A)_{CSG,\delta=0} \approx 1.0 \tag{6.24}$$

If the values of SD(lnA) and CV(A) are higher than those given above then this may be taken as an indication of inhomogeneity of the grain size and/or deviation of the grain shape from a given configuration.

Simple considerations show that for higher values of grain volume variance the following approximate equation can be derived:

$$SD(\ln V) \approx \sqrt{\frac{9}{4} SD^2(\ln A) - 1.0} \tag{6.25}$$

The results obtained for the case of limited resolution are presented in Figure 6.16 against the resolution limit, δ. Figure 6.16a shows that for a given value of SD(lnV) the value of SD(lnA) decreases with an increasing value of the resolution limit δ. Figure 6.16b shows the plot of CV(A) against SD(lnV) and δ. It can be deduced by comparing these two plots that CV(A) is less sensitive to the value of δ. Also it may be noted that for $\delta \geq 0.1$, SD(lnA) becomes considerably lower than predicted for the case of unlimited resolution ($\delta = 0$). For larger values of the resolution limit there is a nearly linear dependence of SD(lnA) on δ.

The results obtained for a bimodal distribution are presented in Figure 6.17 in the form of a plots of SD(lnA) against SD(lnV) and CV(A) against SD(lnV). The results have been obtained for the bimodal distributions obtained by superposition of two log-normal distributions of equal number of grains, ($\theta = 1$), and equal standard deviations ($SD_1 = SD_2 = SD$).

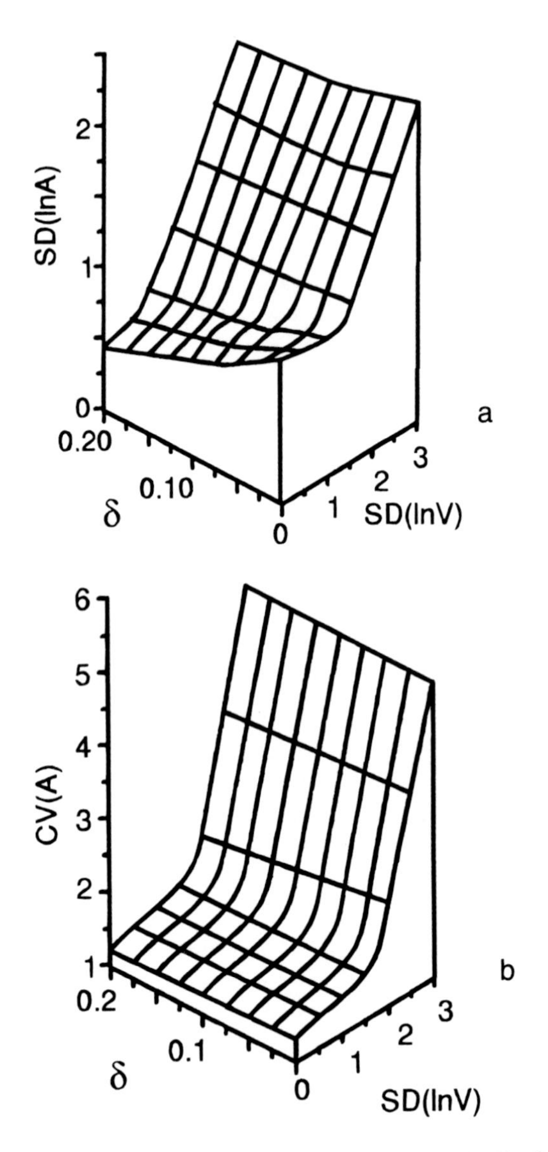

Figure 6.16 Plots of the relations between: (a) SD(lnV) - SD(lnA) - δ; (b) SD(lnV) - CV(A) - δ.

Two important conclusions can be drawn from the plot shown in Figure 6.17. First, for values of SD(lnA) smaller than 1.4, the relationship between SD(lnV) and SD(lnA) depends weakly on the value of SD(lnV). Second, for values of the standard deviation of SD(lnV) higher than about 3, in the case of the SD(lnV)-SD(lnA) relationship, and about 2.0 in the case of the CV(A)-SD(lnV) relationship, a nonmonotonic dependence is observed. There is a range of SD(lnV) values where an increase in the spread in the grain volumes does not lead to an increase in the spread of grain areas. This indicates that

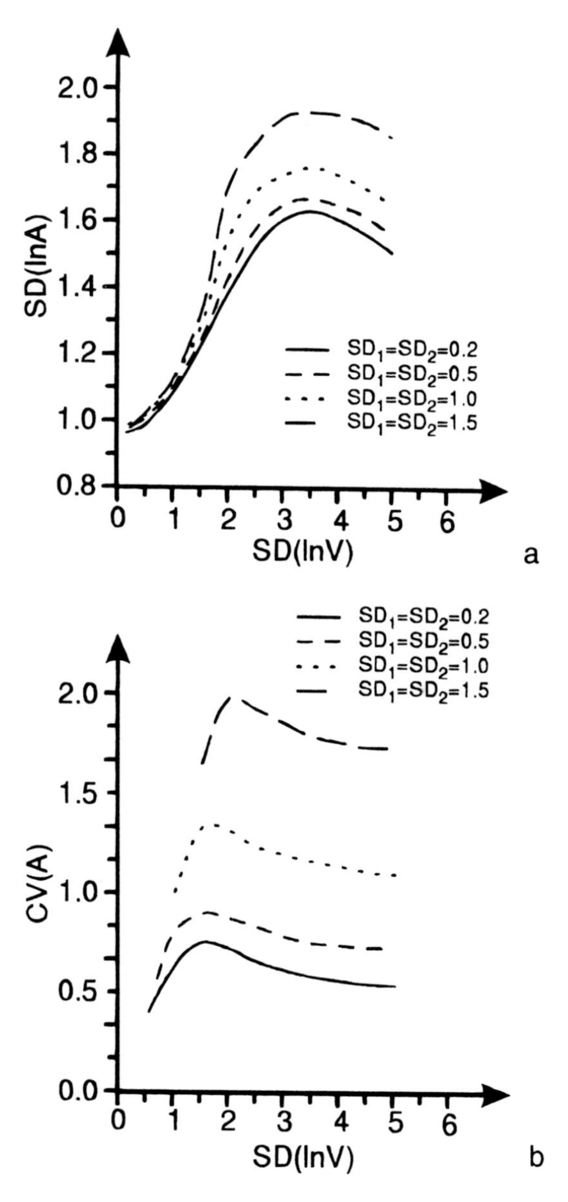

Figure 6.17 Plots of the relations between: (a) SD(lnV) and SD(lnA); (b) SD(lnV) and CV(A). The data have been obtained for a bimodal distribution.

measurements of SD(lnA) and CV(A) by themselves may not always be sufficient for a proper estimation of grain volume spread. It is recommended that such measurements should be supplemented by an examination of the grain area distribution function aimed at checking the possible existence of more than one peak, which would be an indication of a bimodal volume distribution function.

6.7 Mean grain size

For most applications in materials science, the grain size is sufficiently well defined by the value of the mean size of grains, E(X), where X is one of the parameters listed below:

grain volume, V, or equivalent circle diameter, d_3;
grain area, A, or equivalent circle diameter d_2;
intercept length, l.

The value of the mean grain size makes it possible to differentiate between different polycrystals in terms of the average grain size or by the average grain number in a unit volume. To this end, it suffices to use constantly one of the parameters listed above, assuming that it is measured properly according to appropriate procedures. However, the value of the mean size can provide more precise information on the grain geometry forming a polycrystal if its stereological interpretation is known.

Depending on the measurement technique, the mean value of the parameter measured, $E_{meas}(X)$, is the number mean, $E_N(X)$, or weighted mean, $E_W(X)$, of the parameter studied. Currently most of the procedures used lead to weighted mean values of the size parameters. It should be stressed that the weighted means need not be less useful than number means and in fact in some applications are more effective in the sense that they provide numerical answers to important questions from the point of view of understanding properties of the material. This is illustrated Example 6.4.

Example 6.4

Consider a polycrystal that contains grains of volume a^3 and $125a^3$ in the proportion 1:1. The distribution function f(V) is given by the following formula:

$$\begin{cases} f(V) = \frac{1}{2} & \text{for} \quad V = a^3 \\ f(V) = \frac{1}{2} & \text{for} \quad V = 125a^3 \\ f(V) = 0 & \text{otherwise} \end{cases} \tag{6.26}$$

The mean grain volume is given as: $E_N(V) = 63a^3$. This means that in a specimen with a volume $630a^3$ there are on average ten grains. At the same time five large grains occupy over 99% of the specimen volume. This polycrystal might therefore be better characterized by the volume weighted mean volume, $E_W(V)$. This is defined in this case as:

$$E_W(V) = \frac{V_1^2 + V_2^2}{V_1 + V_2} \tag{6.27}$$

which yields: $E_W(V) \approx 124a^3$. The latter value gives a better definition of the grain size if the mechanical behavior of the polycrystal is being considered.

6.7.1 Mean intercept length

The mean intercept length is measured by counting the intersection points of test lines with the grain boundary network revealed on sections of polycrystals. Parallel lines on a random section can be used where the grain geometry is isotropic. It should be stressed, however, that in the case of anisotropic grain structures it is necessary to use a system of lines randomly oriented in 3-D. This means that the lines need to be placed on randomly oriented sections of the polycrystal. Due to the time-consuming aspect of specimen preparation, such an approach is not a common practice in the microscopical characterization of materials. It is simplified by the procedures of modern stereology described in Chapter 3, which require that at least three perpendicular sections be examined. Such sections make it possible to detect the anisotropy and, if carefully directed with respect to the anisotropy axes, provide data that can be used to calculate the mean value of the 3-D random intercepts.

The mean intercept length, E(l), if used for 3-D polycrystals poorly describes the size of individual grains. It can be used as a scale parameter to quantify grain size only in self-similar grain structures, which can be obtained from a "basic pattern" by changing its magnification. However, its importance, apart from simplicity, is related to the stereological equation that relates E(l) to surface-to-volume ratio, S_V:

$$E(l) = \frac{2}{S_V} \tag{6.28}$$

This relationship makes measurements of mean intercept length an important parameter that quantifies the specific surface area of grain boundaries.

The relationship between the mean intercept and specific surface area is valid for any type of grain. However, in the case of equiaxed grains it can be implemented in a much simpler way. Such structures can be studied on a set of parallel sections using grids of parallel lines. One set of parallel sections can also be used in the case of polycrystals with grains elongated along the same axis. It is a requirement then that the measurements be carried out on the sections containing the elongation axis and two sets of parallel grid lines are used perpendicular to each other. The value of S_V is then computed from the equation:

$$S_V = \frac{\pi}{2}(P_L)_p + \left(2 - \frac{\pi}{2}\right)(P_L)_t \tag{6.29}$$

where $(P_L)_p$ and $(P_L)_t$ are the densities of intersection points with the grain boundaries on the lines parallel and perpendicular to the elongation axis, respectively. It should be noted that the density of intersection points with a given set of lines, $(P_L)_n$, is inversely proportional to the intercept length along these lines:

$$(P_L)_n = \frac{2}{E(l_n)} \tag{6.30}$$

In a general case, when no assumption about the shape of grains is justified, the sensible way of drawing the test lines would require pointing them in all directions in space (see Chapter 3). This is implemented in practice by the procedure called vertical sectioning, which is described, together with its modifications, in Chapter 3. In this case sections of material are tested with a system of cycloids aligned perpendicularly to the common direction of the sections. The intersection density, $(P_L)_{cyclo}$, is then used in equation:

$$S_V = (P_L)_{cyclo} \tag{6.31}$$

which is the general form of Equation 6.29.

6.7.2 Mean area

In the case of 2-D polycrystals the area, A, and in turn equivalent area diameter, d_2, can be directly measured from micrographs of the grain boundary network by planimetric methods or by the procedure of counting the number of grains in a unit area, N_A. For a 2-D polycrystal:

$$E_N(A) = \frac{1}{N_A} \tag{6.32}$$

If measured on a section of a 3-D polycrystal, N_A gives directly the weighted value of A, $E_W(A)$. This value can be used to compare the mean size of polycrystals or to retrieve information on the volume of grains.

6.7.3 Mean volume

The number mean volume of grains, $E_N(V)$, can be estimated with the help of the disector method (Chapter 3). This procedure requires, in general, serial sectioning of the material, as its principle is to count the grains below a cross-section instead of the grains cut by the cross-section.

The volume-weighted mean volume of grains, $E_V(V)$, can be determined by the procedure proposed by Jensen and Gundersen.[18] This procedure requires random point sampling of the grains revealed on a cross-section and subsequent measurements of the length intercepts, l_i, drawn through the points, P_i, "thrown" on the image of the grain boundaries. This procedure is shown schematically in Figure 6.18. On the basis of such measurements the mean volume, $E_V(V)$, can be estimated using the formula:

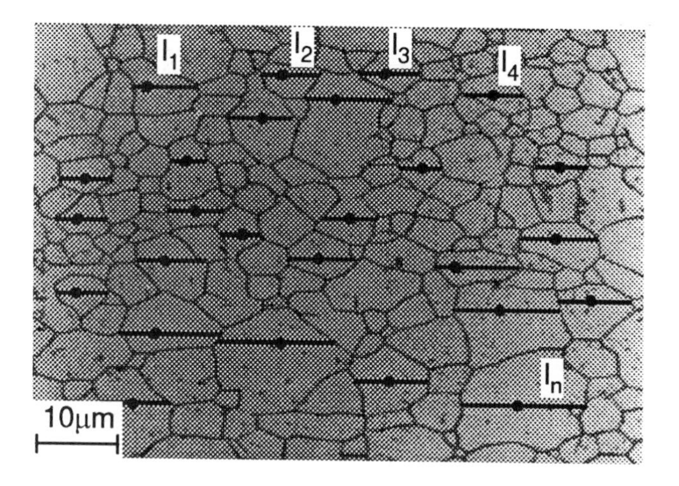

Figure 6.18 Random point sampling and subsequent measurements of the length of intercepts l_i drawn through the points, randomly sampled on a section of a polycrystal.

$$E_V(V) = \frac{\pi}{3} E(l^3) \tag{6.33}$$

6.8 *Relationship between linear parameters of grain size*

Mean area and mean volume can be used to determine mean values of equivalent diameters of circle and sphere, d_2 and d_3, respectively. These two diameters with the mean intercept length, E(l), form a set of three linear measures of the grain size. The values of various parameters describing grain size are related one to the other. For any two parameters, X and Y, a relationship can be proposed in the following form:

$$E(X) = \kappa\, E(Y) \tag{6.34}$$

The value of the proportionality constant appearing in this equation, κ, depends on the grain shape and as a result no universal constant can be given even for such widely used parameters as intercept and area of grains. This is illustrated by the data for the two grain structures shown in Figure 6.19.

6.9 *Degree of grain size uniformity*

Grains in polycrystals differ in their volume and the same mean size can be ascribed to geometrically different populations of grains. This observation rationalizes the concept of grain size homogeneity or uniformity.

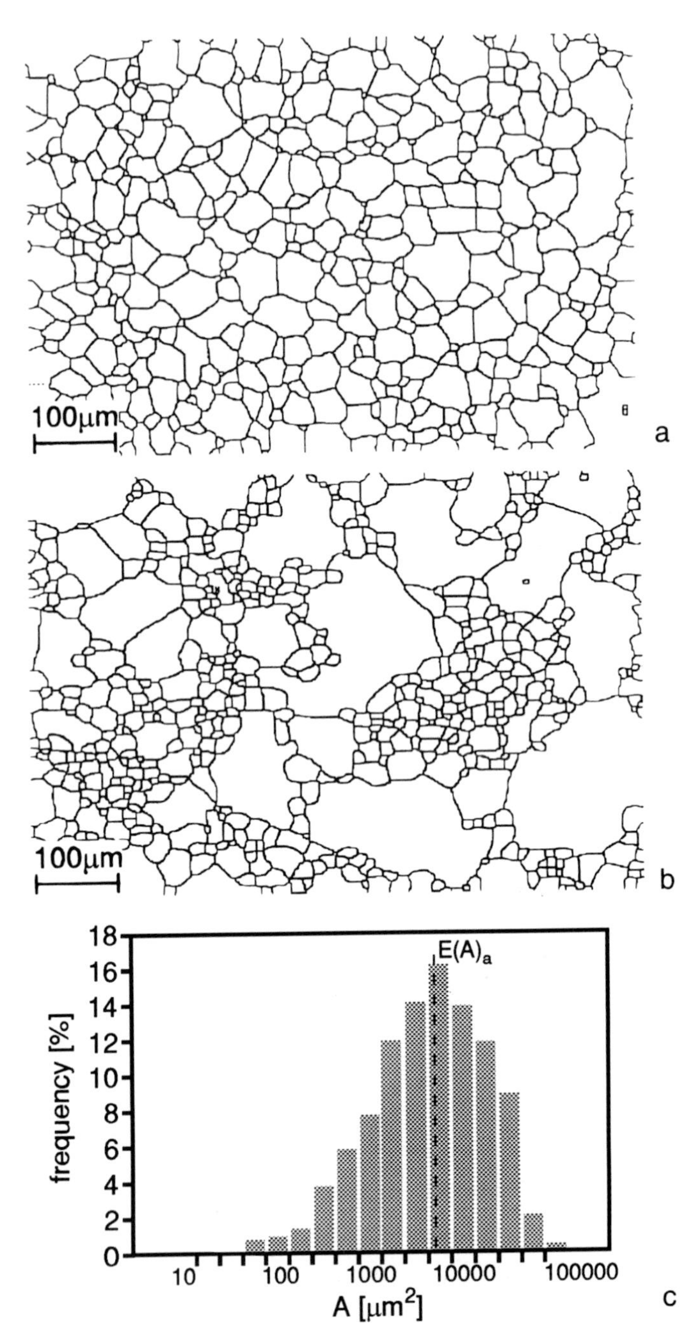

Figure 6.19 An example of two microstructures, (a) and (b) characterized by the same mean grain area, E(A), and therefore the same mean value of the equivalent area diameter, $E(d_2)$. The respective grain area distribution functions are shown in (c) and (d) and intercept length distributions in (e) and (f). The microstructures (a) and (b) yield different values of E(l). (Figure is continued on the facing page.)

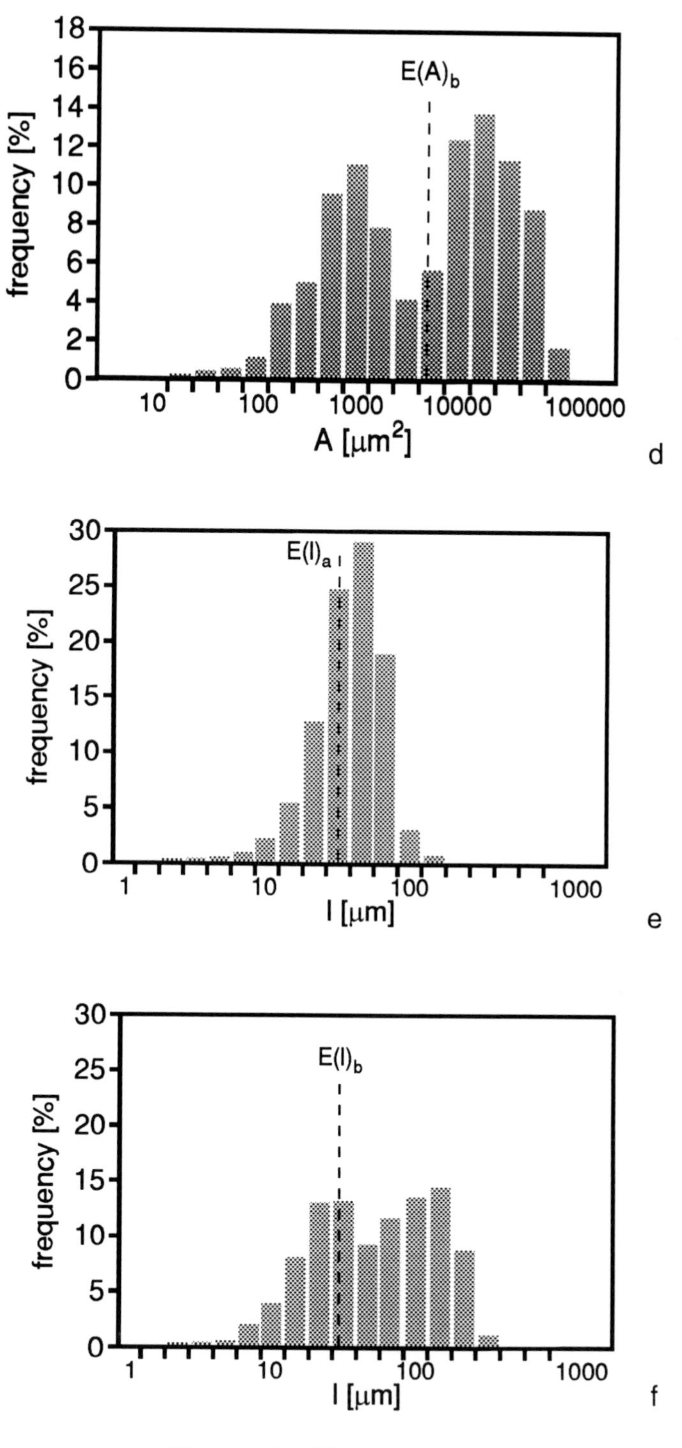

Figure 6.19 *(Continued)*

Example 6.5

Figure 6.19 shows grain boundary networks revealed in polycrystals of different histories. These two micrographs present different grain boundary systems where the main difference is in the size of individual grains. One of the microstructures is less uniform in the sense that the grains show more profound dissimilarity in their size although the mean grain size, measured by mean area, E(A), is approximately the same.

Methods of quantitative stereology make it possible to describe the differences in the degree of the dissimilarity in the grain size in a quantitative way. This problem is discussed under the heading of grain size homogeneity in the literature.

Differences in the grain size may be described in terms of the minimum and maximum, V_{min} and V_{max}, grain volumes. The homogeneity is described by a parameter C_1 defined as:

$$C_1 = V_{max} - V_{min} \tag{6.35}$$

or a parameter C_2 defined as:

$$C_2 = \frac{V_{max} - V_{min}}{E(V)} \tag{6.36}$$

The difference between the two parameters is not only of numerical character. The concept of the dimensionless parameter C_2 assumes that the scale used to measure the grain dimensions is not important. This assumption means that the same degree of grain size uniformity has to be ascribed to polycrystals that are obtained by the mere operation of magnifying some basic microstructure.

It should be noted that any flattening of the grain volume distribution function, f(V), automatically leads to an increase in the value of the parameter C_1. At the same time, the normalized distribution, f(V/E(V)), may remain unchanged as does the value of C_2.

The definition of the C parameters touches on the question of the existence of lower and upper limits of the volume: V_l and V_u. In the general case:

a) $V_l = 0$;
b) $V_u = V_a$—the volume of the artifact.

In practical situations, the minimum volume, V_{min}, is equal to the lower limit of the volume, i.e., $V_{min} = V_l = 0$. This is related to the effect of randomness in grain size in polycrystals as a result of their processing and due to grain boundary rearrangements, such as those leading to growth of grains, activated by temperature (growth of one grain results in shrinking of at least one other). On the other hand, the maximum grain volume in polycrystals can be effectively bounded only by the dimensions of specimens. If no restrictions are placed on the size of the polycrystalline aggregate, the maximum volume, V_{max}, is difficult to define. This problem can be overcome when the degree of uniformity of grain size is defined by the methods of mathematical statistics. In particular, a simple measure of grain size uniformity is the variance of the grain size distribution function, VAR(V), or standard deviation, SD(V), and coefficient of variation, CV(V):

$$CV^2(V) = \frac{VAR(V)}{\left[E(V)\right]^2} \tag{6.37}$$

Another parameter that can be used is the standard deviation of the logarithm of the volume, SD(lnV). The use of this function is rationalized by the fact that the volume distribution function is frequently of a log-normal character.

The point sampling intercept procedure proposed by Jensen and Gundersen[18] can be used to quantify the degree of grain volume uniformity. This is based on the equation that relates volume variance, $VAR_V(V)$, to the mean value of V:

$$VAR_V(V) = E_V(V^2) - \left(E_V(V)\right)^2 \tag{6.38}$$

Jensen and Sorensen[19] have shown that:

$$E_V(V^2) = 4\pi k E(A^3) \tag{6.39}$$

In this relationship A is the area of the grains hit by the P_i points and k is a constant with a value in the range from 0.071 to 0.083. The value of k can be determined either experimentally or by appropriate modelling.

From the last two equations one can obtain the following relationships:

$$VAR_V(V) = 4\pi k E(A^3) - \frac{\pi^2}{9}\left[E(l^3)\right]^2 \tag{6.40}$$

and

$$CV_V^2(V) = \frac{36kE(a^3)}{\pi\left[E(l^3)\right]^2} - 1 \tag{6.41}$$

In practical implementation the procedure based on the use of Equations 6.38 to 6.41 combines standard microscopical measurements such as measurements of intercept length and grain area. These measurements can be conducted with high precision and automated by the use of image analysis systems. In the case of geometrically isotropic grain structures the measurements can be executed on one representative section of the polycrystal, assuming it contains a sufficiently large number of grains.

Experimental data show that the distribution function $f_N(V)$ frequently may be well approximated by a log-normal distribution, $\ln_N(V)$. Such a distribution considerably simplifies the relationships between the weighted and true variances of grain volume and in turn the true and weighted value of CV(V). For a log-normal distribution it has been shown[20] that the following relationship holds:

$$CV_N(V) = CV_V(V) \tag{6.42}$$

This means that for a log-normal volume distribution, $\ln_N(V)$, coefficients of variation for true volume and weighted volume distribution functions are equal. As a result the following relationships are valid:

$$CV_N(V) = \sqrt{\frac{36kE(A^3)}{\pi\left[E(l^3)\right]^2} - 1} \tag{6.43}$$

and

$$E_N(V) = \left(\frac{\pi^4}{972k}\frac{\left(E(l^3)\right)^5}{E(A^3)}\right)^{\frac{1}{3}} \tag{6.44}$$

These two equations can be used to estimate the number mean volume and the coefficient of variation for a log-normal distribution of grain volume from the relatively simple measurement procedures described earlier.

6.10 Data for metallic and ceramic polycrystals

Some data from the literature are given in Table 6.8. These data define the range of values commonly met for the case of metallic and some ceramic materials. It may be noted that typically SD(lnA) is measured to be in the range from 0.9 to 1.4. On this basis one can estimate, that SD(lnV) varies from 0.9 to 1.8 and such a range for SD(lnV) is consistent with the results of grain volume measurements presented in References 21 to 23.

The degree of grain size uniformity has also been evaluated using the method by Jensen and Sorensen.[19] The results of these measurements are shown in Figure 6.20 in log-log coordinates in the form of a plot of the mean grain volume, $E_V(V)$, against the standard deviation, $SD_V(V)$.

6.11 Grain size effect

The grain size of a polycrystalline aggregate has an important effect on its properties. This phenomenon is known in the literature as the grain size effect. As far as the mechanical properties of materials are concerned (such as yield and flow stress, ductility, hardness, and fatigue limit), a refinement of

Table 6.8 Variations of Grain Volume and Grain Area for Some Metallic and Ceramic Polycrystals

Material	SD(lnV)	SD(lnA)	CV(A)	δ*
Aluminum	0.9 – 2.5	Unknown	Unknown	Unknown
Aluminum	1.06 – 1.45	1.1 – 1.35	0.9 – 1.5	<0.03
Alumina	Unknown	0.99 – 1.15	0.96 – 1.2	0.03
Nanocrystals	Unknown	0.70 – 1.05	0.55 – 0.90	0.04
316L	1.03 – 1.47	1.3 – 1.4	1.25 – 1.37	0.01
316L	Unknown	1.04	1.01	0.02
Ti-α	0.95	0.95	Unknown	Unknown
Fe-α (normal)	1.16	0.95	0.88	0.01
Fe-α (selective)	1.82	1.39	3.95	< 0.01

Data collected from References 4, 15, 16, and 17.

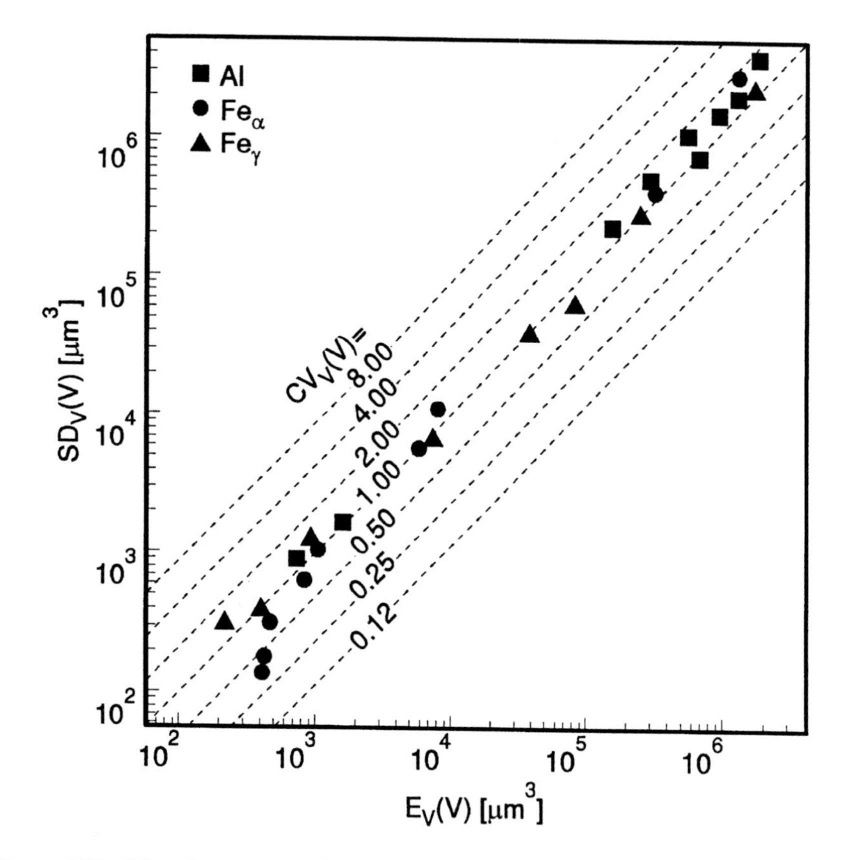

Figure 6.20 Plot of experimental values of $E_V(V)$ and $SD_V(V)$ obtained for aluminum, α-Fe and γ-Fe.

the grain size generally results in an improvement of these properties at low testing temperatures (i.e., a temperature well below 0.5 of the melting temperature of the material). Since the grain boundaries are known to block the movement of crystal dislocations, the grain size effect has been explained in terms of the distance a dislocation can slip. However, there are different specific models of the interaction between dislocations and the grain boundaries in strained polycrystals. They can be broadly divided into:

a) pile ups models;
b) work hardening models;
c) grain boundary sources.

(More information can be found in Reference 24.) All these models lead to a similar equation that was first suggested by Hall and Petch[25] and Armstrong and co-workers[26] and which predicts a linear dependence of the yield or flow stress on (average size of grains)$^{-1/2}$. This dependence is referred to as the Hall-Petch relationship. For example, according to the Hall-Petch relationship the flow stress of polycrystals follows the equation:

$$\sigma = \sigma_o + K d^{-\frac{1}{2}} \tag{6.45}$$

where σ stands either for flow stress at a given plastic strain, ε, or Lüders stress, σ_L, σ_0 and K are constants and d is grain size, typically the mean intercept length.

The constants in this equation can be treated as giving the best fit to the experimental data for the material or, as frequently happens, they are interpreted as the flow stress of the grain interior, σ_o, and the flow coupling effect of the grain boundaries, K.

Figure 6.21 illustrates the effect of the grain size on the flow stress of a 316L austenitic stainless steel. The data are typically plotted in the coordinate system (flow stress – grain size$^{-1/2}$). In the case presented in Figure 6.21 the grain size has been measured in terms of the mean intercept length. This is the most commonly adopted way of measuring the grain size for Hall-Petch relationship applications. The reasons for this are probably twofold. First, this is a relatively simple procedure for grain size measurement. Second, the intercepts can be considered to be equal to the length of dislocation pile-ups. However, no completely formal proof of this has been presented so far.

Equation 6.45 has been successfully tested for a wide range of polycrystalline materials, including metals, ceramics, and intermetallics. At the same time, there is growing evidence that significant departures from Equation 6.45 can be observed, especially in the case of fine- and ultra-fine-grained polycrystals.[27–29]

There are a number of models that give rationalizations of the relationship proposed by Petch and Hall. All of these models are based on the assumption that the grains in polycrystals have the same size and shape. In most of the models these assumptions are not stated in an explicit way; however,

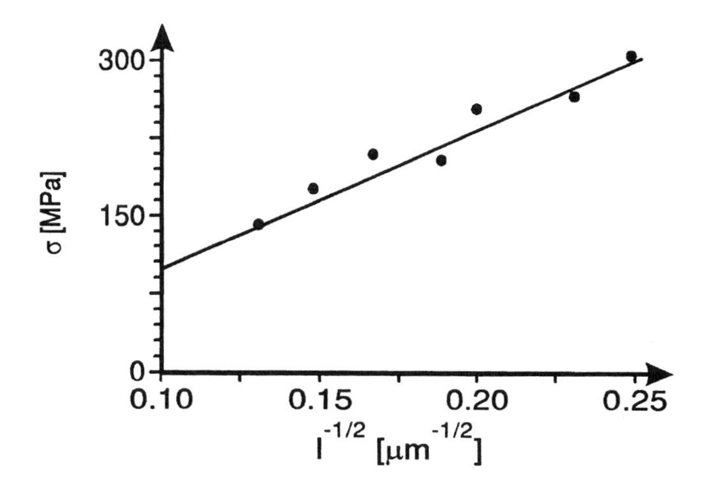

Figure 6.21 Grain size effect on the flow stress for an austenitic stainless steel.

they might be deduced from the fact that no attention is given to a potential diversity in the grain size and/or in the properties of the grain boundaries in the polycrystal.

A new approach was proposed recently[30] that takes into account the statistical nature of grain morphologies. This model follows the lines adopted in the present text and is based on the assumption that a polycrystal can be described by a grain volume distribution function, f(V). It has been further shown[30] that for a polycrystal characterized by log-normal distribution of grain volume, the flow stress can be approximated by the following equation:

$$\sigma = \sigma_o + K_{CSG} \exp\left[-aSD^2(\ln V)\right] d^{-\frac{1}{2}} \tag{6.46}$$

where:

- $SD(\ln V)$ is the standard deviation of the volume logarithm;
- K_{CSG} is the grain boundary coupling constant of the Hall-Petch relationship for CSG polycrystals of a constant grain size;
- σ_o is the friction stress;
- a is a numerical constant.

This modified version of the Hall-Petch equation has been verified[31] on a series of polycrystals formed by a powder-forming process. The powder, with a composition typical of 316L austenitic stainless steel, was sieved in order to produce two fractions, FA and FB, such that:

FA contained particles with a diameter smaller than 44 μm, while
FB contained particles with a diameter larger than 177 μm.

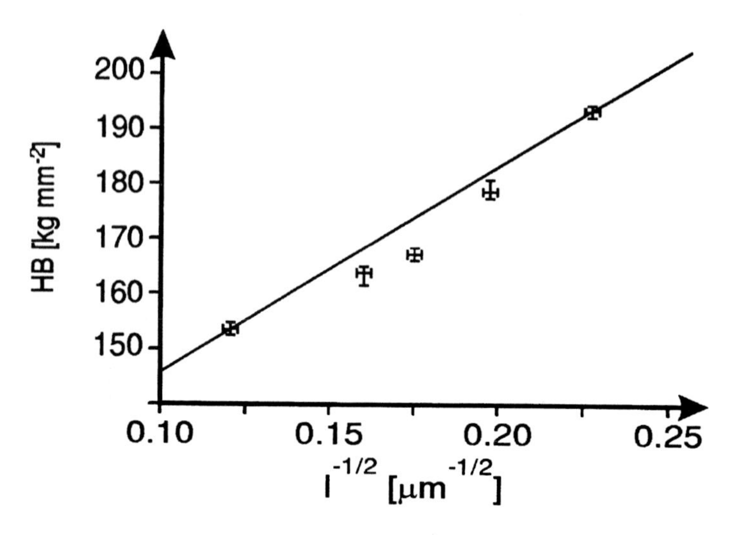

Figure 6.22 Results of hardness measurements against ($l^{-1/2}$) for a series of specimens obtained by a powder forming process.

The two fractions of the particles were later mixed in different proportions to produce a series of specimens with different values of the ratio V_A/V_B specifying the relative volume (weight) of particles from fractions FA and FB, respectively, which resulted in differences in mean grain size as well as in the grain size distribution. Systematic measurements of the grain sizes were performed on these specimens using an automatic image analyzing system. The measurements included grain area, A, and grain intercept length, l.

The results of hardness measurements are plotted in Figure 6.22 against the respective values of ($l^{-1/2}$). It is seen from the plot that the experimental data, in general, confirm a relationship given by the equation:

$$H = H_o + K_H l^{-\frac{1}{2}} \tag{6.47}$$

If the data points are analyzed in total the following values of H_o and K_H were computed:

$H_o = 107$ [kg mm^{-2}], $K_H = 11.1$ [kg mm$^{-3/2}$].

However, upon a closer examination of the experimental data it can be found that the data points systematically deviate from the linear relationship predicted by Equation 6.47.

It has been shown that the data can be explained using the following equation:

$$H = H_o + K_H^* \exp\left[-c_1\left(CV(A) - c_2\right)^m\right] l^{-\frac{1}{2}} \tag{6.48}$$

where K_H^* is the K_H constant for polycrystals built up of grains of constant size, c_1, c_2, and m are numerical constants that depend on the details of the geometry of grains and the resolution limit of the observation technique adopted. The constant c_2 has a value close to 0.7, which is a coefficient of variation of grain area computed for a reference polycrystalline structure containing polyhedra of a constant size.

The above equation can be further simplified by substituting exp(x) with (x + 1) which leads to the following equation:

$$H = H_o + K_H^* \left[1 - c_1 (CV(A) - 0.7)^2\right] l^{-\frac{1}{2}} \tag{6.49}$$

where m = 2 has been assumed and $c_2 = 0.7$. This equation has been applied to the present data. Remarkable agreement with Equation 6.49 has been observed and the following constants were computed:

c_1 = 0.06,
H_o = 110 [kg mm^{-2}],
K_H^* = 11.1 [kg $mm^{-3/2}$].

The results presented here demonstrate that the hardness, and in turn the flow stress, of polycrystals is not only a function of the mean grain size but also of the spread in grain size (degree of grain size uniformity). The Hall-Petch relationship in the classical form of Equation 6.45 should be used in the case of materials showing constant grain uniformity as reflected by a constancy of CV(A) or CV(V). If the equation is applied to data characterized by a variation in the grain size uniformity it may result in a systematic deviation from the expected linear relationship.

Further important modifications of the Hall-Petch formula can be produced if the differences in the properties of grain boundaries are taken into account.[32] It is known for example that low-angle grain boundaries (LAGBs) differ in their properties as compared with general boundaries and strengthen polycrystals in a different manner. In terms of their strengthening role, low-angle grain boundaries may be considered as an element of the grain interior rather than of the grain boundary population. This is rationalized by the fact that low-angle grain boundaries do not act as a complete barrier to dislocation motion but their presence leads to an increase in the value of the friction stress.

The discrimination of LAGBs from the population of grain boundaries in a given polycrystal, implemented in the series of micrographs shown in Figure 6.23, significantly changes the grain geometry. The networks formed by the rest of the grain boundaries (high-angle grain boundaries) are characterized by a higher mean size and a lower degree of uniformity (a higher coefficient of variation).

The fraction of LAGBs controls the variance of the grain size. On the other hand, these boundaries contribute to the strength of the grain surrounded by the higher-angle grain boundaries. This dual function of low-angle

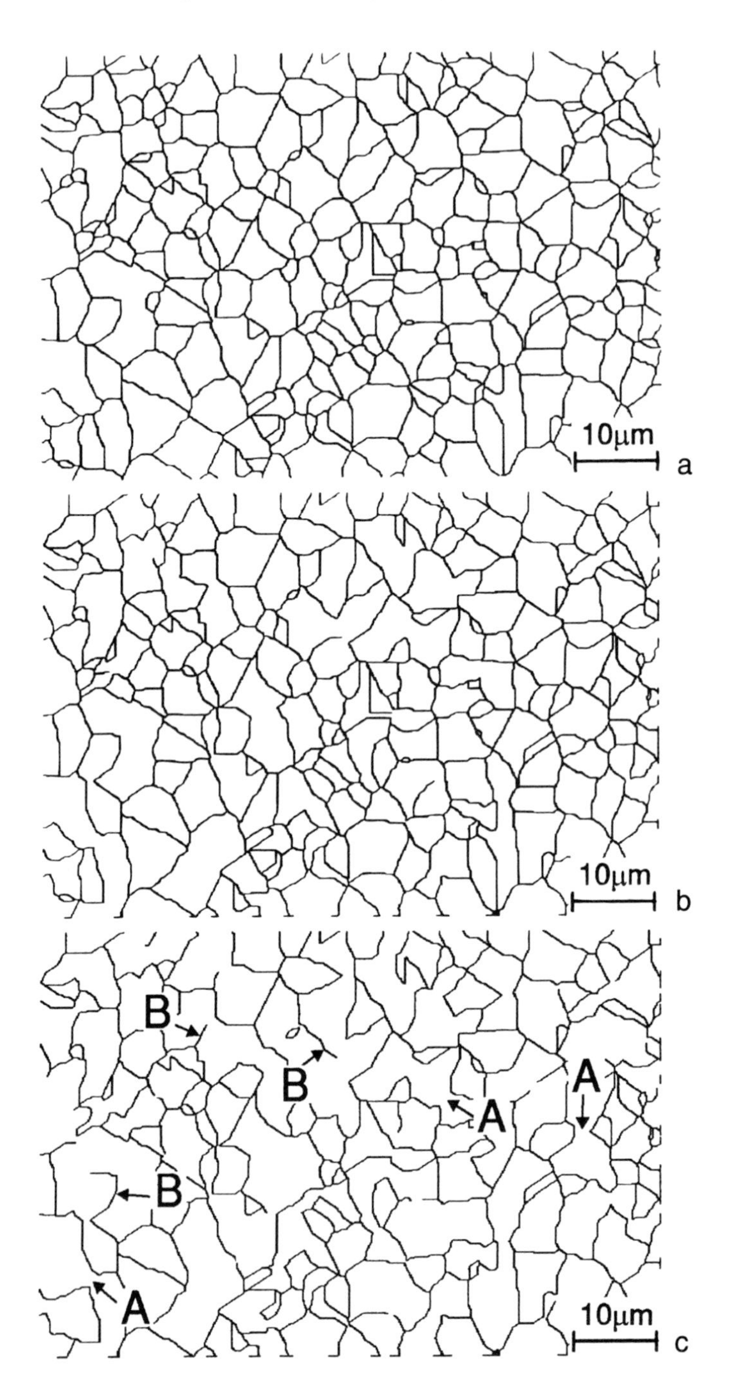

Figure 6.23 A binary image of the grain boundary network revealed in a polycrystalline material (a) and images obtained after removing, from the image of the microstructure, grain boundaries of a misorientation angle smaller than: (b) 5°; (c) 10°. Arrows point to the segments of "ghost" (A) and incomplete (B) grain boundaries.

boundaries is accounted for in the following general formula, which is a modification of Equation 6.45:

$$\sigma = [\sigma_o(f_{LB})] \uparrow + \left\{K \cdot \left[l^{-\frac{1}{2}} g(f_{LB})\right]\right\} \downarrow \tag{6.50}$$

where f_{LB} indicates the fractures of LAGBs.

The arrows in Equation 6.50 show that an increase in the fraction of low-angle boundaries results in higher values of CV(A) thus decreasing in this way the strengthening contribution of the grain boundaries. At the same time, an increased fraction of LAGBs leads to higher values of the friction stress, which partly offsets the decrease in the grain boundary contribution.

6.12 *Geometry of grains and physical properties of grain boundaries*

The geometry of grains in polycrystals is inherently related to the properties of the grain boundaries. As has been discussed in Chapter 5, grain boundaries differ in their structure and properties. Following the recognition of different types of grain boundary, various types of grains can be defined. In the literature the following categories of grains have been discussed:

a) grains bounded by general boundaries—called grains;
b) grains that are bounded by at least one low-angle grain boundary—called subgrains;
c) grains that are bounded by at least one twin boundary—called twins.

The division of grain boundary populations into different types of grain boundary is somewhat arbitrary. One can use as a criterion, the misorientation angle of the grain boundary and then the boundaries with a misorientation less than 10° are typically classified as low angle or subboundaries. On the other hand, the distinction of twin boundaries is usually based on observations of their morphology. The situation is even less clear in the case of the classification of grains since their geometry is defined by a set of grain boundaries, which, on average, contains 14 elements. Further, various definitions for subgrains can be employed that might differ in the critical value for grain boundary misorientation angle as well for the required number of subboundaries per grain.

An important implication of the differences in the properties of grain boundaries is that some of them may not be revealed systematically in the image. A classical example of such a situation is images of grain boundary networks on polished and etched sections of polycrystals. By properly setting the etching conditions it is possible to selectively etch the grain boundaries with specified properties such as: energy higher than a certain threshold value or grain boundaries of a specified chemistry. An example of this shown in Figure 6.24.

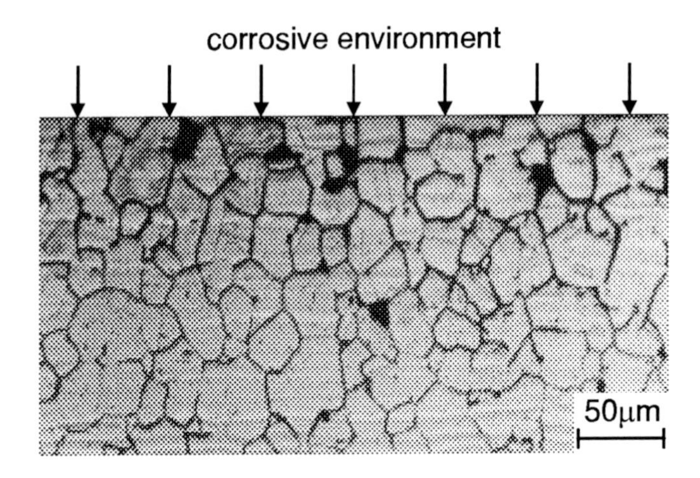

Figure 6.24 An example of a selective etching of grain boundaries of specified properties: corrosion-sensitive grain boundaries in an austenitic stainless steel.

Figure 6.23 shows a series of microstructures obtained from the initial structure in the process of eliminating of grain boundaries with a misorientation smaller than θ_o for θ_o of 5° and 10°. Data from this study is given in Table 6.9, which lists the mean values of the grain area, E(A), intercept length, E(l), and coefficients of variations, CV(A) and CV(l). The microstructures shown in Figure 6.23 contain two characteristic microstructural elements:

a) ghost grain boundaries (marked A);
b) incomplete boundaries (marked B).

Figure 6.25 shows a microstructure revealed in a micrograined austenitic stainless steel sample in the as-recrystallized state. This micrograph illustrates that elements A and B are features of the structures of recrystallized metals.

The data in Table 6.9 show that by discriminating some of the boundaries in the grain boundary network for a given polycrystal, various estimates of the mean grain size can be obtained (as indicated by the E(A) values), as well as different degrees of grain size uniformity (measured in terms of CV(A)).

Selective etching, or more generally imaging, of the grain boundaries changes the observed geometry of grains. Grain sections can be found with grain boundary segments ending inside the grains as well as with two-fold grain edges, which contradicts the commonly accepted views of grain geometry described in the preceding sections. A further consequence of the selective imaging of grain boundaries is that the same microstructure may yield different values of grain shape and size.

Table 6.9 Quantitative Description of the Geometrical Features of Simulated Grain Structures as a Function of the Critical Misorientation Angle θ_0

	Misorientation θ_0 [°]	E[A]	CV(A)	E(l)	CV(l)	N
Microstructure simulated for the assumption of constant rate of grain nucleation	0	1.0	0.82	1.0	0.71	205
	1	1.05	0.64	1.05	0.69	200
	3	1.24	0.87	1.01	0.71	161
	5	1.36	0.99	1.13	0.72	141
	10	1.78	1.18	1.22	0.75	86
Microstructure for the case of grains nucleated at the same time	0	1.0	0.52	1.0	0.66	188
	1	1.07	0.58	1.05	0.65	173
	3	1.25	0.73	1.08	0.69	144
	5	1.48	0.91	1.10	0.71	116
	10	2.09	1.29	1.25	0.72	75

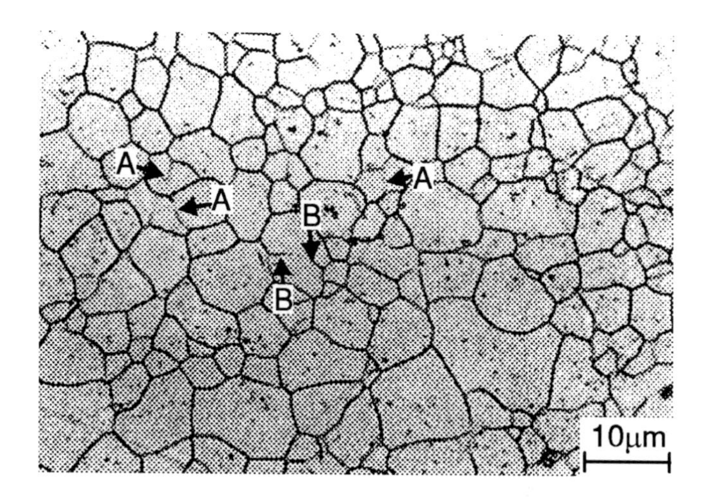

Figure 6.25 Grain boundary networks in recrystallized fine-grained austenitic stainless steel (arrows point to the segments of "ghost" (A) and incomplete (B) grain boundaries.

The problem of selective imaging can be solved by matching the imaging technique with the definition of grains or grain boundaries relevant to the investigation carried out. In other words, the technique used to obtain images of the grain boundaries should, as a rule, reveal all the boundaries of interest. Now, whether or not it should reveal subboundaries, or twin boundaries, will depend on the underlying model for the properties of the material and sometimes the answer is still a matter of controversy. For example, arguments have been given for and against twin boundaries being included in measurements of the grain size. It may be suggested that such boundaries are important elements in some cases of plastic deformation (the large strain regime) and can be neglected in others (the early stages of plastic deformation). However, if the

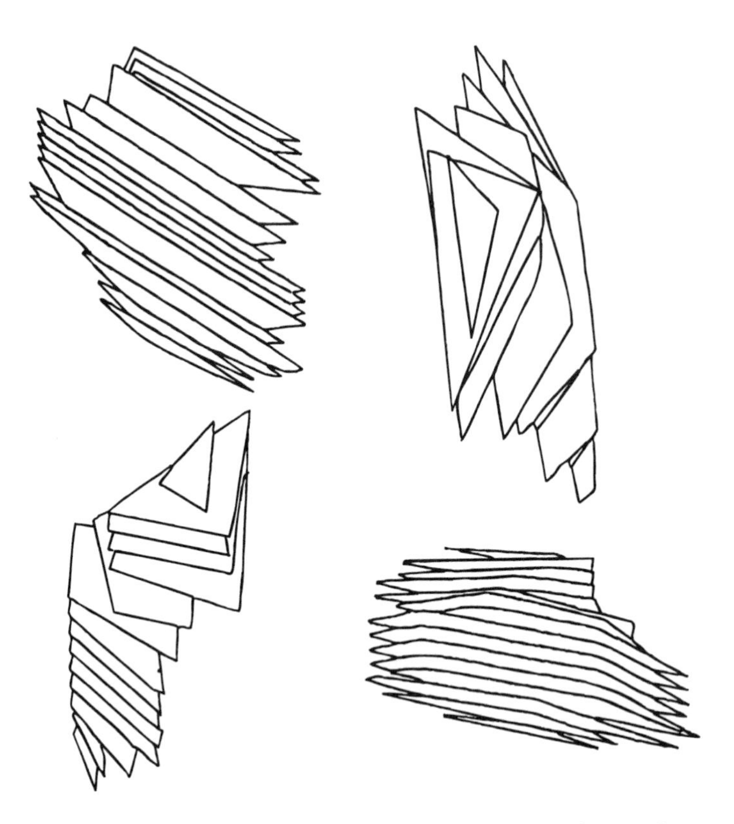

Figure 6.26 Geometry of twin grains in an FCC polycrystal.

twin boundaries are accepted as an element of the grain boundary population, then the description given earlier of grain geometry needs to be modified.

The effect of grain boundary properties on the geometry of grain boundaries is well documented for twin boundaries, which have much smaller energies than the rest of the grain boundary population. Studies using the serial sectioning method have shown that twin boundaries divide general grains into elements (twin grains) of complex geometry. Examples of twin grain geometry are given in Figure 6.26. The analysis of twin grain shape in an FCC alloy lead to the recognition of two twin types:

a) lamellae twins (LTs);
b) edge twins (ETs).

Both twin types occasionally show a step-like structure.

Much smaller differences in energy between grain boundaries, which differ in their microstructure less dramatically, may also have an impact on the geometry of grains as reflected, for example, by the dihedral angles between grain boundaries. It was shown by Smith[8] in the 1940s that dihedral angles may provide useful information on the relative energies of internal surfaces, meeting along a common edge. Since then this idea has been used to develop different techniques to measure the energy of interphase boundaries (see, for example, Reference 33). It has been shown[13] that the population of dihedral angles in a polycrystal changes in a specific way during annealing at high temperatures. Furthermore, Murr[34] argued that at high temperatures the equilibrium configuration is achieved. It has been suggested,[13] therefore, that the changes in the dihedral angle distribution may be used to infer changes in the grain boundary energy distribution function in polycrystals.

A model has been developed for studies of the effect of grain boundary energy effect on the dihedral angles. Three grain boundaries, π_1, π_2, and π_3, meeting along a common edge are assumed to make angles α_1, α_2, and α_3, which are functions of the boundary energies γ_1, γ_2, and γ_3:

$$\alpha_i = f(\gamma_i) \tag{6.51}$$

For an equilibrium configuration the α_i angles can be obtained from the equation:

$$\frac{\gamma_i}{\sin \alpha_i} = const \tag{6.52}$$

The planar dihedral angles β_1, β_2, and β_3 measured on a cross-section cutting the common edge at point O depend on:

a) the values of α_1, α_2, and α_3;
b) the orientation of the edge with respect to the normal to the section (see Figure 6.9).

The relationships were analyzed numerically (details can be found in Reference 33). In the computations it has been assumed that the values of γ are distributed in a normal way around the mean value $E(\gamma)$. The coefficient of variation, described by the value of the ratio $SD(\gamma)/E(\gamma)$, was assumed to be in the range from 0 to 0.4. The results are given in Table 6.10.

Comparing the values of the theoretical estimate with those obtained experimentally, it can be concluded that there is a general agreement between the experimental results and the theoretical values of $SD(\beta)$ obtained for $SD(\gamma)/E(\gamma)$ in the range from 0 to about 0.25.

Table 6.10 Values of $SD(\beta)$ as a Function of $SD(\gamma)/E(\gamma)$ Calculated on the Assumption of a Normal Distribution for the Grain Boundary Energy

$SD(\beta)$ [°]	22.4	22.0	23.8	28.7	33.7	37.5
$SD(\gamma)/E(\gamma)$	0.0	0.02	0.1	0.2	0.3	0.4

References

1. Hatherly, M. and Hutckinson, W. B., *An Introduction to Textures in Metals*, The Institute of Metals, London, Monograph No.5, 1980.
2. Hirth, J. P. and Lothe, J., *Theory of Dislocations*, Johns Wiley and Sons, New York, 2nd ed., 1982.
3. Van Bueren, H. G., *Imperfections in Crystals*, North Holland, Amsterdam, 2nd ed., 1961.
4. Rhines, F. N., *Microstructology, Behavior and Microstructure of Materials*, Rieder-Verlag, Stuttgart, 1986.
5. Cwajna, J., Maliński, M., and Richter, J., Grain-size of hardened high speed steels and nonledeburitic tool alloys, *Acta Stereologica*, 1992 11/Suppl I: 445, Proc. of 8 ICS Irvine, CA, 1991.
6. Rhines, F. N., Craig, K. R., and DeHoff, R. T., Mechanism of steady-state grain growth in aluminum, *Metallurgical Transactions*, 5, 413, 1974.
7. DeHoff, R. T., Quantitative serial sectioning analysis—Preview, *Journal of Microscopy*, 131, 259, 1983.
8. Smith, C. S., Grains, phases and interfaces, an interpretation of microstructure, *Transactions of the Metallurgical Society AIME*, 175, 15, 1948.
9. Liu, G., Yu, H., and Li, W., Efficient and unbiased evaluation of size and topology of space-filling grains, to appear in *Acta Stereologica*.
10. Underwood, E. E., *Quantitative Stereology*, Addison Wesley, Massachusetts, 1970.
11. Kurzydłowski, K. J., McTaggart, K. J., and Tangri, K., On the correlation of grain geometry changes during deformation at various temperatures to the operative deformation modes, *Philosophical Magazine A*, 61, 61, 1990.
12. Rhines, F. N. and Patterson, B. R., Effect of the degree of prior cold work on the grain volume distribution and the rate of grain growth of recrystallized aluminum, *Metallurgical Transactions A*, 13A, 985, 1982.
13. Kurzydłowski, K. J. and Przetakiewicz, W., Transformation of grain boundary structure inferred from the changes in the dihedral angles distribution in $NiMn_2$, *Scripta Metallurgica*, 22, 299, 1988.
14. Liu, G., Applied stereology in materials science engineering, *Journal of Microscopy*, 171, 57, 1993.
15. Okazaki, K. and Conrad, H., Grain size distribution in recrystallized alpha-titanium, *Transactions JIM*, 13, 198, 1972.
16. Conrad, H., Swintowski, M., and Mannan, S. L., Effect of cold work on recrystallization behavior and grain size distribution in titanium, *Metallurgical Transactions*, 16A, 703, 1985.
17. Bucki, J. J. and Kurzydłowski, K. J., Analysis of the effect of grain-size uniformity on the flow-stress of polycrystals, *Materials Characterization*, 29, 365, 1992.
18. Jensen, E. B. and Gundersen, H. J. G., The stereological estimation of movements of particle-volume, *Journal of Applied Probability*, 22, 82, 1985.
19. Jensen, E. B. and Sorensen, F. B., A note on stereological estimation of the volume-weighted 2nd moment of particle-volume, *Journal of Microscopy*, 164, 21, 1991.

20. Bucki, J. J. and Kurzydłowski, K. J., Measurements of grain volume distribution parameters in polycrystals characterized by a log-normal distribution function, *Scripta Metallurgica et Materialia*, 28, 689, 1993.
21. Nunez, C. and Domingo, S., Grain shape and its influence on the experimental-measurement of grain-size, *Metallurgical Transactions A*, 19A, 933, 1988.
22. Takayama, Y., Furushiro, N., Tozawa, T., Kato, H., and Hori, S., A significant method for estimation of the grain-size of polycrystalline materials, *Materials Transactions JIM*, 32, 214, 1991.
23. Kurzydłowski, K. J. and Bucki, J. J., A method for grain size and grain size uniformity estimation—applications to polycrystalline materials, *Scripta Metallurgica et Materialia*, 27, 117, 1992.
24. Armstrong, R. W., The yield and flow stress dependence on polycrystal grain size, *Yield, Flow and Fracture of Polycrystals*, Baker, T. N., Applied Science Publishers, London, 1983, 1.
25. Petch, N. J., The cleavage strength of polycrystals, *Journal of Iron Steel Institute*, 174, 25, 1953.
26. Armstrong, R. W., Codd, I., Douthwaite, R. M., and Petch, N. J., Plastic deformation of polycrystalline aggregates, *Philosophical Magazine*, 7, 45, 1962.
27. Wyrzykowski, J. W. and Grabski, M. W., The Hall-Petch relation in aluminum and its dependence on the grain-boundary structure, *Philosophical Magazine*, 53, 505, 1986.
28. Takeyama, M. and Liu, C. T., Effect of grain-size on yield strength of Ni3Al and other alloys, *Journal of Materials Research*, 3, 665, 1988.
29. Dybiec, H. and Korbel, A., The mechanism of superplastic flow in an Al-Mg alloy, *Materials Science and Engineering*, A117, L31, 1989.
30. Kurzydłowski, K. J., A model for the flow-stress dependence on the distribution of grain-size in polycrystals, *Scripta Metallurgica*, 24, 879, 1990.
31. Kurzydłowski, K. J. and Bucki, J. J., Flow stress dependence on the distribution of grain size in polycrystals, *Acta Metallurgica et Materialia*, 41, 3141, 1993.
32. Kurzydłowski, K. J., The effect of nuclei texture on the geometry of grains and disorientation of grain boundaries in recrystallized polycrystals, 6th International Conference on Intragranular and Interphase Boundaries in Materials, Thessaloniki, Greece, June, 21–26, 1992, *Materials Science Forum*, 126–128, 423–426.
33. Kurzydłowski, K. J., A model for the dependence of dihedral angle distribution on the distribution of grain boundary energy in polycrystalline materials, *Materials Characterization*, 26, 57, 1991.
34. Murr, L. E., Energetics of grain-boundary triple junctions and corner-twinned junctions: transmission electron microscope studies, *Journal of Applied Physics*, 39, 5557, 1968.

chapter seven

Changes in the geometry of grains and grain boundaries under the action of external and internal factors

The microstructure of materials shows dynamical properties in the sense that it changes at a specific rate under the action of external factors, such as applied load, or those changes driven by a tendency to lower the internal energy associated with microstructural elements. In the present chapter, two examples of such evolutions in the microstructure of polycrystals are presented and analyzed. These are changes in the geometry of grains and grain boundaries as a result of plastic deformation and grain boundary migration activated by an elevated temperature.

Plastic deformation is a process that results in a permanent change of shape of an artifact (specimen). The process is defined in terms of three displacement functions:

$$\begin{aligned} u_1 &= u_1\left(x_1, x_2, x_3\right) \\ u_2 &= u_2\left(x_1, x_2, x_3\right) \\ u_3 &= u_3\left(x_1, x_2, x_3\right) \end{aligned} \tag{7.1}$$

which define the displacement at a point with coordinates (x_1, x_2, x_3) along the OX_1, OX_2, OX_3 axes.

If the displacements are large with respect to the dimensions of the body, it is convenient to consider them to be a sum of small increments du_i:

$$u_i = \sum du_i \tag{7.2}$$

Such an approach in most cases is sufficient to avoid complications of nonlinear effects in the description of the deformation.

Since, in general, the displacement field contains a component that is related to translation of the body, the effect of plastic deformation is more precisely described by plastic strain components, ε_{ij}. These components define changes in the shape of volumetric elements of the microstructure in terms of variations in their linear dimensions and change in angles of specified directions in the reference space. In the limit of small increments the components of ε_{ij} are given as:

$$\delta\epsilon_{ij} = \frac{1}{2}\left(\frac{du_i}{dx_j} + \frac{du_j}{dx_i}\right) \tag{7.3}$$

Plastic deformation of materials induced by applied forces is accommodated by a variety of mechanisms operating in the material. These mechanisms are microstructure dependent in the sense that they involve processes and microstructural elements characteristic of the material, such as dislocation slip in crystals and chain molecule rearrangements in polymers. On the other hand, the accommodated plastic deformation changes the microstructure of the material by changing the number and arrangement of the microstructural elements. Plastic deformation of materials is therefore a complex phenomenon that can be analyzed at different levels of precision: on the macro-, mezo-, and microscopic scales.

In a macroscopic approach the deformed artifact is treated as a totally homogeneous body or, in other words, the material is assumed not to contain any specific microstructural elements pertinent to the process of plastic deformation. In this approach the displacements and plastic strains are "inherently uniform" and depend only on the applied load. They are designated in the rest of this text by the superscript "mac", for example, $(\varepsilon_{ij})^{mac}$. Since the macroscopic strain depends on the applied load and shape of the artifact and is not related to its microstructure, $(\varepsilon_{ij})^{mac}$ can be used to characterize the external conditions of the deformation imposed.

Macroscopic strains present an exact description of a plastic deformation process only in the case of a few "microstructure-less" materials such as viscous liquids. Most materials accommodate plastic deformation in an inherently nonuniform, localized way. Depending on the details of the microstructure of a material, the plastic strain can be specifically distributed on a mezo- or microscopic scale.

The deformation of polycrystals is an example which defies the simple concept of macroscopic scale. Polycrystals are aggregates of a large number of grains that are distinguished by different plastic properties and it is logical to assume that on a finer scale their deformation is nonuniform. In order to describe the distribution of plastic deformation in polycrystals, it is convenient to consider an aggregate of N nonoverlapping elements of the volume dV_i. Depending on the deformation conditions and the size of these elements they can be assumed to be fractions of grains, grains, or groups of grains.

The total deformation of a polycrystalline specimen results in a specific deformation, ε_{ij}, accommodated in the elements dV_i. There are in general two specific cases of the relationship between the macroscopic deformation ε_{ij}^{mac} and the local deformation $\varepsilon_{ij}(dV_i)$ accommodated in the elements:

1. uniform deformation characterized by compatible deformation of the specimen and of all of the elements, i.e, by the condition:

$$\underset{k}{\wedge} \epsilon_{ij}\left(dV_k\right) \approx \epsilon_{ij}^{mac} \tag{7.4}$$

where the symbol $\wedge$ means for any integer $k \leq N$;

2. nonuniform deformation with specific elements (dV_i^*) that, disregarding their positions within the specimen, systematically absorb different deformation levels than the other elements:

$$\epsilon_{ji}\left(dV_i^*\right) \neq \epsilon_{ij}\left(dV_k\right) \tag{7.5}$$

The nonuniform distribution can be further broken down into more specific situations such as:

a) distributed "hard and soft" elements;
b) statistically compatible deformation;
c) deformation bands (of various types);
d) deformation localized to interfaces.

In a number of materials it is possible to distinguish those elements of their microstructure that show systematically different resistances to plastic deformation. They are termed here hard and soft elements and designated dV_i^h and dV_i^s respectively. For any soft elements dV_k^s ($k = 1, 2, 3, \ldots, n < N$) and any hard elements dV_l^h ($l = 1, 2, 3, \ldots, m < N$):

$$\epsilon_{ij}\left(dV_k^s\right) > \epsilon_{ij}\left(dV_l^h\right) \tag{7.6}$$

In an extreme case the hard elements may remain undeformed:

$$e_{ij}\left(dV_l^h\right) = 0 \tag{7.7}$$

This is accompanied by a strain concentration effect in the soft elements, which absorb plastic deformation to a greater extent than the macroscopic plastic strain:

$$\epsilon_{ij}\left(dV_k^s\right) \geq \kappa \epsilon_{ij}^{mac} \tag{7.8}$$

where κ is termed strain concentration constant, and is larger than 1.0.

The hard and soft elements situation applies to a number of two-phase materials including composites. An example can also be given of plastic deformation of a single-phase polycrystalline aggregate with the plastic strain nonuniformly apportioned to the larger grains. This nonuniformity is of special importance in the early stages of plastic deformation when the resistance to plastic deformation can be controlled by the fraction of large grains in the volume of the polycrystal.

However, it should be noted that, in general, plastic deformation of volumetric elements in the specimen volume is not expected to differ too much by comparison to the deformation of their neighbors, as otherwise it would lead to material decohesion. On this basis, the distribution of plastic strains among the soft and hard elements of the material is anticipated to be statistically compatible, in the sense that the differences in the strain accommodated in different elements does not exceed some critical value ε_o:

$$\sum_i \sum_j \left| \epsilon_{ij}\left(dV_k^s\right) - \epsilon_{ij}^{mac}\left(dV_l^h\right) \right| \leq \epsilon_o \tag{7.9}$$

In particular, the constant ε_o can be defined as a fraction of the macroscopic plastic strain. In this case the statistically compatible deformation requires the following:

$$\left| \epsilon_{ij}\left(dV_k^s\right) - \epsilon_{ij}^{mac} \right| < \chi\, \epsilon_{ij}^{mac} \tag{7.10}$$

where χ is a constant.

This situation is expected to be found in the case of large deformations of some polycrystalline metallic materials deformed at low temperatures.

An important exception to the rule of small differences in the deformation of soft and hard elements is the case of a deformation band. In this case, the soft elements are not distributed randomly and are arranged in the form of a band that extends throughout the volume of the specimen dividing it into separate parts. An example in this case would be the so-called Lüders bands observed in some alloys.[1]

Another possibility of a nonuniform deformation in polycrystals is the case of deformation that is absorbed by a system of soft elements that have a significantly smaller thickness than that of the hard elements.[2] In the extreme case, the soft elements can be considered to be reduced to the interfaces between the hard elements. In this situation, as a result of plastic deformation, the hard elements change their positions with respect to their neighbors through the process of sliding. In such a situation, the hard elements dV_l^h absorb little plastic strain:

$$\epsilon_{ij}\left(dV_l^h\right) \approx 0 \tag{7.11}$$

and the deformation is predominantly absorbed in the soft surface elements S_m between the elements:

$$\epsilon_{ij}\left(S_m\right) \gg 0 \tag{7.12}$$

Different basic variants of the distribution of plastic strains are schematically shown in Figure 7.1.

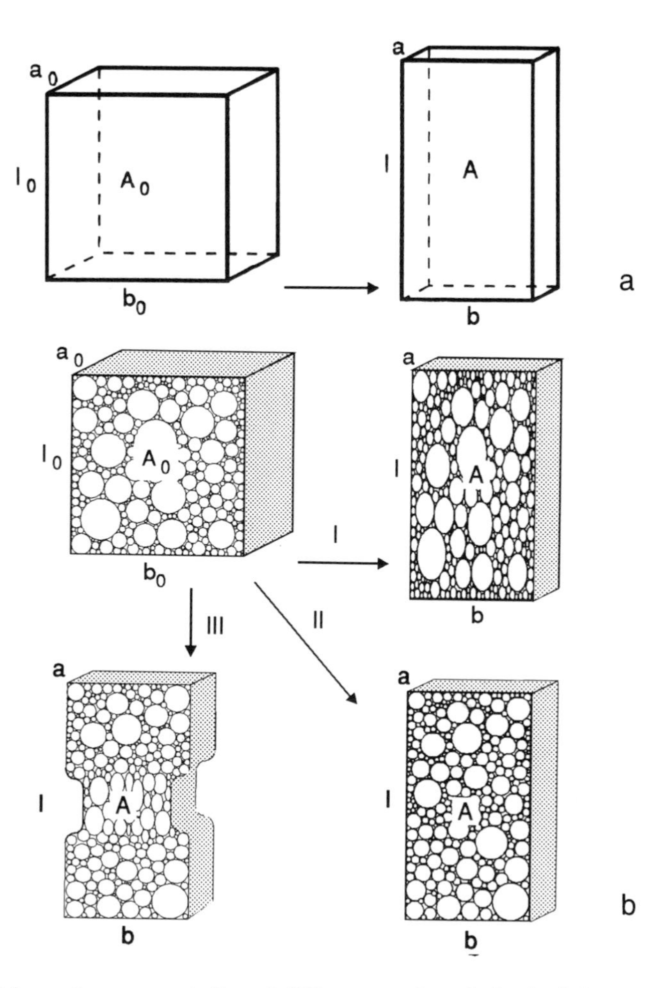

Figure 7.1 Schematic representation of different modes of plastic deformation in a specimen of a complex microstructure: (a) a specimen of the initial dimension $l_0 \times a_0 \times b_0$ is deformed to assume the dimension $l \times a \times b$; (b) the macroscopic change of shape can be brought about, on the microscale level, by:

I—uniform deformation,
II—sliding over the surfaces of the elements shown in the initial structure,
III—a localized deformation band.

7.1 Tensile test

A tensile test is one of the most commonly used methods of examining the mechanical properties of materials.[3,4] It provides information on elastic and plastic characteristics, and on the fracture mode of the material tested. Specimens used for such a test are conventionally flat or cylindrical, although no restriction on the shape of the specimens is imposed. During the test, deformation is accommodated in the gauge section of the specimens whilst their shoulders remained undeformed.

In a macroscopic description, the displacement field in the gauge section is uniform. In the coordinate system for a fixed point in the section the displacement functions are linear and dependent only on the respective spatial coordinates:

$$u_i\left(x_1, x_2, x_3\right) \propto x_i \tag{7.13}$$

Such a displacement field predicts that on a macroscopic scale all sections of the strained specimen remain flat and sections perpendicular and parallel to the axis of straining preserve their orientation.

The displacement along the axis of straining, X_3 in Figure 7.2, is directly controlled by the conditions of the test and is usually measured in terms of the instantaneous gauge section length, l, which is initially equal to l_o. On this basis, the macroscopic linear engineering strain, ε_l, is defined:

$$\epsilon_l = \frac{l - l_o}{l_o} \tag{7.14}$$

The macroscopic true strain of the gauge section in the direction of tension can be obtained by integrating the following expression:

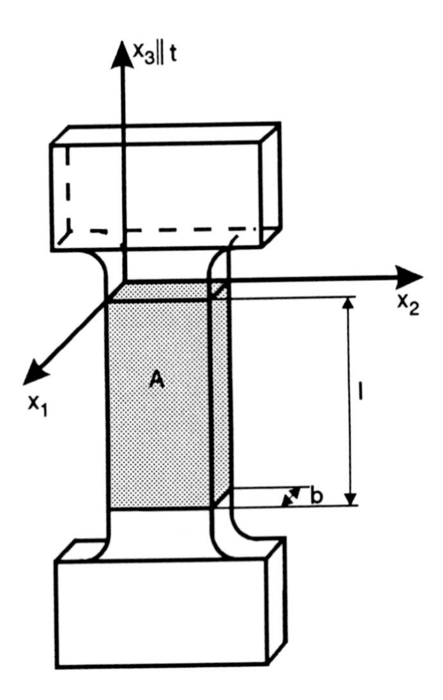

Figure 7.2 Schematic representation of the geometry and the coordinate system adopted in the analysis of tensile deformation.

$$\delta\epsilon_{33}^{mac} = \frac{\delta l}{l} \tag{7.15}$$

This leads to the formula:

$$\epsilon_{33}^{mac} = \ln\left(\frac{l}{l_o}\right) \tag{7.16}$$

The macroscopic linear nominal strain of the specimen differs from the component ε_{33} of the physical strain tensor, which relates the relative displacement to the instantaneous length change in the course of deformation.

For uniaxial straining these two are related by the following formula:

$$\epsilon_{33}^{mac} = \ln\left(1 + \epsilon_l\right) \tag{7.17}$$

Macroscopic strain in directions perpendicular to the axis of tension, ε_{22} and ε_{11}, cannot be treated in the same way as they are not controlled by the test conditions. The contraction of the specimen cross-sections, or in other words, its reaction to imposed macroscopic elongation, depends on the anisotropy of its plastic properties. These two components are equal for isotropic materials:

$$\epsilon_{22}^{is} = \epsilon_{11}^{is} = -\frac{1}{2}\epsilon_{33} \tag{7.18}$$

For anisotropic materials these two components may differ significantly with the smaller approaching zero and the larger the value of the plastic strain along the axis of straining. This can be described in terms of the anisotropy ratio, f_a, such that:

$$\max\left[\epsilon_{11}, \epsilon_{22}\right] = f_a \min\left[\epsilon_{11}, \epsilon_{22}\right] \tag{7.19}$$

This ratio varies from 1 to infinity.

The strain components in the directions perpendicular to the axis of straining describe changes in the cross-section, A, of a strained specimen. Since, on the macroscopic scale, the volume of material strained usually remains constant, the following general relationship can be derived:

$$\frac{A}{A_o} = \frac{l_o}{l} = \exp\left(-\epsilon_{33}^{mac}\right) \tag{7.20}$$

On this basis it is convenient to define the macroscopic reduction of the specimen transverse section, ε_r:

$$\epsilon_r = \frac{A - A_o}{A_o} = \exp\left(-\epsilon_{33}^{mac}\right) - 1 = \frac{-\epsilon_l}{1 \rightarrow \epsilon_l} \tag{7.21}$$

One of the major consequences of deformation by tensile testing is the fact that the specimen, and possibly the microstructure, acquires specific anisotropy as illustrated in Figure 7.3. Depending on their orientation with respect to the axis of straining some of the specimen sections increase their area while others are reduced in area. The largest reduction occurs for the case of the section perpendicular to the axis of straining; the largest increases in those sections parallel to the straining direction. Also, with the exception of these two special orientations, the sections rotate towards the axis of straining.

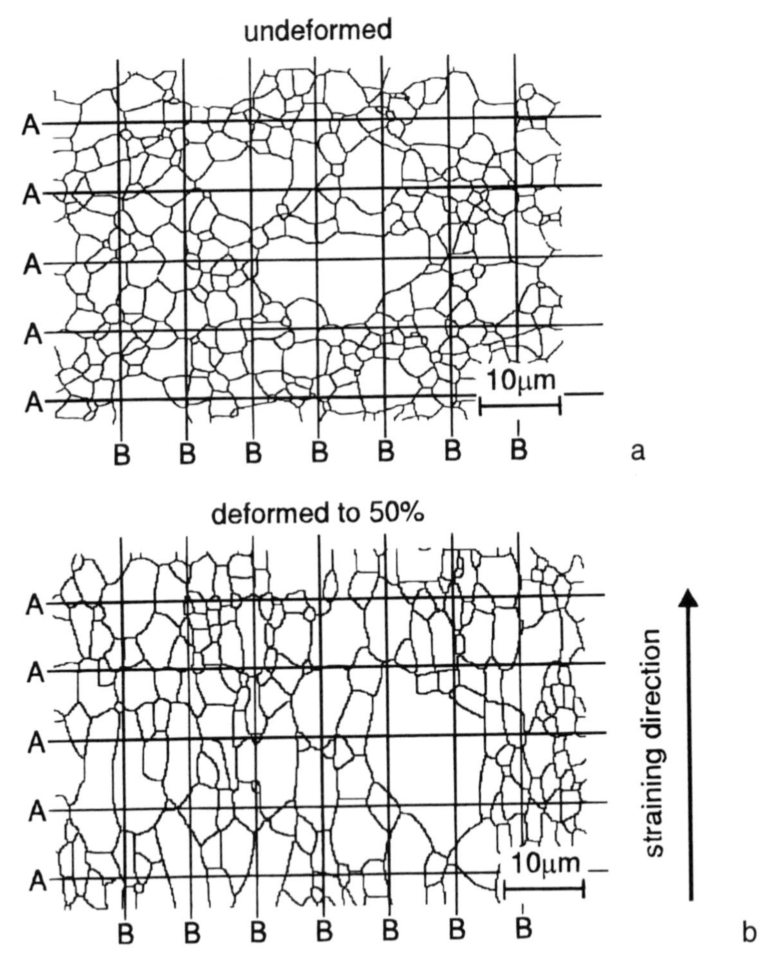

Figure 7.3 System of grain boundaries revealed in: (a) annealed and (b) deformed polycrystal. In both cases the grain boundaries have been revealed on sections parallel to the deformation/specimen axis. Mean intercept length, E(l), was measured along the lines A and B. The following values, $E_A(l)$ and $E_B(l)$, were obtained for the as-annealed polycrystal: $E_A(l) = E_B(l) = 11$ μm.

For the strained specimen the following values were found: $E_A(l) = 9$ μm, $E_B(l) = 12$ μm. The difference in the values $E_A(l)$ and $E_B(l)$ is an indication of the anisotropy of the strained microstructure.

Due to the anisotropy induced by tensile straining, the results of microstructural observations of strained materials usually depend on the orientation of the sections with respect to the axis of straining, **t**. The following orientations of sections are of special importance:

a) longitudinal sections with their normal, **n**, perpendicular to **t**;
b) transverse sections with their normal, **n**, parallel to **t**.

The longitudinal sections can be further specified by the angle their normals make with some reference direction **p** perpendicular to **t**. Thus, a given section used for observations can be described in terms of two angles its normal, **n**, makes with **t** and **p**:

$$\varphi = \measuredangle(\bar{n}, \bar{t}) \quad \psi = \measuredangle(\bar{n}, \bar{p}) \tag{7.22}$$

These two angles can be used to define uniquely a given section of the strained material from the point of view of the macroscopic description of the deformation.

7.2 Changes in the geometry of grains in strained polycrystals—uniform deformation approximation

Within the framework of the uniform distribution of plastic strain approach, each subelement of the material, also each grain, absorbs the same amount of deformation. This means that each grain elongates by ε_l along the same direction in the reference system, X_i. As a result, the grains gradually become elongated in the direction **t** and the grain structure develops a textured grain shape (the grain geometry becomes of anisotropic character). This effect can be analyzed quantitatively with some more specific assumptions regarding the shape of grains, or numerically studied for the case of a stochastic grain geometry.

7.3 Spherical grains approach

Changes in the geometry of equiaxed grains as a result of uniaxial tension can be approximated by an evolution in the geometry of deformed spheres. The process of elongation changes a sphere of radius R into an ellipsoid (Figure 7.4). Due to the applied deformation by uniaxial straining, a spherical grain of radius R is elongated along the straining axis **t** by ε_l while its dimensions in the direction perpendicular to **t** reduce. Let us assume that the direction of the maximum reduction is defined by unit vector **p** perpendicular to **t**. In this case the dimensions of the body, initially equal to R, change in the following way:

a) along **t**: R into $R(1 + \varepsilon_l)$,
b) along **p**: R into $R(1 - \varepsilon_p)$
c) along $\mathbf{p} \times \mathbf{t}$ (in the direction perpendicular to **t** and **p**)
R into $R(1 - \varepsilon_m)$
where $\varepsilon_p \geq \varepsilon_m \geq 0$.

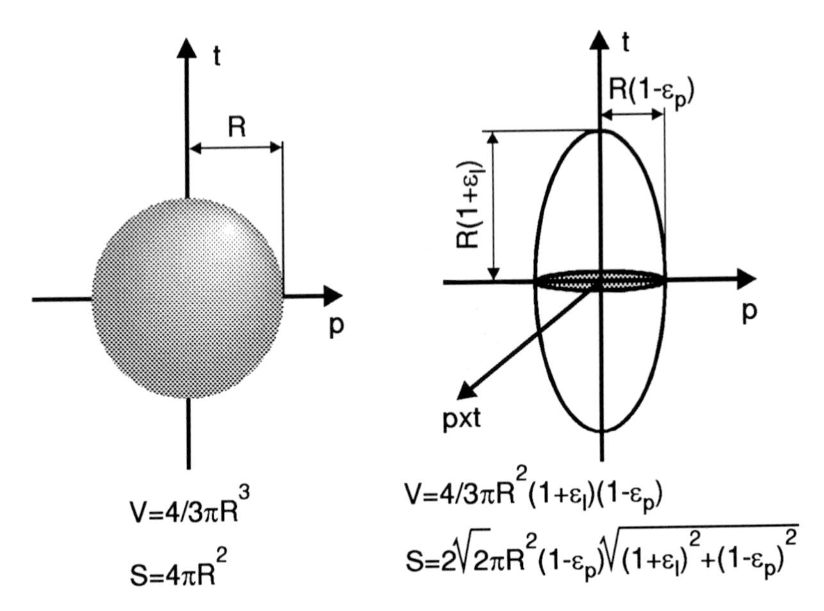

Figure 7.4 Schematic description of the changes in the geometry of spheres of an isotropic material deformed by uniaxial straining.

The values of ε_p and ε_m depend on the degree of anisotropy in reduction in the two directions perpendicular to the direction of straining. If $\varepsilon_m = 0$, this means in one direction the dimensions remain preserved, then the condition of constant volume requires that:

$$\epsilon_l = \frac{\epsilon_p}{1-\epsilon_p} \tag{7.23}$$

On the other hand, for the case of isotropy in the directions perpendicular to the axis of straining (axial symmetry with $\varepsilon_p = \varepsilon_m$) the following formula is valid:

$$\epsilon_p = \epsilon_m = \frac{\sqrt{1+\epsilon_l}-1}{\sqrt{1+\epsilon_l}} \tag{7.24}$$

Transformation from a sphere (equiaxed shape) into an ellipsoid of the same volume increases the **surface area** of the grain and surface-to-volume ratio, S_V. In the case of axial symmetry, the ratio of the final to initial surface areas is given by the equation:

$$\frac{S(\epsilon_l)}{S_o} = \frac{1}{2}\left(\frac{1}{1+\epsilon_l} + \sqrt{1+\epsilon_l}\,\frac{1}{c}\sin^{-1}c\right) \tag{7.25}$$

where

$$c = \sqrt{1 - \frac{1}{\left(1+\epsilon_l\right)^3}} \tag{7.26}$$

The function given by Equation 7.25 is plotted in Figure 7.5.

The elongation of grains also changes the area of their sections revealed in the cross-sections of given orientations. In particular, grain sections observed on the cross-section perpendicular to the axis of straining reduce their area. This is a consequence of the fact that homogeneous deformation preserves flatness of the sections and that the number of grains cut by the cross-sections must remain constant. The mean grain section area, $E_p(A)$, of grains cut by a perpendicular cross-section is a function of the applied elongation, ε_l. The ratio $[E_p(A)/E_p(A_o)](\varepsilon_l)$, (where $E_p(A_o)$ is the mean grain section area in undeformed specimens) in the case of a uniform distribution of plastic strain is given by:

$$\left[\frac{E_p(A)}{E_p(A)_o}\right](\epsilon_l) = \frac{1}{1+\epsilon_l} \tag{7.27}$$

The decrease in the area of grain sections in cross-sections perpendicular to **t** is accompanied by an increase in the mean area of grain sections revealed in cross-sections parallel to **t**. In the general case, the changes in area depend on the relative orientation of the section to the **p** axis. For axial symmetry the following formula applies:

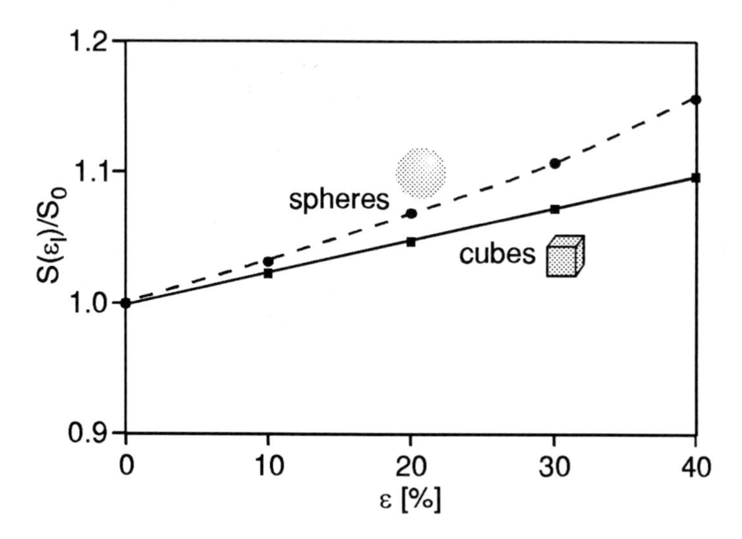

Figure 7.5 A plot of $S(\varepsilon_l)/S_0$ ratio for initially spherical and cubic grains strained uniformly as a function of the elongation ε_l.

$$\left[\frac{E_t(A)}{E_t(A)_o}\right](\epsilon_l) = \sqrt{1+\epsilon_l} \tag{7.28}$$

The same arguments can be used to extend the analysis to the case of the grain section distribution function, f(A). Prior to the deformation, the areas of grain sections revealed in a cross-section of a given orientation form a population defined by the distribution function $f(A)_o$. As a result of straining this function changes into $f(A)_\varepsilon$:

$$f(A)_o \text{ transforms into } f(A)_\varepsilon$$

or shortly:

$$f(A)_\varepsilon = G[f(A)_o].$$

The function $G[f(A)_o]$ depends on the orientation of the cross-section of interest. In particular, grain sections revealed in a longitudinal cross-section of a tensile specimen deformed to an elongation ε_l (measured along the axis of straining) will be extended in a direction parallel to the axis of straining. As the plastic deformation of the tensile specimen to a strain ε_l will be accompanied by a contraction, say ε_p, in a transverse direction, the grains will also decrease their dimensions in the transverse direction. Since in the case of homogeneous deformation the sections remain flat, the area of the section of grain i, A_i, assumes an area $A_i(\varepsilon_l)$ given by:

$$A_i(\epsilon_l) = A_i\left[1+(\epsilon_l)\right]\left[1-(\epsilon_p)\right] \tag{7.29}$$

This is a linear function of A_i, which implies that the transformation of the distribution function discussed is essentially a rescaling of the variable A. Consequently, the normalized grain area distribution f(A/E[A]) determined on undeformed and deformed cross-sections will remain unchanged, as will the higher-order moments of the normalized distribution function such as the standard deviation, SD, skewness, SK, and kurtosis, K. It has been mentioned already that in some applications it is more convenient to use as a variable the logarithms of grain sections areas, ln(A), instead of A. In this case the population of grains revealed in a cross-section is described by the distribution function f[ln(A)]. By taking logarithms of two sides of the equation that determine the changes in the area of a grain section, the following formula can be obtained:

$$\ln\left[A_i(\epsilon_l)\right] = \ln\left[A_i\right] + \ln\left[1+(\epsilon_l)\right] + \ln\left[1-(\epsilon_t)\right] \tag{7.30}$$

This means that the distribution function f[ln(A)] for longitudinal cross-sections, as a result of straining, is shifted to the higher values of the ordinate axis without changing the shape of the function.

Example 7.1

Measurements of changes in the distribution of grain area, A, were performed on two commercial batches of a 316L austenitic stainless steel.[5] The material was cold rolled to 90% reduction in thickness. Tensile specimens with a gauge length of 20 mm and a cross-section of 1.1 × 7.0 mm were machined from the rolled plate. Specimens were annealed at temperatures of 900°C and 1000°C. The specimens were strained at room temperature to total elongations ε_l = 0.24 and ε_l = 0.50. The gauge and shoulder sections representative of deformed and as-recrystallized states were mounted and polished using standard techniques. In all cases a layer of material of thickness greater than several tens of grain diameters was removed from the surface during polishing. The changes in the specimen dimensions resulting from the plastic deformations were carefully measured using a micrometer.

Metallographic observations were conducted both with a scanning electron microscope operating with an accelerating potential of 20 kV and with an optical microscope. Photomicrographs were taken randomly of selected fields from each specimen. Each field contained approximately 50 to 100 grains. In order to increase the precision of the measurements, line tracings of the grain boundary networks were prepared. Two-dimensional grain boundary networks characteristic of the recrystallized and deformed specimens are shown in Figure 7.3a and b, respectively.

Plots of the normalized cumulative frequency vs. the grain area divided by the mean grain area are given in Figure 7.6. The values of the mean, standard deviation, skewness, and kurtosis for the distribution function f(lnA) are given in Table 7.1.

It should be noted that the results obtained for the specimens from different batches and annealed at different temperatures show significant degrees of consistency. The initial structures are characterized by the same values of

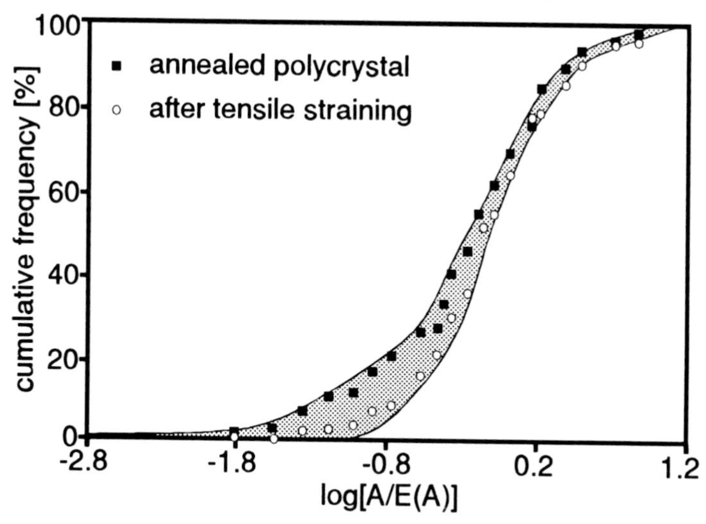

Figure 7.6 Cumulative frequency for the logarithm of normalized grain section area, log[A/E(A)], for an annealed polycrystal and the same polycrystal after tensile straining at a low temperature.

Table 7.1 The Values of the Standard Deviation, Skewness, and Kurtosis for the ln(A) Distribution Functions for the As-Annealed Material and for the Material Deformed at a Low Temperature.

Parameter	$\varepsilon = 0$	$\varepsilon = 0.24$	$\varepsilon = 0.50$
SD[lnA]	1.15	0.97 ± 0.14	1.04 ± 0.14
SK[lnA]	0.55	0.07	0.06
K[lnA]	3.06	3.47	3.00

Note: The measurements of the grain section area were carried out on longitudinal cross-sections of specimens of an austenitic stainless steel strained at room temperature.

SD[lnA] and similar values of higher order moments (SK,K) of the grain area distribution functions. A chi-square test was used to prove that there is good agreement of the experimental distributions with a lognormal distribution.

Comparison of the data for recrystallized and deformed specimens indicates that while there is both an increase in the grain areas and scatter in the data with increasing plastic strain, the size and shape of the logarithm of grain area distribution remains constant.

Grains that are elongated as a result of uniaxial tension change their **shape**. Within the limits of the assumption of a spherical grain shape they transform into ellipsoids. Such a transformation does not change the shape of grain sections on the cross-section perpendicular to the axis of straining. On the other hand, profound changes in the shape of grain sections occur in sections parallel to **t** where grains are expected to transform from circles into ellipses as schematically shown in Figure 7.4. The change in the shape of grain sections is accompanied by variations in the values of shape factors. In particular the ratio $(d_{max})_p/d_2$ varies with the amount of elongation according to the formula:

$$\frac{\left(d_{\max}\right)_p}{\left(d_2\right)} = \left(1 + \epsilon_l\right)^{\frac{3}{4}} \tag{7.31}$$

where $(d_{max})_p$ is the maximum chord length measured in the section parallel to the direction of elongation and d_2 is the equivalent circle diameter.

A similar formula for perimeter p to equivalent diameter d_2 ratio, p/d_2, measured for grain sections revealed in a longitudinal cross-section is given below:

$$\frac{p}{d_2} = \frac{\pi}{\sqrt{2}}\sqrt{\left[\left(1 + \epsilon_l\right)^{\frac{3}{2}} + \left(1 + \epsilon_l\right)^{-\frac{3}{2}}\right]} \tag{7.32}$$

Plots of the functions given by Equations 7.31 and 7.32 are given in Figure 7.7. It can noted that in both cases the change in the shape factors does not depend on the grain size.

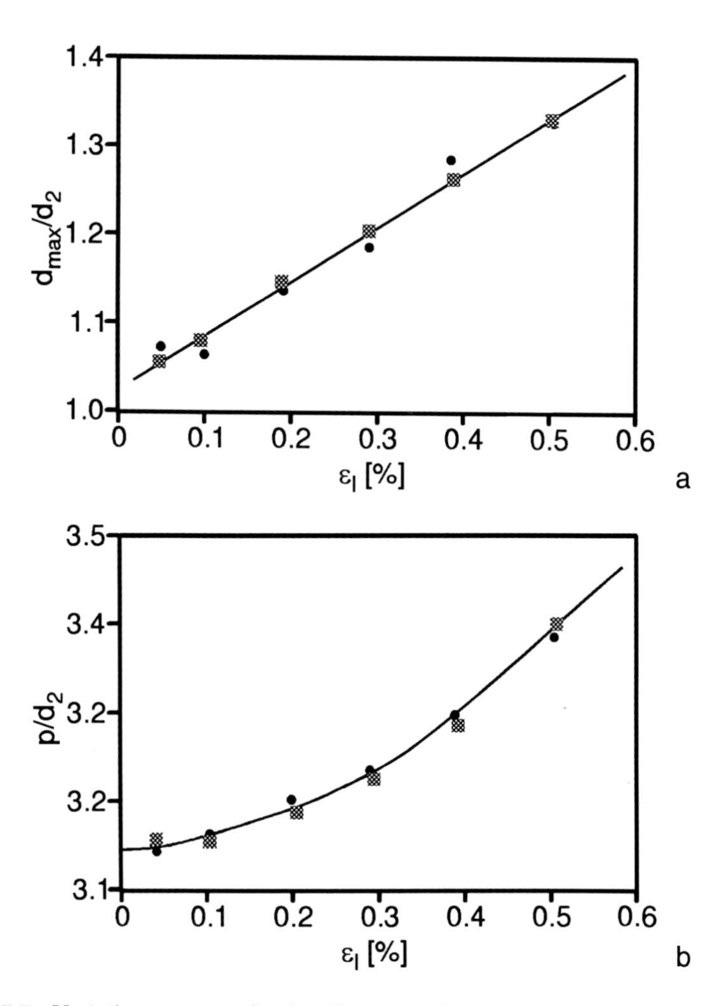

Figure 7.7 Variation, as a result of uniform tensile strain, in the factors describing the shape of grain sections: (a) the shape factor d_{max}/d_2 and (b) the shape factor p/d_2 dependence on the plastic elongation ε_l. The lines represent analytical solutions for spherical grains. The points designated by ● were measured for polycrystals strained at a low temperature; the points designated ▒ were obtained by computer modelling of homogeneous deformation.

In concluding this discussion of the changes in grain shape, it should be noted that simple arguments can be used to show that homogeneous deformation does not change the connectivity of grains and, in particular, the distribution function f(n), which defines the number of sides of a grain section. This is true for any cross-section of a uniformly strained polycrystal.

It has been stated earlier that grain size is an important parameter that has an impact on the properties of the material. Properties of strained polycrystals change significantly as a function of plastic strain. A question may arise as to what extent these changes are contributed to from the variation in the grain size and to what extent they are related to the changes in other microstructural elements, for example, an increase of the dislocation density.

The grain size in polycrystals can be measured using different parameters (see Chapter 6). Depending on the theoretical model for the properties of the polycrystal one can use parameters such as:

a) grain volume, V;
b) grain area, A;
c) intercept length, l.

Since plastic deformation by definition, preserves the volume of the deformed body, it may be concluded that the mean volume of grains, E(V), as well as the volume distribution function remains unchanged during the straining:

$$f_o(V) = f_\varepsilon(V).$$

On the other hand, the other parameters listed above change with the course of deformation in terms of their mean values measured on cross-sections of particular orientations and in terms of the respective distribution functions.

Changes in the area of grain sections have been described already for two particular orientations of the cross-sections: (a) longitudinal and (b) transverse sections. The results of this analysis can be generalized by the equation:

$$f_{\varphi\psi}(A) \xrightarrow{\epsilon} f_{\varphi\psi}(\alpha A) \tag{7.33}$$

with the constant α being either positive or negative depending on the values of the angles defining the orientation of the sections.

For the variation in the intercept length, l, it should be recalled that straining induces a specific anisotropy in the microstructure of a deformed specimen. There are a number of important implications of this.[6] First, intercepts that are measured on sections of different orientations are likely to show different mean values and distribution functions. In particular, if the measurements are performed on longitudinal cross-sections parallel to the elongation axis, the mean intercept length, $E_l(l)$ shows little variation with ε_l:

$$\frac{d}{d\epsilon_l} E_l(l) \approx 0 \tag{7.34}$$

At the same time the mean intercept length measured with test lines placed on perpendicular sections yields the mean intercept length, $E_p(l)$, which decreases with increasing elongation:

$$\frac{d}{d\epsilon_l} E_p(l) < 0 \tag{7.35}$$

In view of a systematic variation in the mean intercept length as a function of the section orientation, the term *mean intercept length* should be understood as the mean over all possible section orientations taken with the same probability. Experimental procedures for taking such sections are, in general, quite complicated. However, the mean intercept length and its variation can be estimated from the formula that links E(l) to S_V, i.e., density of grain area in the volume of a polycrystal. According to the basic relationship:

$$E(l) = \frac{2}{S_V} \tag{7.36}$$

An increase in the S_V value results in a decrease of the mean intercept length. On the other hand, it has been already noted that S_V increases with increasing specimen elongation due to the departure from equiaxial to elongated geometry. It can be concluded, therefore, that the mean intercept length decreases with increasing elongation:

$$\frac{d}{d\epsilon_l} E(l) < 0 \tag{7.37}$$

The function that relates S_V and ε_l can be theoretically determined for some simple shapes of grains. In the case of spherical grains the formula given earlier can be used. The results are plotted in Figure 7.8. An experimental approach to changes in S_V values is discussed in Example 7.2.

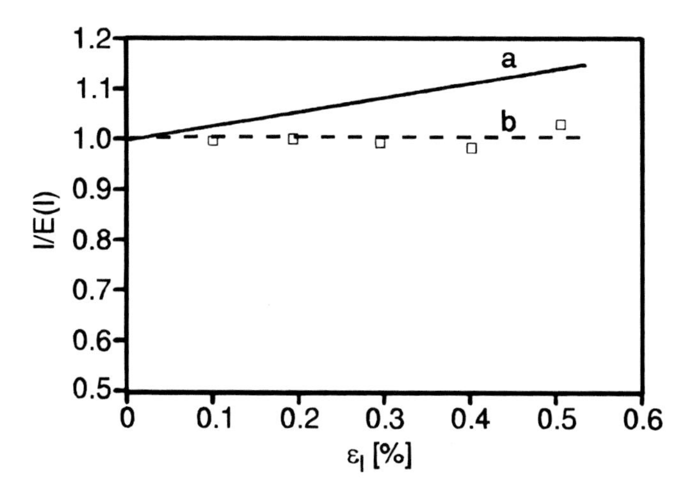

Figure 7.8 Variations in the normalized mean intercept length as a function of plastic elongation, ε_l, in a tensile test. The line (a) represents an estimate based on the assumption of a spherical grain shape and (b) was obtained from numerical analysis. The experimental points were obtained for aluminum.

Example 7.2

From an experimental point of view the easiest way to estimate the mean intercept length is probably via measurements of S_V. To this end one can use the serial sectioning procedure described in Chapter 3. Another possibility is to use a longitudinal section and intercepts parallel and perpendicular to the axis of straining. Consider the microstructures shown in Figure 7.9. By orienting the probing lines along the axis of straining it is possible to estimate the surface to volume ratio of the grain boundaries of a "random" orientation (see Chapter 2), $(S_V)_r$:

$$\left(S_V\right)_r = 2\left(P_L\right)_t \tag{7.38}$$

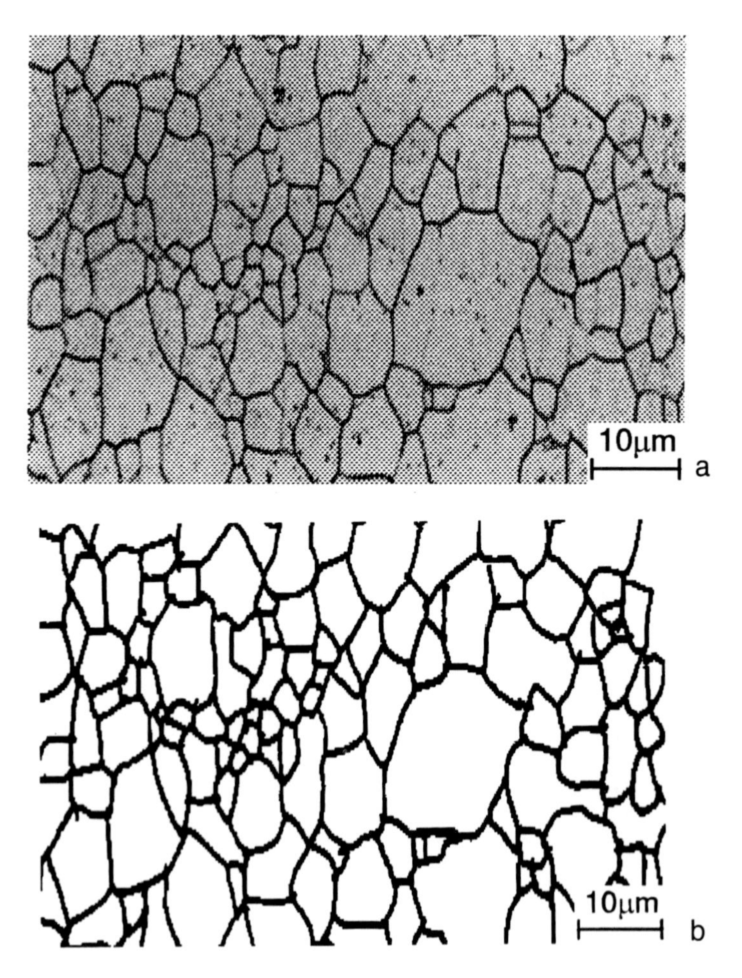

Figure 7.9 An image characteristic of grain boundaries revealed in a longitudinal section of a polycrystal strained at a low temperature (a) and its binary representation (b).

For the microstructure in Figure 7.9 the following values were measured: $2(P_L)_t = (S_V)_r = 0.6\ \mu m^{-1}$.

In the subsequent step the system is probed with lines perpendicular to this axis and the number of intersections per line length $(P_L)_p$ determined ($(P_L)_p = 0.9\ \mu m^{-1}$ for the microstructure studied). The difference between these two numbers of intersections is a measure of the linear anisotropy of the surface orientations. More detailed considerations show that the total surface-to-volume ratio, S_V, is given by the following formula:

$$S_V = \frac{\pi}{2}\left(P_L\right)_p + \left(2 - \frac{\pi}{2}\right)\left(P_L\right)_t \tag{7.39}$$

In the present case Equation 7.39 yields $S_V = 1.7\ \mu m^{-1}$. The dimensionless ratio of the grain boundary anisotropy, which is defined as:

$$\Omega_{fiber} = \frac{\left(P_L\right)_p - \left(P_L\right)_t}{\left(P_L\right)_p + \left(\frac{4}{\pi} - 1\right)\left(P_L\right)_t} \tag{7.40}$$

is in this case equal to 0.6.

Example 7.2 and the preceding discussion underlines the fact that the mean intercept length alters as a result of changes in the grain geometry induced by plastic deformation. These changes in the mean intercept length are unlikely to explain all aspects of the variation in the properties of strained polycrystals. However, they should be accounted for in the models describing this phenomenon.

7.4 Stochastic geometrical modelling

The analysis carried out so far was based on the assumption that the grain shape is approximately spherical. On the other hand, it is known that grains in polycrystals have more complicated shapes and their geometry is of a stochastic nature. The results obtained within the limits of a spherical grain shape model can be used only as a first approximation.

A more precise description of the changes in grain geometry, and other defects, as a function of the homogenous deformation of polycrystals may be obtained if the initial geometry is characterized by distribution functions, $f_o(x)$, of some geometrical parameter, x, such as grain elongation, grain section area, etc. determined for unstrained material.

These distribution functions of geometrical parameters change as the result of the distortion introduced by plastic deformation. A given function, $f_o(x)$, transforms into $f_\varepsilon(x)$:

$$f_o(x)dx \xrightarrow{\epsilon} f_\epsilon(x)dx \tag{7.41}$$

The transformed distribution function, $f_\varepsilon(x)$, is related to the initial one through the function G, such that:

$$f_\epsilon(x) = G\left[f_o(x)\right] \tag{7.42}$$

In general, the analysis of the function G, would prove to be mathematically complicated. However, this analysis may be simplified significantly with the help of computer modelling of the effects of homogeneous deformation.

Example 7.3

Grain structures representative of well-annealed polycrystals were chosen and subjected to measurements of some geometrical parameters. The following parameters were examined:

a) grain area, A;
b) d_{max}/d_2;
c) p/d_2;
d) angle ω of the maximum chord to a reference axis;
e) F_x/F_y;
f) intercept length, l.

Some of these parameters are schematically explained in Figure 7.10. The measurements were carried out with the help of an automatic image analysis system and yielded the initial distribution functions $f_o(x)$ with x being one of the parameters a) to f).

In the next step of the analysis, the initial microstructure was digitized and uniformly deformed by a linear transformation applied to the digitized im-

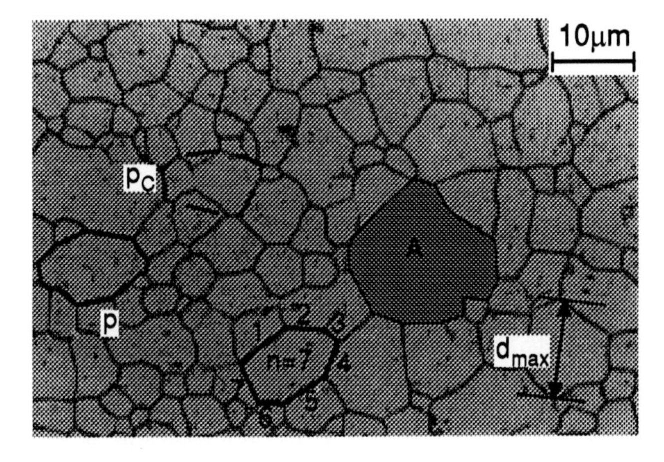

Figure 7.10 Schematic definition of the parameters used to define geometry of grains: A is grain section area, p—perimeter, p_c—Cauchy perimeter, d_{max}—maximum chord, n—number of edges.

ages. In this transformation, each point P(x,y) of the image is transformed into $P_\varepsilon = P'(x', y')$ such that:

$$x' = x(1 + \varepsilon_l),$$
$$y' = y(1 - \varepsilon_t).$$

This transformation can be used to model the effect of tension, along the X axis, in the cross-section parallel to the direction of straining and can readily be modified to describe the deformation in any other section. The microstructures, distorted as a result of the transformation, of digitized images were printed and the measurements of the parameters (a) to (f) performed. However, in this case the measurements yielded the distribution functions for the deformed structures, $f_\varepsilon(x)$. The results of the measurements are illustrated by plots in Figure 7.11 and are summarized in Table 7.2.

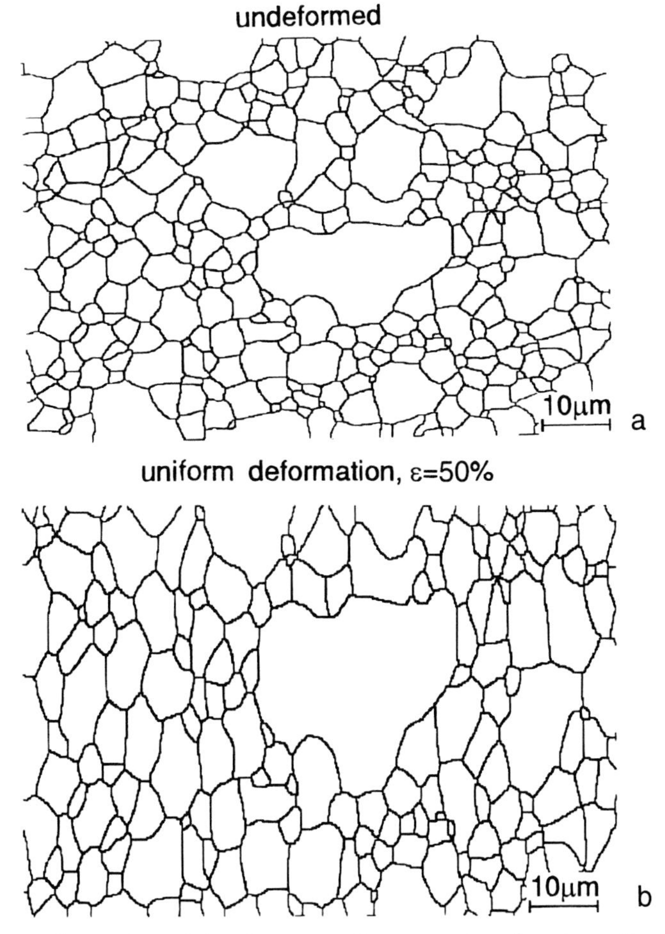

Figure 7.11 Illustration of the changes in the geometry of grain sections in an annealed polycrystal (a) as a result of uniform tensile deformation (b). The microstructure in (a) is representative of an FCC metal; (b) was obtained by deformation of the microstructure exemplified by (a); (c) (on next page) shows the distribution function of d_{max}, $f(d_{max})$, before and after distortion to 50%.

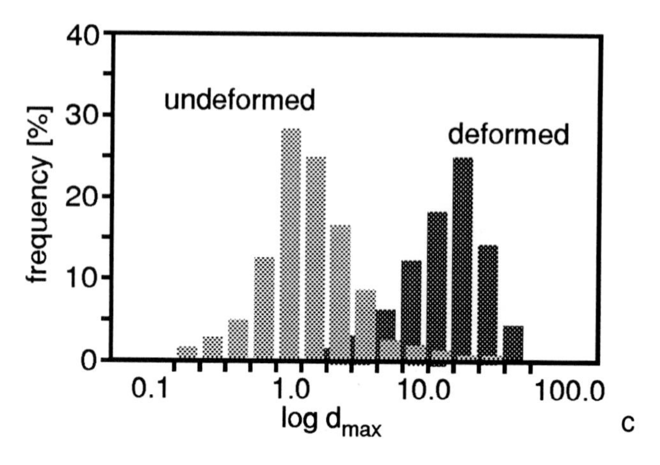

Figure 7.11 (*Continued*)

Table 7.2 Changes in the Computed Values of the Grain Boundary Specific Area, S_V, the Mean Value, E(), of d_{max}/d_2 and p/d_2 for the Model Geometry of Grains in the Case of Macroscopic Deformation of 50%

Parameter	Uniform deformation, ε = 50%	
	Spheres	Stochastic geometry
$S_V/(S_V)_o$	1.24	1.20
$E\left(\frac{d_{max}}{d_2}\right) : E\left(\frac{d_{max}}{d_2}\right)_o$	1.35	1.23
$E\left(\frac{p}{d_2}\right) : E\left(\frac{p}{d_2}\right)_o$	1.09	1.10

7.5 *Statistically compatible deformation*

The statistically compatible deformation of polycrystals can be studied experimentally on polycrystals strained at low temperatures. At a temperature below about 0.2 T_m (T_m—the melting point) a number of metals deform via intragranular slip of dislocations on slip systems of an appropriate orientation with respect to the applied load. As a result, in crystalline structures with five independent slip systems each grain of the aggregate tends to accommodate to a similar plastic strain. The grains deform in a similar but not identical manner. There are some differences in the deformation of individual grains due to the differences in their crystal orientations and variation in their shape and neighbors. Individual grains deform statistically in a manner compatible with the macroscopic straining of the aggregate.

The changes in the distribution functions of geometrical parameters, which take place in a statistically compatible deformation of grains, are exemplified by the results obtained in investigations performed on an austenitic stainless steel strained at a room temperature.

The same parameters were measured as in Example 7.3. The results are summarized in Figure 7.12 and Table 7.3.

7.6 Localized deformation

In the case of localized deformation a strained specimen can be divided broadly into two nonoverlapping and fragmented parts of the volume, V_1 and

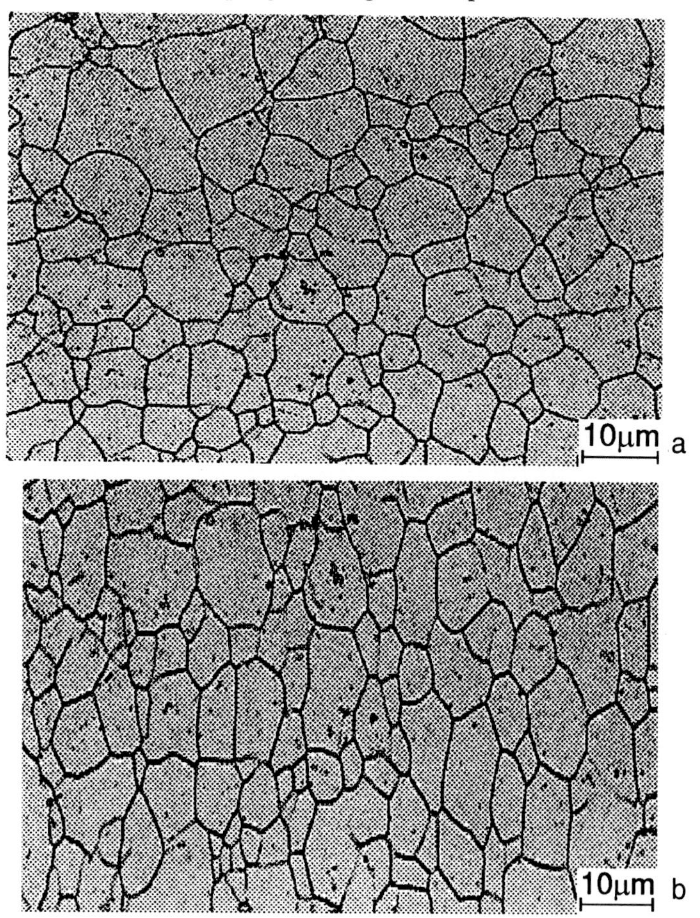

Figure 7.12 Illustration of the changes in the geometry of grain sections in an annealed polycrystal (a) as a result of geometrically compatible tensile deformation (b). The microstructures in (a) and (b) are representative for an austenitic stainless steel in the as-annealed state and after tensile strain at room temperature, respectively; (c) (on next page) shows distribution function of d_{max}, $f(d_{max})$, before and after distortion to 50%.

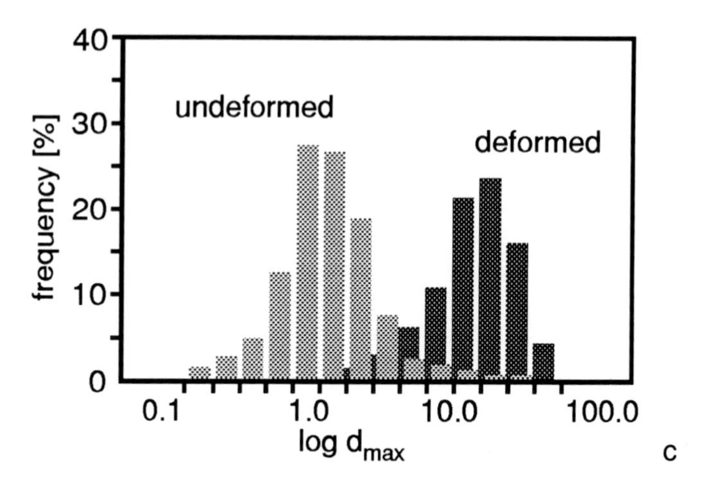

Figure 7.12 (*Continued*)

Table 7.3 Changes in the Measured Values of the Grain Boundary Specific Area, S_V, the Mean Value, E(), of d_{max}/d_2 and p/d_2 for the Geometry of Grains Revealed in a Single Phase Polycrystal Strained at a Low Temperature to a Macroscopic Deformation of 50%

Parameter	Statistically compatible
$S_V/(S_V)_o$	1.15
$E\left(\frac{d_{max}}{d_2}\right):E\left(\frac{d_{max}}{d_2}\right)_o$	1.10
$E\left(\frac{p}{d_2}\right):E\left(\frac{p}{d_2}\right)_o$	1.03

V_2. One of these, say V_1, does not absorb any plastic strain or absorbs a significantly smaller amount than the second part, V_2. From this definition of the two regions the analysis of a plastically strained specimen can be effected, in general along the same lines previously given; with the exception that the results are applied to the deformed volume of the material, V_2. One may assume for example a uniform or statistically compatible distribution of the plastic strain within V_2 and use the results of the preceding sections. However, it should be noted, that the macroscopic strain for the volume V_2 differs from the macroscopic strain for the specimen as schematically explained in Figure 7.1. There is an effect of strain magnification and the following relationship can be used to estimate the macroscopic strain ε^{mac} in V_2, $\varepsilon^{mac}(V_2)$:

$$\epsilon^{mac}\left(V_2\right) = \epsilon^{mac}\,\frac{V_2}{V_1 + V_2} \qquad (7.43)$$

For the case of a small deforming volume within the specimen, the strain accommodated might be much higher than the macroscopic strain of the specimen.

Plastic strain localization is accompanied by a specific deformation in the regions adjacent to the one absorbing the strain. This effect has received a great deal of interest in the case of grain boundary sliding. Since the grain boundaries do not form a flat surface, sliding over them requires accommodation at the points of change in the grain boundary curvature. This in turn results in deformation of parts of the grain interiors. Such deformation is controlled by the condition of sliding and is expected not to alter the shape of those grains that are equiaxial, as in superplastic deformation.

An example of localized deformation via grain boundary sliding is shown in Figure 7.13 where a micrograph of a fine-grained polycrystal strained to a total elongation of 25% ($\varepsilon_l = 0.25$) is presented. This micrograph is representative of grain sections revealed in the cross-section containing the axis of straining, **t**. It may be noted that, despite a significant macroscopic elongation of the specimen, the grain sections revealed do not show significant elongation. This observation is substantiated by the results of measurements shown in Figure 7.14 and Table 7.4.

7.7 *An application to studies of plastic deformation mechanisms*

As a result of plastic deformation the distribution functions of geometrical features are transformed according to the scheme:

$$f_o(x) \rightarrow f_\varepsilon(x).$$

As has been discussed in the preceding sections, this transformation is a characteristic for plastic deformation processes taking place in the material. Apart from the applied load, which is an "external factor", it depends on:

a) the operating mechanism of plastic deformation;
b) interactions between the grains;
c) the shape and size of a given grain.

On a microscopic scale, plastic deformation of polycrystals involves different processes of straining and recovery. At this level there are a number of physically distinguishable modes of deformation, among them are the following:

a) intragranular dislocation slip (IS);
b) intergranular sliding, sometimes called grain boundary sliding (GBS);
c) diffusional flow (DF).

The way in which strained materials respond to plastic deformation depends on the macroscopic strain induced by the applied load, the temperature, and their microstructure. In a tensile test the resistance to plastic

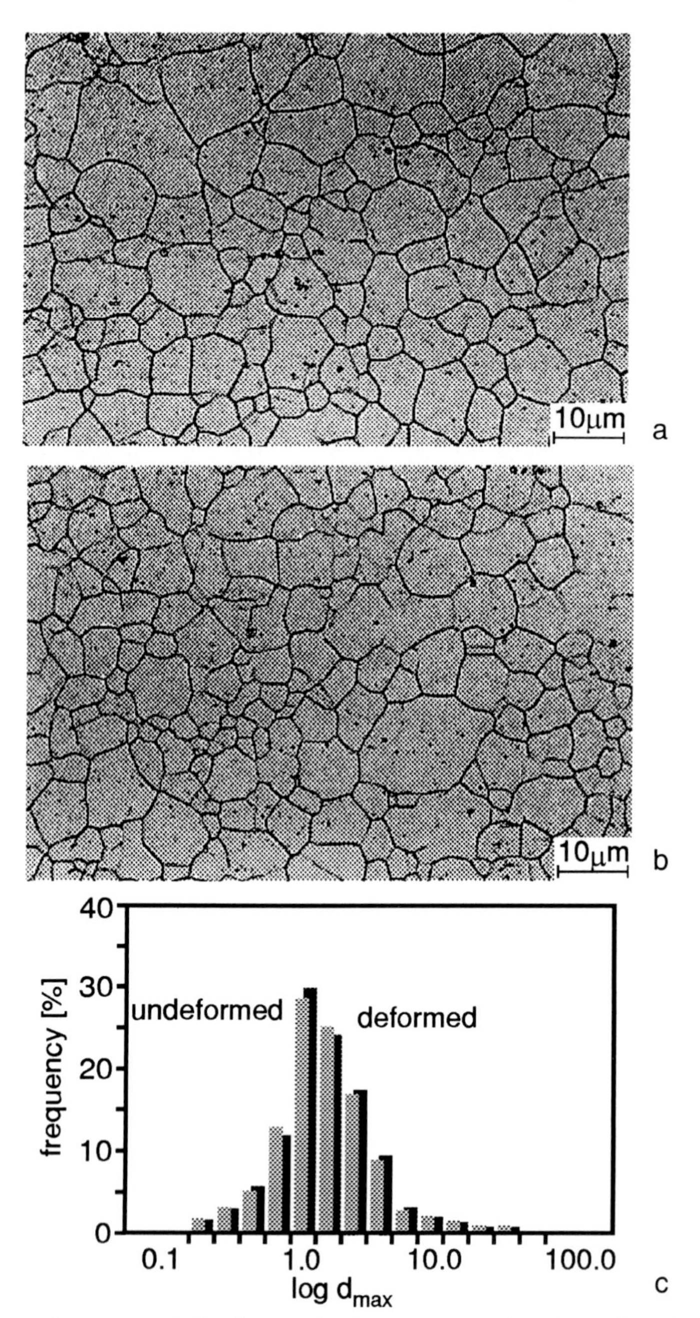

Figure 7.13 Illustration of the changes in the geometry of grain sections in an annealed polycrystal (a) as a result of plastic deformation predominantly localized in the regions adjacent to the grain boundaries (b). The microstructures in (a) and (b) are representative for a fine-grain sized austenitic stainless steel in the as-annealed state and after tensile strain at 400°C, respectively; (c) shows distribution function of d_{max}, $f(d_{max})$, before and after deformation to 25%.

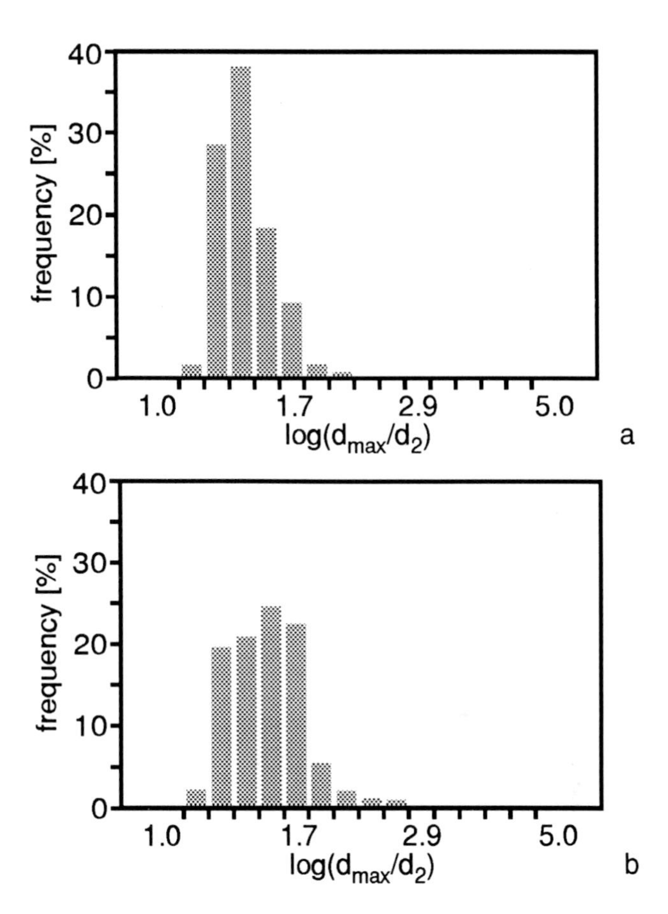

Figure 7.14 Histogram of d_{max}/d_2 ratio for longitudinal sections of: (a) as-annealed specimen; (b) specimen strained under the conditions resulting in the plastic deformation being predominantly localized in the regions adjacent to the grain boundaries.

Table 7.4 Changes in the Measured Values of the Grain Boundary Specific Area, S_V, the Mean Value, E(), of d_{max}/d_2 and p/d_2 for the Geometry of Grains Revealed in a Single-Phase Polycrystal Strained to a Macroscopic Deformation of 25% Under the Conditions where the Plastic Deformation is Preferentially Accommodated in the Regions Adjacent to the Grain Boundaries

Parameter	Deformation localized at the grain boundaries
$S_V/(S_V)_o$	1.05
$E\left(\frac{d_{max}}{d_2}\right):E\left(\frac{d_{max}}{d_2}\right)_o$	1.02
$E\left(\frac{p}{d_2}\right):E\left(\frac{p}{d_2}\right)_o$	1.01

deformation is measured by the flow stress, σ. The flow stress of a polycrystal is the lowest stress σ, such that:

$$\sum a_k \frac{d}{dt}\epsilon_{ij}(\sigma) = \frac{d}{dt}\left(\epsilon_{ij}\right)^{mac} \tag{7.44}$$

where a_k (k = 1, 2, . . . ,m) are relative contributions of the m possible mechanisms of plastic deformation.

The relative contributions of different deformation mechanisms to the total plastic strain of a specimen are functions of the test conditions and of the mutual relationships between the mechanisms.[7,8] In the case of tensile tests, test conditions are specified by the value of the applied strain rate, ε, and by the test temperature, T. The flow stress, σ, recorded during a tensile test is a function of these two parameters. On the other hand, in so-called creep tests the material is subjected to a constant load or stress at a given temperature and the strain rate is measured as a reaction of the material to the imposed conditions.

The mechanisms of plastic deformation may operate independently or may be of synergistic character. Independent mechanisms of plastic deformation are additive and their contribution to the total strain depends on their stress-strain rate characteristics. In the case of synergistic mechanisms, the law of addition does not hold and the relative contributions, a_i, depend on the way the mechanisms are coupled. If the mechanisms are required to take place simultaneously (parallel mechanisms) then $a_i = a_j$ and the flow stress is controlled by the slower mechanism.

The existing data on the mechanisms operating in a wide range of materials have been collected and depicted in the form of so-called deformation maps.[7] These maps are usually drawn in (σ, T) space and are divided into areas distinguished by a dominance of one particular mechanism of plastic deformation. Despite the great value of such maps for mechanical engineers, it should be noted that they may provide less accurate information if used for test conditions represented by points falling near the borders of different regions in the map and, in general, are less precise for high strain rate and high-stress regions. In the latter case, a number of mechanisms might be operating at the same time and an estimation of their contributions is of fundamental importance to understanding the behavior of the material. It has been shown recently[8] that the data relevant to the contributions of different plastic deformation and recovery modes may be obtained from a stereological analysis of metallographic measurements on plane sections.

Polycrystals of single-phase materials strained over a wide range of temperature show varying dependence of the flow stress, σ, measured at a constant total elongation, ε_l, on the test temperature. The question arises as to what extent the dependence $\sigma(T)$ reflects the temperature dependence of a particular mechanism of plastic strain. There may be a direct relationship between $\sigma(T)$ and the kinetics of the one dominant mechanism. However, the dependence $\sigma(T)$ might just reflect a variation in relative contributions of a number of operating mechanisms.

Example 7.4

Studies of operating plastic deformation mechanisms have been carried out on an austenitic stainless steel.[2] Gauge (deformed) and shoulder (undeformed) sections of these specimens deformed at various temperatures in the range 0.2–0.8 T_m and to a total plastic strain of 25% were polished and etched to reveal the grain boundaries. The observations were carried out on the cross-sections parallel to the axis of straining **t**. Metallographic measurements were made, using an image analyzer, on images obtained from scanning electron (magnification 2500×) and optical micrographs. Over 500 grains were selected randomly and for each individual grain the parameters listed earlier have been measured. Studies of the correlation between the values of d_2 vs. d_{max} for individual grains from the specimens tested at various temperatures showed a linear correlation between these two parameters in the form:

$$d_{max} = C\,d_2 \tag{7.45}$$

Changes in the values of the mean value of d_{max}/d_2 obtained by the least-squares method are given in Table 7.5. It can be concluded that straining at low temperatures results in a statistically significant increase in the value of the ratio. Straining by 25% increased the mean value of d_{max}/d_2, $E(d_{max}/d_2)$, from 1.24 to 1.50. Straining at intermediate temperatures, in the range 0.3 to 0.6 T_m, does not significantly change the mean value of the ratio in this micrograined material.[2] This observation suggests that different mechanisms of plastic deformation control the properties of the material. At low temperatures the dominant mechanism is trans-crystalline sliding of dislocations. At intermediate temperatures the contribution of grain boundary sliding becomes important.

Table 7.5 Changes in the Mean Value of the Ratio d_{max}/d_2, $E(d_{max}/d_2)$, as a Function of the Test Temperature for Specimens of an Austenitic Stainless Steel with an Initial Grain Size Yielding a Mean Intercept Length, E(l), equal to 3.1 μm

T	R.T.	600 K	800 K	1000 K
$E\left(\frac{d_{max}}{d_2}\right):E\left(\frac{d_{max}}{d_2}\right)_{\epsilon=0}$	1.10	1.02	1.03	1.07

7.8 Grain geometry evolution under the action of internal factors

Plastic deformation of polycrystals is an example of the situation when grains change their geometry under the action of external factors. However, at higher temperatures the grain boundaries become sufficiently mobile to be able to change their geometry under the action of internal factors, which are related to the microstructure of the material. In particular, boundaries may migrate, i.e., move in a direction along their normals, as a result of:

a) the dependence of grain boundary energy on curvature;
b) the dependence of grain boundary energy on the orientation of the boundary plane;
c) interactions of the boundaries along their common edges;
d) interactions with other defects in the material, for example, dislocations.

The changes in the geometry taking place under the action of internal factors, in general, can be analyzed along the same lines as the modifications in the geometry induced by external factors. However, they are subject to different rules since here the evolution is frequently accompanied by a decrease in the number of grains and an increase in their mean size—a process termed grain growth.

Grain growth has an important effect on the properties of polycrystalline materials. For low-temperature applications usually fine-grained materials are required with the mean grain size, measured by mean intercept length, below 20 μm. On the other hand, materials designed to sustain creep conditions at high temperature may require large grain sizes in excess of 100 μm. As a result, grain growth has attracted considerable attention in the past and is still the subject of studies, both of a theoretical and applied nature.[9,10]

One way of looking at the phenomenon of grain growth is to analyze its microstructural aspects. Such an analysis typically takes into account the properties of grain boundaries, solutes, dispersoids, textural factors, and local orientations of the grains.[9–18] Another way of investigating grain growth concentrates attention on the changes in the size of grains and their relationship to the process conditions. This approach was in the past limited to studies of the mean or maximum size of grains. However, it has been recognized that the process of grain growth may follow different patterns leading to drastically different microstructures (for example, Reference 15). Simple mean size measurements are not sufficient to differentiate precisely between variants of the grain growth processes that may occur. However, this can be achieved if the analysis measures the grain size distribution functions and their evolution with the time of annealing or as a function of the mean size of grains.

Grains in polycrystals form populations that can be described by distribution functions, f(x)dx, which define the relative number of grains of size in the range (x, x + dx) where x, depending on the dimensionality of the polycrystal and the demand of the theoretical considerations usually is

a) volume V, (3-D polycrystals);
b) area A (2-D polycrystals, thin layers, etc.);
c) intercept l (1-D polycrystals, bamboo structures, etc.).

The process of grain growth results in a transformation of f(x) with annealing time t:

$$f(x,t=0) \Rightarrow f(x,t) \tag{7.46}$$

The general condition for grain growth to occur requires that the mean grain size increases. This is equivalent to the condition:

$$\frac{d}{dt}\left[\int_0^\infty xf(x,t)dx\right] > 0 \tag{7.47}$$

Assuming some initial distribution function, f(x,0), which yields the initial mean size, $E(x)_0$, there are a number of different specific time evolutions of f(x, t) that lead to a given mean size, $E(x)_t$, after time t. Among these, special interest is attached to normal or steady-state grain growth, which is characterized by constancy of the function of normalized size:

$$\frac{d}{dt}\left[f\left(\frac{x}{E(x)},t\right)\right] = 0 \tag{7.48}$$

Normal grain growth can therefore be characterized by an invariant function of the normalized size distribution f(x/E(x)). Such a process assumes constancy in the degree of grain size homogeneity measured in terms of CV(x) values:

$$\frac{d}{dt}\left(CV(x,t)\right) = 0 \tag{7.49}$$

Also, higher-order moments of the size distribution function, such as skewness and kurtosis, are constant. However, they are less interesting at the moment since they have little meaning to the current theories of grain growth.

The changes in the size distribution function as a result of normal grain growth are shown schematically in Figure 7.15 for different coordinate systems including f(x, t), f(x/E(x, t)), and f(lnx, t). Grain area, A, is used as a measure of grain size (x = A). It may be noted that in the case of normal grain growth the normalized size of grains is preserved while the distribution function of lnx is shifted parallel along the size axis.

It should be also noted that in certain analyses it may be more convenient to replace the number distribution functions, f(x), with weighted distribution functions, $f_w(x)$, such that:

$$f_w(x) = \frac{x}{E(x)} f(x) \tag{7.50}$$

The number distribution functions define the number of grains of size x, while the weighted functions define the fraction of space occupied by those grains. If x stands for volume V, such a function determines the fraction of the specimen's volume that is filled with grains of volume V. In the case of grain growth studies it is frequently desirable to focus attention on the largest grains, which by the nature of the formalism make insignificant contributions to the population in terms of the number of objects but occupy a significant fraction of the volume, area, etc. of the material studied.

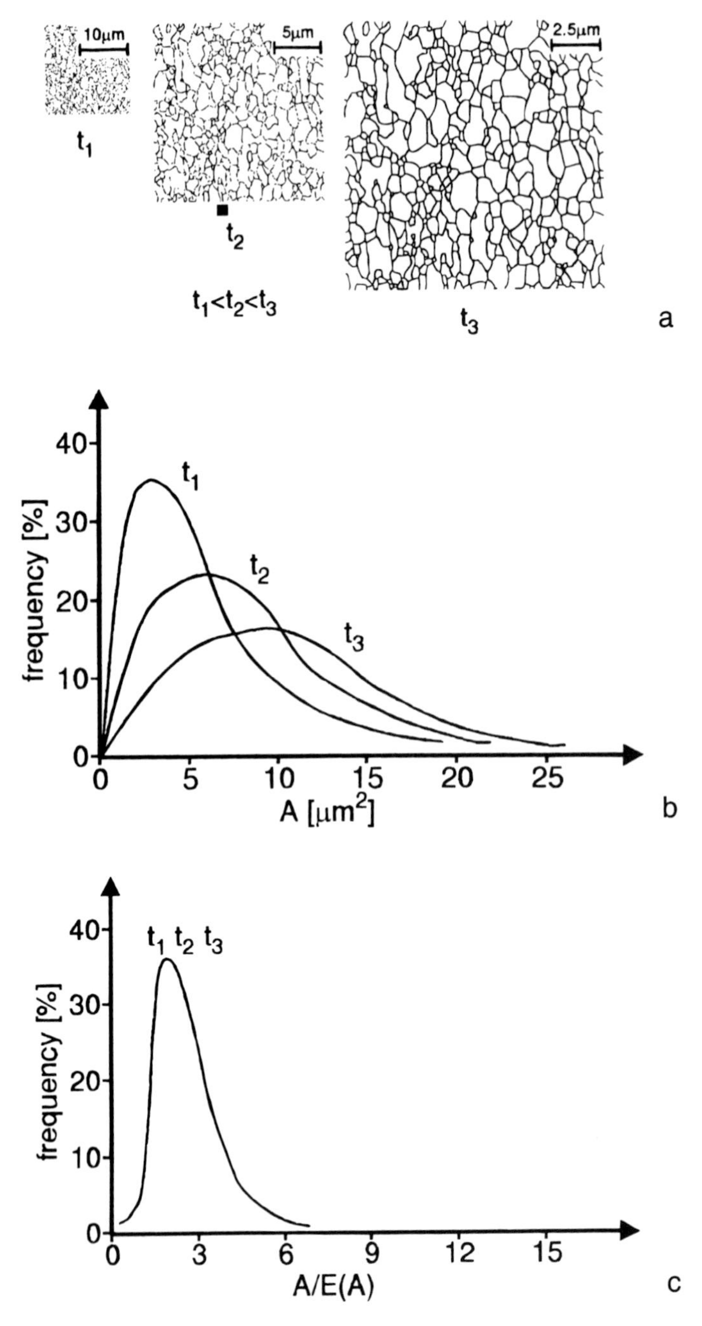

Figure 7.15 Changes in the geometry of grain sections and the distribution function of the grain section areas for the case of normal grain growth: (a) microstructures after the annealing time t_1, t_2, and t_3; (b) the distribution functions of grain area, A, in linear scale; (c) the distribution function of the normalized grain area A/E(A); (d) (facing page) the distribution functions of ln(A). (Figure is continued on the facing page.)

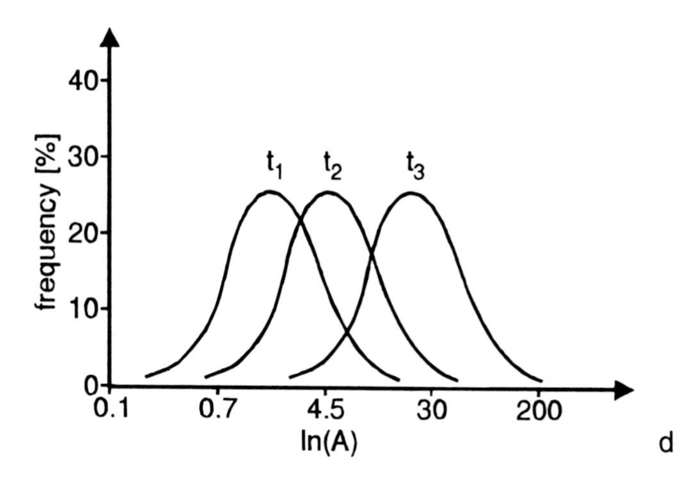

Figure 7.15 (*Continued*)

The process of normal grain growth has been studied theoretically in a number of papers (for example, References 9, 10, and 16). The grain structures obtained in the process of steady-state grain growth differ only in the scale of the grain size dimensions. Such grain structures remain self-similar, i.e., they can be made to coincide statistically by mere adjustments in the magnification or change in the scale.

The steady-state approach to grain growth restricts the scope of the analysis to the special experimental conditions (long annealing time) and special grain size distributions. In practice, the steady-state situation is rarely observed. In most practical situations grain growth is characterized by changes in the normalized distribution function of grain size.[17] A special case of this deviation from steady-state growth (normal growth) gives selective growth of some part of the initial grain population and is often termed secondary recrystallization (anomalous or abnormal grain growth). This process is characterized by a change in the grain size distribution function from unimodal to bimodal due to an intensive growth of a few grains at the expense of the rest of the grains, which statistically preserve their volume and are gradually consumed by the growing giants. An example of selective grain growth is given Figure 7.16. The data plotted in this figure (7.16b) concerns the weighted distribution functions of the grain area, which are more sensitive to abnormal growth.[19]

From the above discussion, it follows that a given process of grain growth is defined by a function F_t that transforms the distribution function from the initial one to the final one after time t:

$$F_t : f(x,t) = F_t[f(x,0)] \tag{7.51}$$

In experimental studies of grain growth, the function F_t is characterized in a descriptive way via classification of the changes in the distribution functions. Different methods that can be used to estimate experimentally the

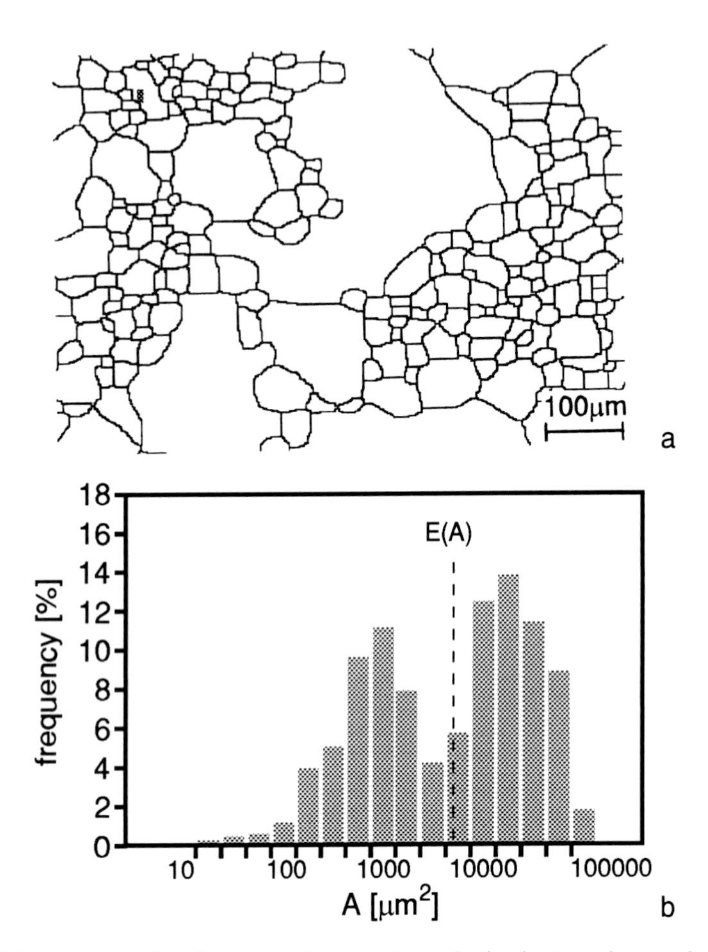

Figure 7.16 An example of a microstructure typical of selective abnormal grain growth, (a) and the corresponding distribution function of grain area (b).

distribution functions f(x) for various parameters x have been discussed and can be found in the preceding chapter. However, it may be noted that in practical situations the distributions functions are sufficiently well described by two parameters such as E(x) and CV(x) to which eventually more can be added, such as SK(x) and K(x)* (see Chapter 2). With this simplification, F_t is a vector function of time defined in a multivariable space of the parameters listed above. Such a function can be plotted as a line in a multidimensional space of these parameters. In particular, the growth function in 2-D space E(x)–CV(x) is shown schematically in Figure 7.17.

Various types of grain growth processes that do not preserve the normalized volume distribution function can be generally termed unsteady or abnormal. From a practical point of view, such processes of grain growth can

*SK(x) and K(x) are the coefficient of skewness and kurtosis and are defined in Chapter 2.

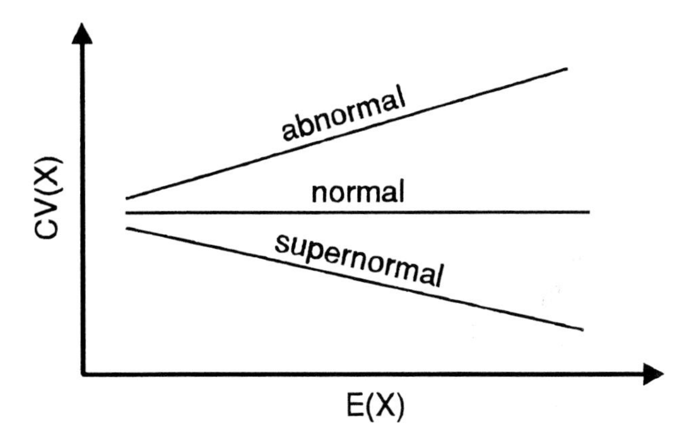

Figure 7.17 Schematic representation of the grain growth processes in the space of the coordinates defining the mean size, E(x), and coefficient of variation, CV(x), of the grain size.

be divided into two groups depending on whether or not they lead to a widening of the grain size distribution. As the "width" of the grain size distribution function can be characterized by the coefficient of variation, CV(x), the variants of grain growth can be characterized based on the way they change the relative degree of grain size homogeneity expressed trough CV(x).

If CV(x) decreases with time:

$$\frac{d}{dt}CV(x) < 0 \tag{7.52}$$

the process can be termed supernormal. Consequently if CV(x) increases with time:

$$\frac{d}{dt}CV(x) > 0 \tag{7.53}$$

the process can be called abnormal. These variants are schematically shown in Figure 7.17.

The line obtained for a given F_t function determines a path of grain growth in the space of relevant parameters that describes the properties of the grain size distribution—effectively the mean grain size and the degree of grain size homogeneity.

Depending on the parameter chosen to describe grain size, the grain growth path can be plotted in the space E(V)–CV(V), E(A)–CV(A), or $E(d_2)$–$CV(d_2)$. It should be noted that if the grain area is used merely as a result of difficulties involved in measurements of grain volume, a true path in the space of E(V)–CV(V) can be obtained by the methods described earlier.

Abnormal grain growth has been studied in a number of polycrystalline metals and alloys: Al, α-Fe, γ-Fe (316L type steel), and Ni-2% Mn alloys (Table 7.6), which are characterized by significantly different values of stacking fault

Table 7.6 Chemical Compositions in weight percentages, (except for Al) of the Materials Used in Studies of Grain Growth in Polycrystalline Metals

I. Fe-γ (316L type)

C	Si	Mn	P	S	Cr	Ni	Mo	N	Fe
0.02	0.84	1.6	0.01	0.015	17.8	13.2	2.2	0.001	balance

II. Fe-α (Armco)

C	Mn	P	S	Fe
<0.04	<0.1	<0.02	<0.025	balance

III. Al (ppm)

Si	Fe	Cu	Zn	Ba	S	P	Ca	Cr	Mn	Pb	K	Al
24	24	23	19	17	12	8	2	<1	<1	<1	<1	balance

IV. NiMn alloy

Mn	Fe	Si	O	H	S	N	C	Ni
2.1	0.01	0.001	0.0045	0.0008	0.002	0.0015	0.068	balance

energy. A detailed description of the changes in grain geometry has been obtained for specimens that were recrystallized and then annealed over a wide range of temperatures.

These specimens were recrystallized and annealed at various temperatures for a variety of times so as to obtain significantly different ratios of the final to the initial mean grain size measured in terms of the mean values of equivalent area diameters $d_2/(d_2)_0$. Grain size measurements were performed using an image analysis system with specialized software that enabled both measurement and a statistical analysis of grain area, A, grain equivalent diameter, d_2, and shape factors to be made. The results obtained are presented in Figure 7.18 in the form of plots of the coefficient of variation of the equivalent diameters $CV(d_2)$, against the mean values, $E(d_2)$. It can be seen that although the data points, in general, show a considerable scatter in the materials studied, the coefficient of variation correlates with the mean value of the equivalent diameter. In the case of aluminum it is observed that the variation in the diameter decreases with increasing mean grain size. An opposite trend is noticed for the Ni-2% Mn alloy. In the other two materials studied one can argue that both tendencies appear in different ranges of the mean grain sizes.

The results presented in Figure 7.18 demonstrate that the grain growth observed in commercial alloys is characterized by significant changes in the grain size distribution function and does not meet the steady-state conditions assumed frequently in the theoretical models. The data presented show a significant variation that cannot be explained in terms of experimental errors, which for all estimated values of CV(A) and E(A) were smaller than 3% (more than 350 grains were selected for each specimen studied).

The specimens for the measurements reported were heat treated in a conventional way that involved recrystallization and further annealing at

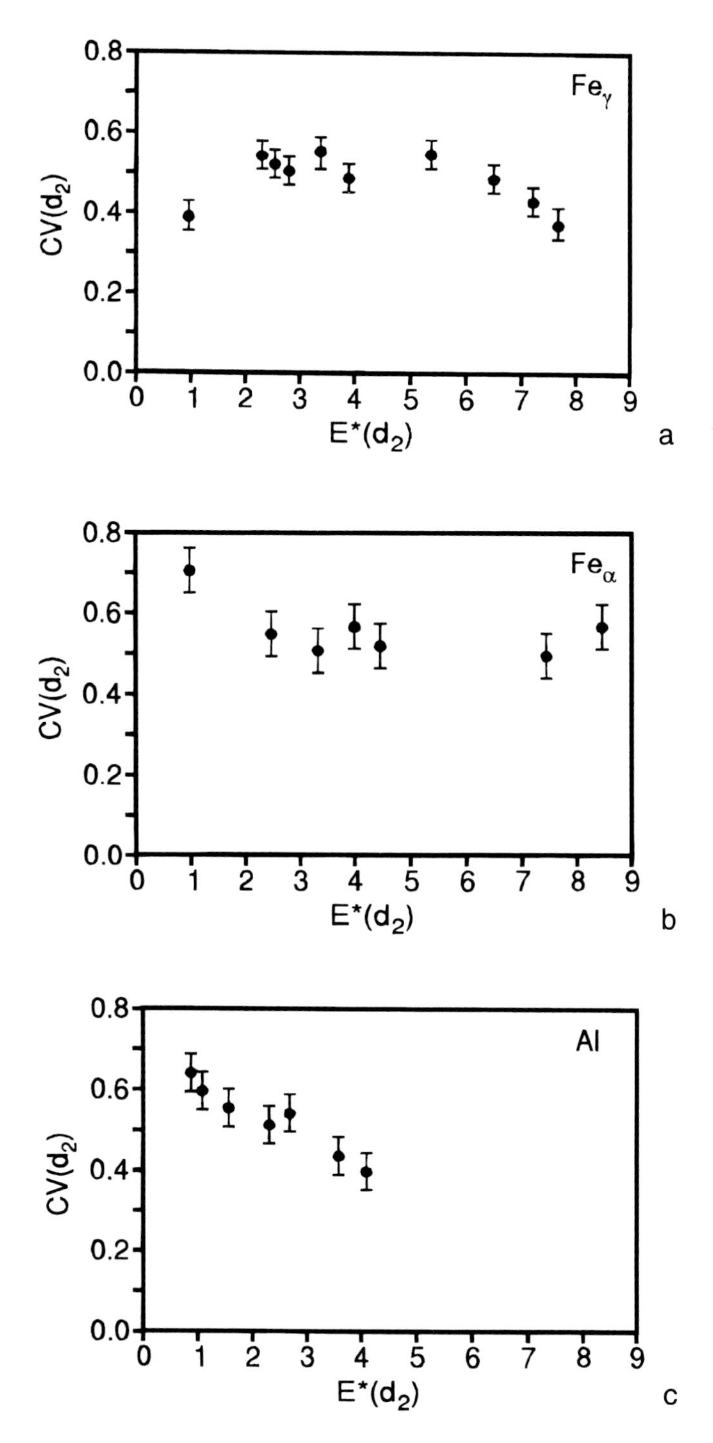

Figure 7.18 Experimental plots obtained in studies of grain growth of selected metallic materials Fe_γ (a), Fe_α (b), Al (c), and Ni-2%Mn (d) (on next page).

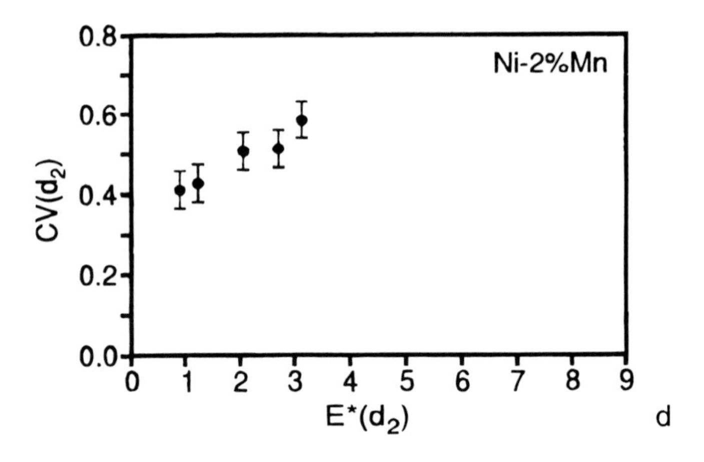

Figure 7.18 *(Continued)*

temperatures in the range 0.6 to 0.8 T_m for annealing times up to 8 h. The conclusion can be drawn that the process of grain growth in these commonly employed conditions upon closer examination shows deviations from the normal type of growth.

The deviations from normal grain growth are expected to be related to the microstructural changes in a material. Grain growth is known to change the crystallographic texture of a material and also the characteristics of the grain boundaries. Such microstructural changes in polycrystals affect the processes of grain growth and lead to variations in the volume distribution functions.

The complexity demonstrated in geometrical changes related to grain growth calls for a more precise description of the process. It has been shown that grain growth can be more precisely described in an analysis that takes into account a number of parameters describing the grain size. For convenience, such an analysis can be performed in a 2-D space, E(A)–CV(A) or $E(d_2)$–$CV(d_2)$. These two parameters are relevant from the point of view of physical metallurgy as they define the mean grain size and the degree of grain size homogeneity. In this coordinate system the data points for a given path of grain growth are represented by lines that indicate changes in the normalized spread of the grain size distribution function. The lines determine the growth path in this space of geometrical parameters. If necessary, this 2-D space may be extended by adding more parameters such as the mean shape factors.

The four materials studied have been found to follow different patterns/paths of grain growth over the range of experimental conditions covered in the present studies. In α-Fe, Al, and to lesser extent γ-Fe, the growth proceeds along a path that can be termed a supernormal grain growth, i.e., the one which results in an increase of grain size uniformity (with a decrease in the coefficient of variation of the grain area CV(A)). On the other hand, the grain growth in Ni-2% Mn alloy is of abnormal character, i.e., it results in a widening of the normalized grain size distribution function (a decrease of grain size uniformity). The reason for such behavior of this material was a 4% prestrain imposed on the specimens after recrystallization (more details can be found in Reference 18).

The results of the present studies give an indication that the process of grain growth is accompanied by variations in the degree of grain size homogeneity. These variations are expected to be linked to the microstructural processes taking place in the materials. It should be noted also that most of the CV(d_2) values fall into the range from 0.6 to 0.3. This seems to be a characteristic range for CV(d_2) values in homogenous polycrystals to which the smaller and larger values converge as demonstrated in the case of Al and Ni-2% Mn. However, as yet no such "stable" value of CV(A) has been justified on theoretical grounds and to proceed further more data need to be collected.

7.9 Annealing of deformed materials

Grain growth is commonly studied in annealed materials. In this case the migration of grain boundaries is driven by the surface tension of the grain boundaries. Migration of grain boundaries also takes place under more complicated circumstances where two or more driving forces exist. An example of this is migration of grain boundaries through a deformed grain interior or, in the extreme case, through a matrix of heavily deformed polycrystals.

Figure 7.19 schematically shows changes in the mean grain size, E(A), reported for a number of alloys subjected to plastic deformation in the range of 4 to 9% strain and subsequently either annealed for various times at a constant temperature or annealed at different temperatures for a fixed time. The variations in the value of mean grain size are characterized by the existence of an "incubation period" marked in Figure 7.19. During the incubation period the mean grain size remains constant. However, this constancy does not imply lack of grain boundary migration and constancy of shape of grains.

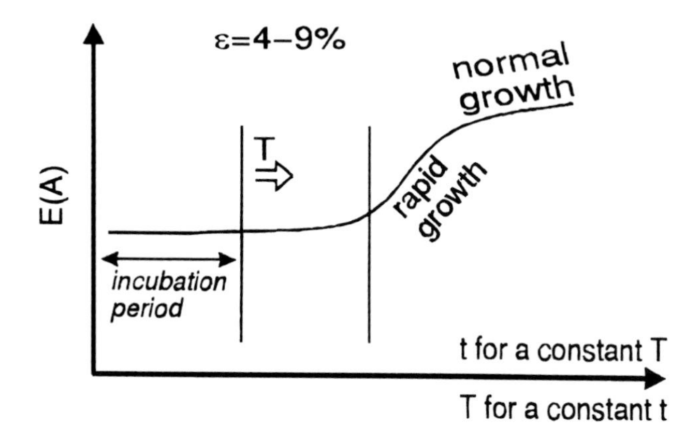

Figure 7.19 Schematic representation of the changes in the mean size of grains induced by prestraining and subsequent annealing.

Example 7.5

In order to prove grain boundary activity during the incubation period, studies were conducted on a Ni-2% Mn alloy with the chemical composition given in Table 7.6. The material, described also in References 5 and 6, exemplifies a low stacking fault energy, single phase FCC metal. Cylindrical specimens were preannealed at 1000°C, i.e., 0.8 T_m, for 1 h and subsequently strained in tension to a total plastic strain of $\varepsilon_t = 0.07$. The strained specimens were later annealed for 1 h at various temperatures in the range from 0.3 to 0.8 T_m. Observations of the grain boundary network were conducted on longitudinal cross-sections. Examples of grain boundary binary images revealed on the sections of the specimens annealed at different temperatures and annealing times are given in Figure 7.20. The measurements and analysis covered the following parameters:

1. grain area, A, and equivalent diameter, d_2 ($d_2 = (A/4\pi)^{1/2}$);
2. grain perimeter, p;
3. maximum dimension, d_{max};
4. shape factor, d_{max}/d_2;
5. shape factor, p/d_2.

As discussed earlier, grain area, A, and equivalent diameter, d_2, can be used directly to characterize the **size of individual grains**. On the other hand, the two ratios describe the **shape of individual grains**. Measurements of A, d_{max}, and p, and calculations of the ratios were performed for at least 300 grains randomly selected from the population of grains revealed on cross-sections of the specimens. The results of the measurements were analyzed both in terms of the mean values, E(), and standard deviations, SD(), of the parameters given above and their distribution functions.

The plots of E(A) and grain size coefficient of variation CV(A) against the annealing temperature T_a are shown in Figure 7.21. The dependence of $E(d_{max}/d_2)$ and $E(p/d_2)$ on T_a for a constant annealing time and on time, t_a, for a constant temperature are illustrated in Figure 7.22. The results show the existence of an incubation period observed for annealing temperatures lower than or equal to 0.7 T_m and in the case of annealing at 0.7 T_m for times longer than 20 min.

The results of these observations lead to the conclusion that the range of annealing times classified as an incubation period is associated with changes in the grain geometry. These changes are indicated by variations in the value of the d_{max}/d_2, p/d_2 ratios, grain shape factors, and CV(A) (the grain size coefficient of variation).

The results presented in Example 7.5 underline the complexity of the phenomena of recrystallization and grain growth. The data show that the size of individual grains may change without a change in the mean size value. In other words, the size distribution function under certain conditions may change its shape without changing its mean value. Obviously, such a change is reflected by an alteration in the coefficient of variation.

As said before, grains in a polycrystalline aggregate strained at a low temperature tend to change their geometry in a manner similar to the overall

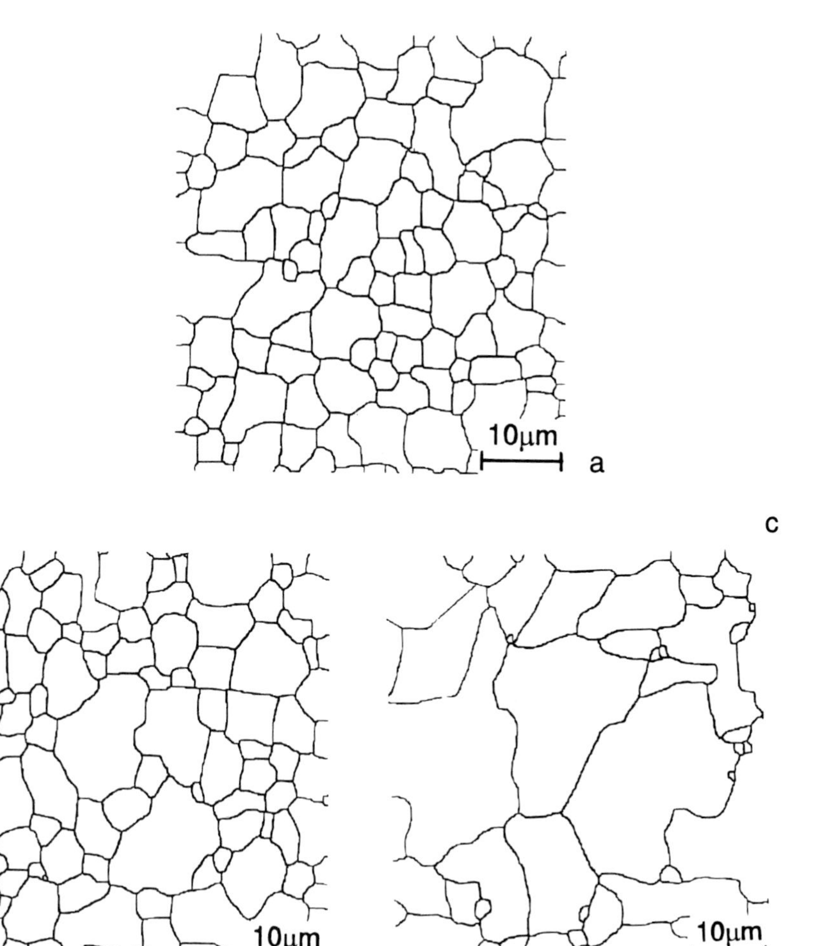

Figure 7.20 Binary images of the grain boundaries in the material after: (a) prestrain, (b) prestrain and subsequent annealing at 0.5 T_m, (c) prestrain and subsequent annealing at 0.7 T_m.

changes in the geometry of the deformed specimen. In the case of a tensile test they gradually elongate in the direction of straining, while in a compression situation they become elongated in the directions perpendicular to the axis of straining. In both cases, profound changes are observed in the arrangement of the grain boundaries with a clear anisotropy in the grain boundary spatial distribution, which may often be related to the anisotropy of the properties of the material.

Upon annealing the deformed microstructure tends to release stored energy and to equilibrate its microstructure. This is usually accompanied by a reestablishment of some of the properties, the phenomenon being termed

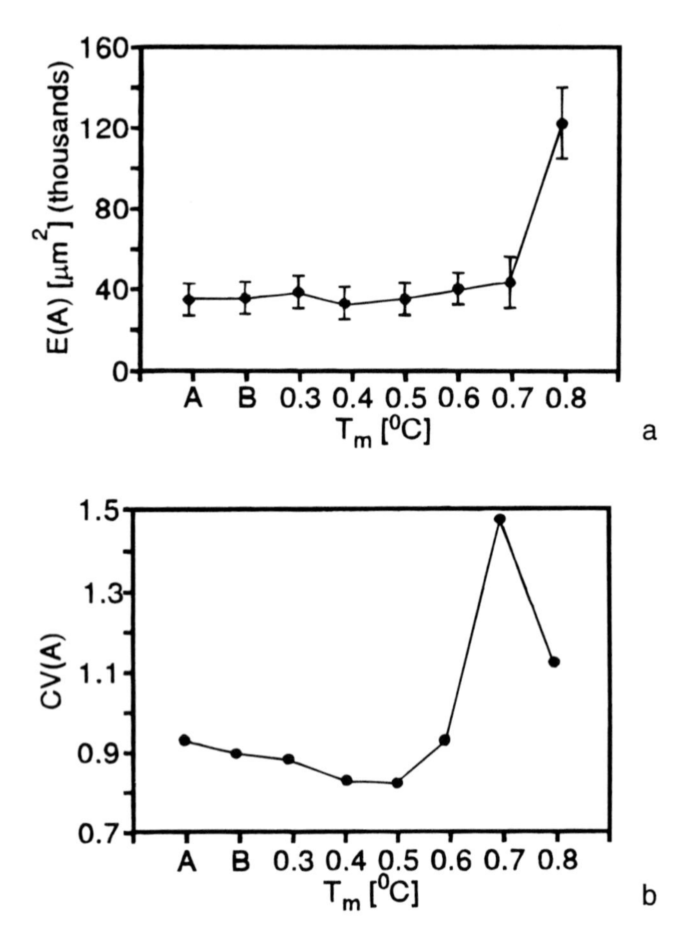

Figure 7.21 Mean area E(A) (a) and coefficient of variation CV(A) (b) as functions of T_a. (A, B on the T_a scale indicate "as-annealed" and "prestrained", respectively.)

static recovery. The term *static* underlines that recovery does not act concurrently with the deformation process, which it does in the "warm" range where dynamic recovery occurs.

It is commonly appreciated that recovery is characterized by a decrease in the density of point defects (vacancies and self-interstitial atoms) and a rearrangement of dislocations, which reform into more stable configurations in the form of dislocation walls. However, in the case of polycrystals an important role may also be played by processes taking place at grain boundaries. Grain boundaries are a major strengthening factor of the microstructure of polycrystals and their nonequilibrium geometry and arrangement as a result of plastic deformation contributes to work hardening of the material. Upon annealing of deformed polycrystals, the rearranged grain boundaries are

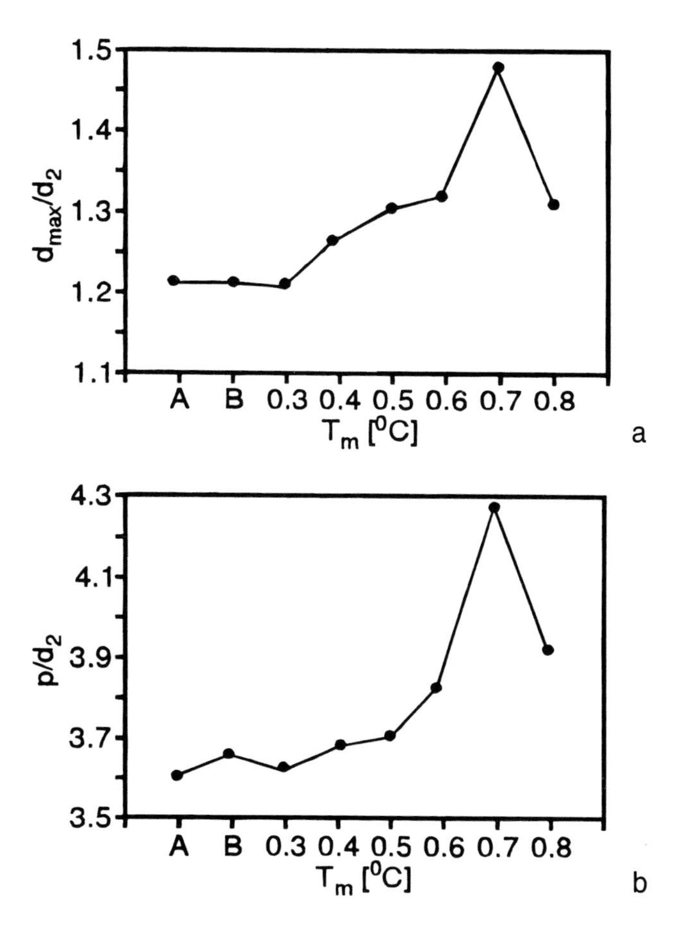

Figure 7.22 Dependence of the shape factors: (a) d_{max}/d_2 and (b) p/d_2 on the annealing temperature T_a.

expected to migrate to more equilibrium positions, accelerating the annihilation of dislocations. Studies of the grain boundary rearrangement in a deformed polycrystal are described in the following example.

Example 7.6

Studies were performed on α-Fe (Armco) with an initial grain size of approximately 30 μm. Specimens in the form of rods were compressed at room temperature to give a total reduction in thickness of 50%. The deformed material was isothermally annealed at 400°C for annealing times from 0.5 to 8 h. In order to establish a reference level for the properties of the material, some specimens were also annealed in the range 600 to 800°C and for 1 to 3 h. These specimens were found to be fully recrystallized with a grain size significantly smaller than the initial one. The specimens used for the quantitative characterization of the microstructures were prepared by vertical sectioning (see Chapter 3). This method assumes that the observation planes are parallel to a specified direction, and in this case they were parallel to the direction of deformation. Specimens were mechanically polished and electrolytically etched to reveal the grain boundary networks according to standard metallographic procedures.

In the observations of the microstructure in the specimens annealed under different conditions, attention was focused on the description of the overall grain boundary geometry. This geometry has been described through measurements of grain boundary traces revealed on the cross-sections of the specimens. The following parameters were used:

1. S_V, grain boundary area per unit volume;
2. A_i, area of an individual grain section;
3. d_2, equivalent diameter of grain sections computed as the diameter of the same area circle;
4. d_{max}, maximum chord of the grain section;
5. p, grain section perimeter.

The measurements of these geometrical parameters were made with the help of an automatic image analysis system. The measurements of S_V, due to the anisotropy of the microstructure, were carried out using the procedure based on vertical sectioning and employing cycloids (see Chapter 3).

Figure 7.23 shows micrographs characteristic of microstructures of the material after 50% compression and the subsequent anneals. It may be noted that the sections of the grains in the deformed material are clearly elongated in the direction perpendicular to the axis of straining. The microstructures revealed on section of specimens after annealing at 400°C (for 0.5, 1, 3, and 8 h) show successive stages of grain boundary rearrangement in the process of annealing. After 30 min annealing the grain boundaries lost their preferential orientation along the direction perpendicular to the axis of straining. The shape factor d_{max}/d_2, plotted against annealing time in Figure 7.24, shows a drop in value from 1.6 to 1.3, the latter of which is typical of annealed α-Fe. This value remains effectively unchanged with longer annealing times. At the same time the grain boundaries revealed in cross-sections become more curved and lose their predominantly convex shape. These changes are reflected by increasing values of p/d_2 (see Figure 7.24).

Annealing the deformed material at 400°C also results in a change in the total grain boundary area as shown in Figure 7.25. The grain boundary area per unit volume, S_V, in the beginning increases in value by approximately 5% and then gradually drops by 20% after an 8-h anneal. As a result, the mean intercept length decreases in the first stage of the recovery and then steadily increases as recovery/continuous recrystallization proceeds. The observed rearrangement of the grain boundaries is accompanied by a systematic increase in the spread in the grain size distribution as indicated by increasing values of CV(A) and CV(d_2) depicted in Figure 7.26.

The results presented in Example 7.6 show that the processes that occur when deformed polycrystals are annealed are complex. It is usual to term the

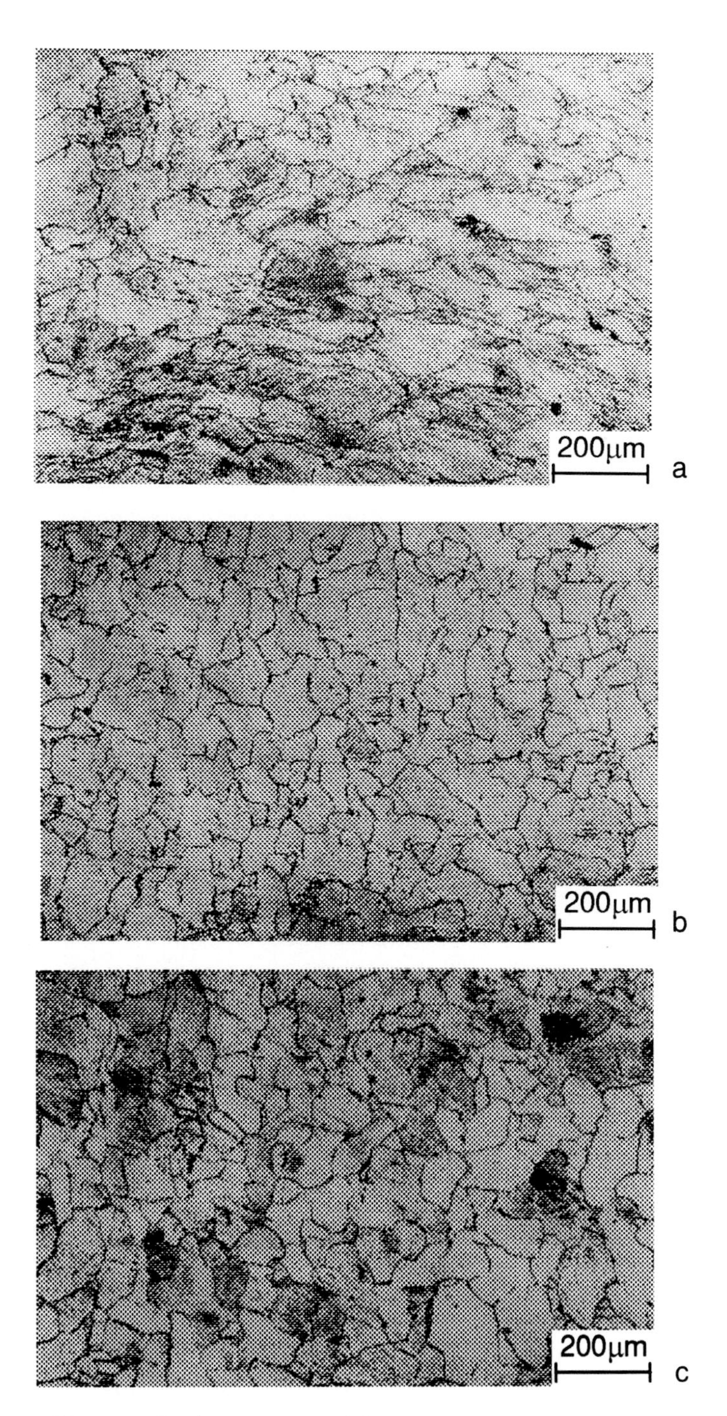

Figure 7.23 Micrographs characteristic of the microstructures of α-Fe after: (a) compression; (b) subsequently annealed at 400°C for 30 mins; (c)–(e) subsequently annealed at 400°C for 1, 3, and 8 h, respectively. (Figure 7.23d and e are on the next page).

Figure 7.23 (Continued)

Figure 7.24 Plot of shape factors, d_{max}/d_2 and p/d_2, against the annealing time at 400°C.

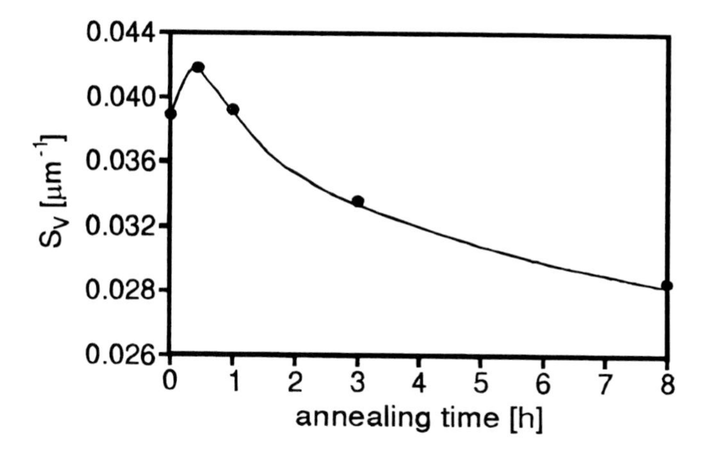

Figure 7.25 Variation in the density of the grain boundaries, S_V, as a function of annealing time at 400°C.

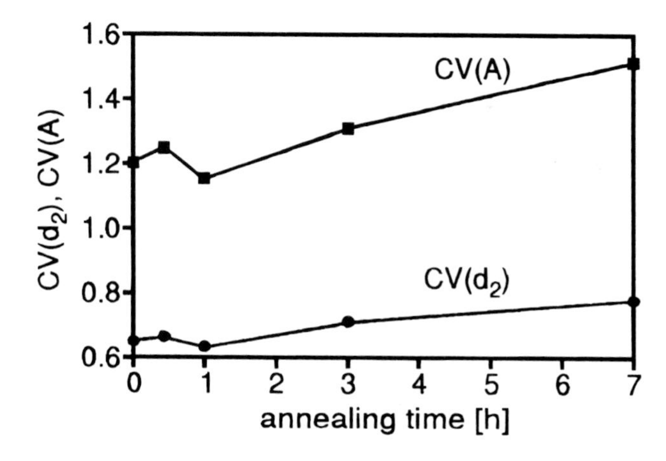

Figure 7.26 Variation in CV(A) and CV(d_2) as functions of annealing time at 400°C.

processes occurring below the recrystallization temperature under the overall title *recovery* where the grain pattern is largely unchanged. However, the study reported here shows that grain boundaries can play an important part in annealing processes below the recrystallization temperature and this can lead to major alterations in the grain boundary pattern.

The examples presented illustrate that a quantitative description of the changes in the grain size distribution function and monitoring variations in the grain shape might provide useful information about processes taking place in the material and in turn lead to a better understanding of, and control over, its properties.

References

1. Kurzydłowski, K. J., Varin, R. A., and Tangri, K., The effect of boron on discontinuous yielding of ultrafine-grained austenitic stainless steel, *Zeitshrift fur Metallkunde*, 80, 469, 1989.
2. Kurzydłowski, K. J., McTaggart, K. J., and Tangri, K., On the correlation of grain geometry changes during deformation at various temperatures to the operative deformation modes, *Philosophical Magazine A*, 61, 61, 1990.
3. Dieter, G. E., *Mechanical metallurgy*, McGraw-Hill Book Company, New York, 1986.
4. Meyers, M. A. and Chawla, K. K., *Mechanical Metallurgy—Principles and Applications*, Prentice-Hall Inc., New Jersey, 1984.
5. Kurzydłowski, K. J., McTaggart, K. J., Łazęcki, D., and Tangri, K., The use of grain area measurements to determine the distribution of plastic deformation in polycrystals of austenitic stainless steel, *Scripta Metallurgica et Materialia*, 26, 1377, 1992.
6. Kurzydłowski, K. J., The changes in the mean intercept length distribution as a function of strain and their implications to the Hall-Petch analysis, *Materials Science and Engineering Letters*, 9, 788, 1990.
7. Ashby, M. F. and Frost, H. J., *Deformation—Mechanism Maps, the Plasticity and Creep of Metals and Ceramics*, Pergamon, Oxford, 1982.
8. Kurzydłowski, K. J. and Bucki, J. J., The effect of grain size on the plastic deformation at elevated temperatures and its implications to theory of athermal flow stress of polycrystals, *The Bulletin of Polish Academy of Sciences*, 39, 463, 1991.
9. Ralph, B., Grain growth, *Materials Science and Technology*, 6, 1139, 1990.
10. Randle, V., Ralph, B., and Hansen, N., Grain growth in crystalline materials, *Proc. 6th Int. Risø Symposium* (Annealing), 123, 1986.
11. Grant, E., Porter, A. J., and Ralph, B., Grain-boundary migration in single-phase and particle-containing materials, *Journal of Materials Science*, 19, 3554, 1984.
12. Randle, V. and Ralph, B., Local texture changes associated with grain-growth, *Proceedings of the Royal Society of London*, A415, 239, 1988.
13. Tweed, C. J., Hansen, N., and Ralph, B., Grain-growth in samples of aluminum containing alumina particles, *Metallurgical Transactions A*, 14A, 2235, 1983.
14. Haroun, N. A., Theory of inclusion controlled grain growth, *Journal of Materials Science*, 15, 2816, 1980.
15. Kurzydłowski, K. J. and Tangri, K., An analysis of the kinetics of stochastic grain growth, *Scripta Metallurgica*, 22, 785, 1988.
16. Hillert, M., On the theory of normal and abnormal grain growth, *Acta Metallurgica*, 13, 227, 1965.
17. Łazęcki, D., Bystrzycki, J., Chojnacka, A., and Kurzydłowski, K. J., Grain geometry evolution during grain growth in polycrystalline materials: variation in the degree of grain size uniformity, *Scripta Metallurgica et Materialia*, 29, 1055, 1993.

18. Kurzydłowski, K. J., Grain geometry evolution in the process of grain growth induced by temperature and plastic deformation—the problem of the incubation period, *Scripta Metallurgica et Materialia*, 27, 871, 1992.
19. Tweed, C. J., Hansen, N., and Ralph, B., Methods of assessing grain size distribution during grain growth, *Metallography*, 18, 115, 1985.

chapter eight

Particles, pores and other isolated volumetric microstructural elements

Particles, pores and precipitates are examples of 3-D, or in other words volumetric elements of the microstructure of materials. They are characteristic features of a large group of steels and metallic alloys, as well as of ceramics, composites and concretes, which are formed either as a result of careful microstructural design (as, for example, precipitates in aluminum alloys) or due to shortcomings in manufacturing (as happens in the case of inclusions in some grades of steels). For simplicity all these different examples of volumetric elements will be called particles in the text that follows.

In most cases these particles are present in the material to improve properties. For this function usually it is required that the particles be harder than the matrix in which they are embedded. Such particles contribute to a so-called particle-strengthening effect. However, hard particles might also be sites of crack generation and a reason for a decrease in toughness of the material. As a result, particle-strengthening mechanisms can be used extensively only if the role of the particles is fully understood. This in turn requires precise information about their nature (chemical composition and microstructure) and a description of their geometrical features. The physical properties of particles in different materials are the remit of physical metallurgy, phase transformations, and chemistry. In the present text, attention is focused on a description of their geometrical properties. This geometrical approach means that the solutions and methods presented here are of general character and do not depend on the nature of the particles studied. The results can be applied to any system of particles; to nanometre size precipitates as well as to millimetre size inclusions in cast iron.

Some examples of microstructures containing volumetric elements are shown in Figure 8.1. These are

a) nonmetallic inclusions in a low-alloy steel (SEM);
b) carbides in an austenitic stainless steel (TEM);
c) particles of sand in a concrete (optical microscopy);
d) pores in sintered tungsten.

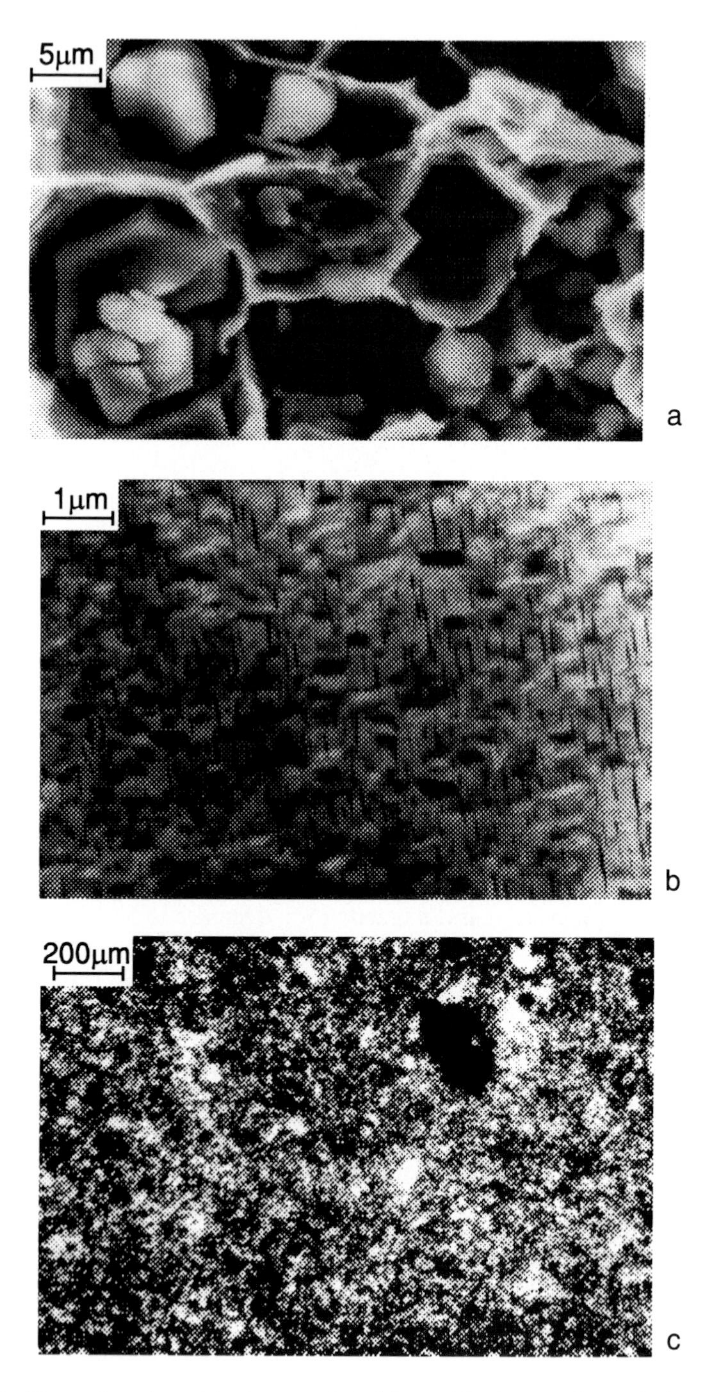

Figure 8.1 Examples of particles in materials: (a) particles in a steel revealed on a fracture surface; (b) transmission electron microscopy image of carbides in an austenitic stainless steel; (c) optical microscopy of particle sections revealed on a cross section of a concrete; (d) (facing page) pores revealed on a section of sintered tungsten.

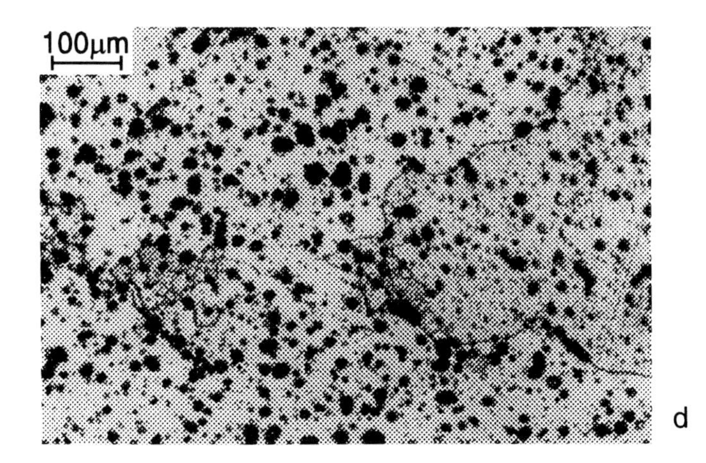

Figure 8.1 (*Continued*)

These examples illustrate important aspects of particles in materials: they differ among themselves in the details of their geometry. In other words, particles form populations. A population in a given material consists of all particles $V_1, V_2, \ldots$, or in shorthand notation V_i, with i ranging from 1 to N_T, the total number of particles.

A description of the geometrical features of the particles, in general, consists in characterizing their:

a) content;
b) shape;
c) size;
d) spatial distribution;
e) location with respect to other defects;
f) orientation.

Such a characterization can provide information on the whole system of particles without recognizing individual properties of the particles. This approach leads to such parameters as:

V_V—volume fraction;
N_V—mean density,

which are discussed in the further text.

The individual parameters, X_i, imply the use of distribution functions, $f(X_i)$. These distribution functions provide information on the relative number of particles distinguished from the rest of the population by certain specific values of X. In materials science applications such a description might be simplified to include information on mean properties of the particles. The mean value, E(X), is given by:

$$E(X) = \frac{1}{N_T}(X_1 + E_2 + \ldots + X_{N_T}) \tag{8.1}$$

In this case the properties are averaged over all the particles in the material. This approach leads to parameters such as:

E(V)—mean volume;
E(S)—mean surface area.

In some cases a more detailed description is required that includes higher-order parameters such as variation VAR(X):

$$VAR(X) = \frac{1}{N_T}\left[(X_1 - E_X)^2 + (X_2 - E_X)^2 + \ldots + (X_{N_T} - E_X)^2\right] \tag{8.2}$$

where $E_X = E(X)$. Alternatively, the standard deviation SD(X):

$$SD(X) = \sqrt{VAR(X)} \tag{8.3}$$

or coefficient of variation CV(X):

$$CV(X) = \frac{SD(X)}{E(X)} \tag{8.4}$$

can also be used. These three parameters provide information on variability in the property X of the particles. They are all equal to zero for a uniform population of particles in which all the elements show the same property (X = constant)—see also Chapter 2.

8.1 Particle sampling and choice of parameters

Particles, like other components of the microstructure, usually appear in large numbers, N_T, unless the object of the investigation has an extremely small volume. As a result, in the process of microstructural characterization, observations are made on a set of particles sampled from the population in the material studied. There are four general methods for sampling particles that can be used to obtain quantitative information on their properties:

a) sampling with points, (grid of points Q_i);
b) sampling with lines, (grid of lines L_i);
c) sampling with planes (sections C_i);
d) sampling with disectors (slicing pair of sections C_1, C_2).

Among these, sampling with sections C_i is of special importance for the examination of nontransparent materials. The different sampling procedures can also be used in a combination of two or more methods, and then the sectioning is usually the first one in a sequence that may contain further sampling by points and/or lines.

These sampling procedures make it possible to select some smaller number of objects representative of the microstructure studied, on which the further measurements are carried out. These procedures are not equivalent in the sense that each generates different descriptions of the material.

In the case of point sampling with a 3-D set of points of uniform density, the particles are sampled with a probability proportional to their volume, V_i. As a result particles with a larger volume are sampled more frequently than the smaller ones, even if they appear in equal numbers per unit volume. This sampling with points generates volume-weighted estimates of microstructural parameters. These volume-weighted estimates of particle volumes, $E_V(V)$, are given as:

$$E_V(V) = \frac{1}{K}\left(V_1^h + V_2^h + \ldots + V_K^h\right) = \frac{1}{N}\frac{\left(V_1^2 + V_2^2 + \ldots + V_N^2\right)}{\left(V_1 + V_2 + \ldots + V_N\right)} \tag{8.5}$$

where the summation is carried out over K particles hit by a grid of points.

In the case of probing with lines, the probability of hitting a given particle is proportional to its surface area, S_i. If a system of lines is drawn through a population of particles, a mean volume, $E_S(V)$, can be calculated by averaging the volume of particles hit by the lines:

$$E_S(V) = \frac{1}{K}\left(V_1^h + V_2^h + \ldots + V_K^h\right) = \frac{1}{N}\frac{\left(S_1V_1 + S_2V_2 + \ldots + S_NV_N\right)}{\left(S_1 + S_2 + \ldots + S_N\right)} \tag{8.6}$$

It can be shown on the other hand that random sections of the material reveal the particles with a probability proportional to their height h_i in the direction perpendicular to the section:

$$E_h(V) = \frac{1}{K}\left(V_1^h + V_2^h + \ldots + V_K^h\right) = \frac{1}{N}\frac{\left(h_1V_1 + h_2V_2 + \ldots + h_NV_N\right)}{\left(h_1 + h_2 + \ldots + h_N\right)} \tag{8.7}$$

Distributed in the volume of a material, particles are sampled with the same probability regardless of their size and shape only in the case of volume sampling by means of a disector. This procedure, described in Chapter 3, assumes that a slice of the material is cut, which is bounded by two cross-sections, and particles are accepted that do not emerge on one of the sections.

The particles sampled with the disector yield a number mean volume:

$$E_N(V) = \frac{1}{N}\left(V_1 + V_2 + \ldots + V_N\right) \tag{8.8}$$

These different procedures of sampling the particles are discussed in Example 8.1.

Example 8.1

The system shown in the Figure 8.2 contains two types of particle: equiaxed and extended. These particles, appearing with *the same density per unit area*, have the following properties:

Equiaxed $A = 30\ \mu m^2$, $L = 20\ \mu m$
Extended $A = 20\ \mu m^2$, $L = 100\ \mu m$

Measurements of particle area have been carried out on the particles sampled in three different ways: by the 2-D disector, by a grid of lines, and a grid of points. In the case of the disector the particles are sampled that are located entirely below the look-up section and have at least one point above the reference section. In the case of sampling with lines, the particles are taken into account that are hit by the lines and in the case of the point grid—those hit by the points.

Depending on the way the particles are sampled, the following mean values can be obtained for the studied system of particles (the mixture of equiaxed and extended particles):

a) the disector (particles are sampled with a probability that disregards their size)

$$E(A) = (30 + 20)/2 = 25\ \mu m^2;$$

b) with a grid of lines (particles are sampled in proportion to their perimeter)

$$E(A) = (30 + 5 \times 20)/6 = 21.6\ \mu m^2;$$

c) with a grid of points (particles are sampled in proportion to their area)

$$E(A) = (3 \times 30 + 2 \times 20)/5 = 26\ \mu m^2.$$

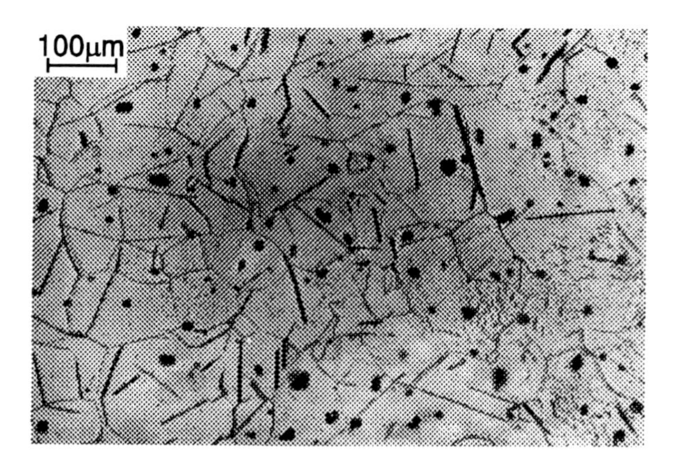

Figure 8.2 A section through a system containing two types of particles: spherical and plate-like ones. These two types of particles cover approximately the same fraction of the cross-sectional area.

The sampling procedures are frequently used as a combination of two or more:

Sampling I–Sampling II–Sampling III–. . .

For surface type observations, e.g., by reflected light microscopy, one usually starts by making a cross-section:

Sampling I = Sectioning–Sampling II–Sampling III–. . .

The subsequent sampling may involve point probing, linear sampling, or both in this order:

Sampling II = Point sampling–Linear sampling–. . .
Sampling II or III = Linear sampling–. . .

In fact, this combination of sectioning, point sampling, and linear sampling provides estimates of the volume-weighted volume of particles described later in the text (see also Chapter 3). A typically used combination of sampling techniques is illustrated in Figure 8.3.

Example 8.1 shows that a set of particles in a given material can be characterized by different parameters obtained through different sampling procedures. A natural question then arises: What is the best way of describing the properties of the system of particles? Answers to this question can be formulated from different points of view. In terms of a mathematical statistics approach it might be desirable to use the number distribution function, or its mean value and higher moments (such as the standard deviation), as they can be transformed simply into weighted distributions. However, in studies of the microstructure of materials estimating the number distribution functions is a most complicated task. On the other hand, the parameters used to characterize microstructures should suit those theoretical models that explain the properties of materials. From this point of view weighted distributions, which are more accessible by experimentation, might be found to meet better the needs than the more fundamental number distributions.

It should be remembered also that frequently the objective of microstructural characterization is to detect differences between two or more materials or samples of the same material. In this case, it is possible to use any parameter that is expected and found to be sensitive to those elements/aspects of the microstructure that are studied. As a result, sampling procedures may be chosen in a less rigorous way, as long as the same procedure is applied to each specimen, which is particularly easy to achieve if the specimens have the same geometry and the materials studied have the same texture. Detecting differences essentially means a descriptive approach to the microstructure and the numbers that are obtained as a result of these measurements, for example, the number of particles on a longitudinal section of a specimen has only relative meaning. They correlate properties of the microstructures studied by adopting a particular procedure.

If the description of the microstructure is intended to provide a quantitative description in terms of some physical quantities (such as density of

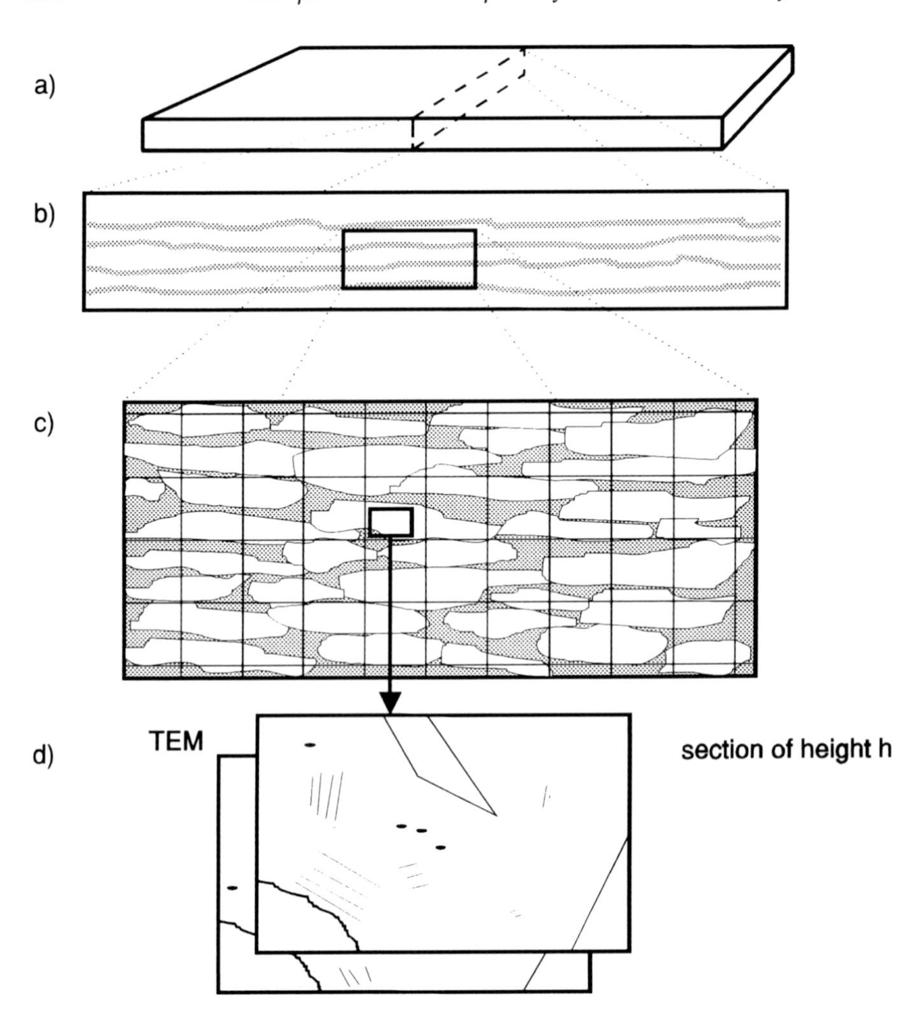

Figure 8.3 Illustration of the different ways of sampling: (a) sectioning; (b) selection of a larger field on a section of the material; (c) selection of a smaller observation field; (d) selection of an observation field for transmission electron microscopy.

particles), the procedures for particle sampling must be carefully designed so as to ensure unbiased estimates of the parameters required. Particles of a given i-type can be sampled proportionally to their:

	Parameter	Method
a)	number, N_i,	disector;
b)	height, h_i,	section;
c)	surface, S_i,	lines;
d)	area of section, A_i,	points on a section;
e)	volume, V_i,	points in volume.

Each of these sampling procedures leads to a different estimate of a given parameter for the same set of particles. It should be stressed that as long as the procedures used are in accordance with the principles of stereology, all the estimates are formally correct and should not be differentiated in terms of their correctness. Their effectiveness can only be judged in terms of applications to some theoretical models of the material's properties. Such models may require a specific estimate of the mean volume in terms of the number of particles of a given volume, $E_N(V)$. These models may as well dictate the use of volume-weighted mean volume, $E_V(V)$. In the first case all particles, small and large, are assumed to be basically of the same importance while in the second, by contrast, the bigger particles have a larger effect on the estimate.

The existing theories of the properties of materials as yet are not very formal and do not clearly indicate how the particles should be sampled in order to provide an appropriate description. As a result, in most experimental studies the intention is to sample particles in proportion to their number. However, this is rarely achieved, due to difficulties involved in such a sampling procedure. On the other hand, weighted estimates of some parameters can be much more easily obtained and, even more important, certain weighted values seem to make more sense from the materials science point of view. For example, the volume-weighted mean volume of particles, $E_V(V)$, can be a more effective description of the particle size than $E_N(V)$ in studies of the strengthening mechanisms in multiphase materials. This is due to the fact that the weighted value takes into account not only the number of particles but also the volume occupied by them.

Here it is suggested that in studies of particles and other 3-D microstructural elements, the following rules be obeyed:

1. Before starting to make measurements one should decide on the theoretical grounds which make a parameter appropriate or acceptable.
2. In the same way, one should make a decision with regard to the proper sampling procedure(s).
3. If there is no theoretical guidance, one should use simple sampling procedures and not try to avoid weighted estimates.

8.2 *General characteristics of particle populations: volume, surface area, curvature and number*

A system of particles can be characterized generally in terms of their total volume, V, surface area, S, curvature, M, and number, N. These parameters depend on the geometry of the individual entities forming the population of the particles studied. However, this is a "one way" functional dependence, in the sense that for given values of V, S, M, and N, there are different possible distribution functions of particle volumes, surface area, etc. In the approach adopted in materials science, these quantities are usually referred to unit volume of the material. This is due to the fact that the microstructure of materials is usually sufficiently homogenous to show volume-extensive properties. If referred to unit volume of a material these parameters define the:

a) volume fraction, V_V;
b) specific particle interface, S_V (interphase boundary per unit volume);
c) specific particle curvature, M_V;
d) particle density, N_V.

These parameters form basic, general descriptions of the particles. They can be used to explain those properties of materials that do not depend on the distributions of particle size, shape, etc. An example of such a property is material density, which, in sintered materials, is a linear function of the volume fraction of pores:

$$\rho = \rho_t\left(1 - V_V\right) \tag{8.9}$$

where ρ is the density, ρ_t is the theoretical density, and V_V is the volume fraction of the pores.

8.2.1 *Volume fraction*

The volume fraction, $(V_V)_\alpha$, of α particles is one of the fundamental quantities used to characterize multiphase materials. Its applications range from the theory of composite materials to phase transformations. The volume fraction defines the fraction of the volume filled with particles. For a specimen of total volume V_T the volume fraction is defined as:

$$\left(V_V\right)_\alpha = \sum_{i=1}^{N} \frac{\left(V_i\right)_\alpha}{V_T} \tag{8.10}$$

where this summation is over all N particles in the volume of the specimen, V_T.

According to the basic stereological relationships, the volume fraction can be estimated by probing the microstructure with either sections, lines, or points using the following set of equations (see also Chapter 2):

$$\left(V_V\right)_\alpha = \left(A_A\right)_\alpha = \left(L_L\right)_\alpha = \left(P_P\right)_\alpha \tag{8.11}$$

In these equations, the symbols have the following meanings:

$(A_A)_\alpha$ the area fraction occupied by particle sections on randomly cut cross-sections;
$(L_L)_\alpha$ the length of the intercepts inside particles as a fraction of the total length of random lines in three dimensions;
$(P_P)_\alpha$ the fraction of points hitting the particles from a system of points randomly distributed in space.

Application of these equations requires random sectioning of the specimens and either area measurements or further sampling with points or lines. In this context, the term *random* can be replaced by *systematic*, assuming that all possible orientations are taken with the same probability.

There are two special situations that should be distinguished:

1. particles are isotropic;
2. particles are anisotropic, either in their orientation or distribution.

Example 8.2

The microstructure of the two-phase material shown in Figure 8.4 was found to be isotropic. A section of the material was made and the particles revealed by an etching process. The volume fraction of particles was estimated using:

a) particle area measurements;
b) test lines;
c) grid of points.

These three methods yielded similar values of the volume fraction, V_V = 14%.

If the particles are isotropic, in principle all measurements can be performed on one representative cross-section of the material. However, it is strongly recommended to use more sections. Measurements on three mutually

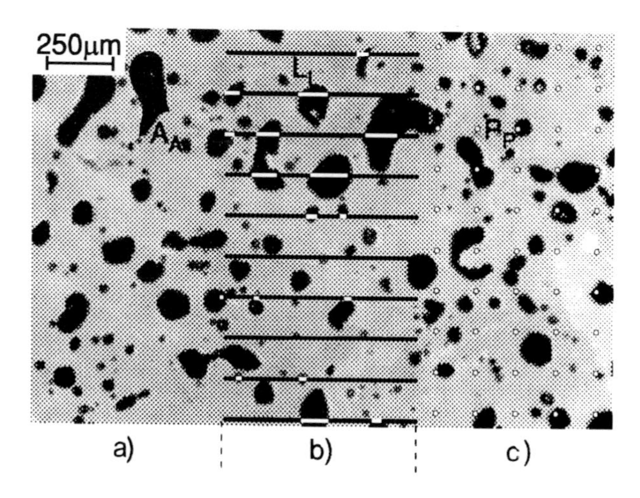

Figure 8.4 An example of a microstructure of an isotropic two-phase material. A system of test lines (b) and points (c) are shown that have been used to estimate the volume fraction of the particles.

perpendicular sections provide better estimates as well as an opportunity to test the isotropy of the particles (see Chapter 3).

The measurement of particle sections, A, can be made with the help of an image analysis system. The same applies to measurements of intercept length (the intercept lines may form a grid of lines parallel to a given direction). On the other hand, simple counting can be carried out if a grid of points is used, although this again can be effected using an image analysis system. It should be noted that the precision of estimation by point counting is no worse than the other two methods.[1]

If the particles are anisotropic, it is recommended that the sections be random in three dimensions. This might be difficult to implement and as an alternative vertical sections can be made. More information about vertical sectioning can be found in Chapter 3.

In a simplified version three sections can be studied inclined at angles of 60 and 120°. If the volume fraction has been estimated by measurements on N sections, it is taken as the average value; for example:

$$V_V = \frac{(A_A)_1 + \ldots + (A_A)_N}{N} \tag{8.12}$$

(See also Example 8.3.)

Example 8.3

A two-phase material of possibly anisotropic properties was sectioned in the way shown in Figure 8.5a. Three sections made angles of 60° (120°) with each other. Particle area measurements yielded the following values of the area fraction:

$$(A_A)_a = 15\%,\ (A_A)_b = 20\%,\ (A_A)_c = 16\%.$$

The average area fraction is equal to 17% and this can be used as an estimate of the volume fraction.

8.2.2 *Specific interface/surface area*

Specific interface/surface area, or in other words, surface-to-volume ratio, S_V, is another elementary parameter that describes some properties of a system of particles. It can be used generally in any studies related to interfacial phenomena such as:

—phase transformations;
—particle growth;
—interfacial cracking.

In its basic form, it defines the total area of the surface separating particles from the matrix in which they are embedded. Based on the results of stereology, this parameter can be estimated using a system of random lines in three dimensions. If such lines are drawn throughout the volume V_o and hit N_l particles per unit length, the total particle surface area can be estimated from the relationship:

$$S = 4 N_l V_o \tag{8.13}$$

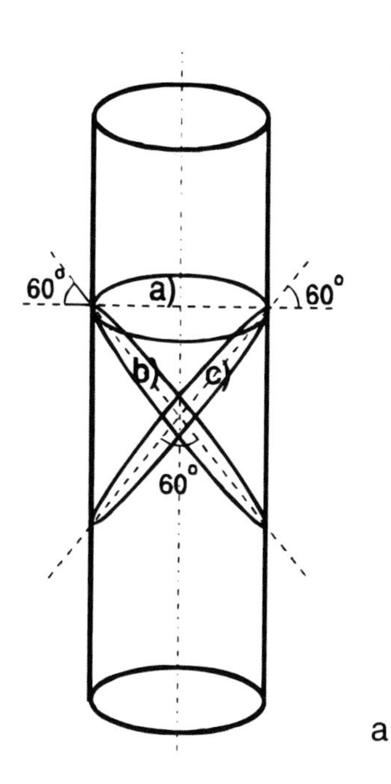

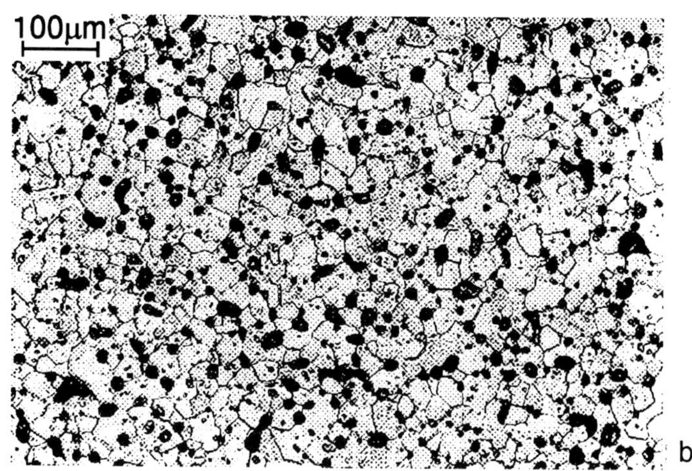

Figure 8.5 Images of particle sections revealed on different cross-sections of a possibly anisotropic two-phase material. Microstructures shown in (b)–(d) show the same microstructure [a, b, and c in (a)] sectioned at a different angle to the axis of the specimen. (Figure 8.5c and d can be found on the following page.)

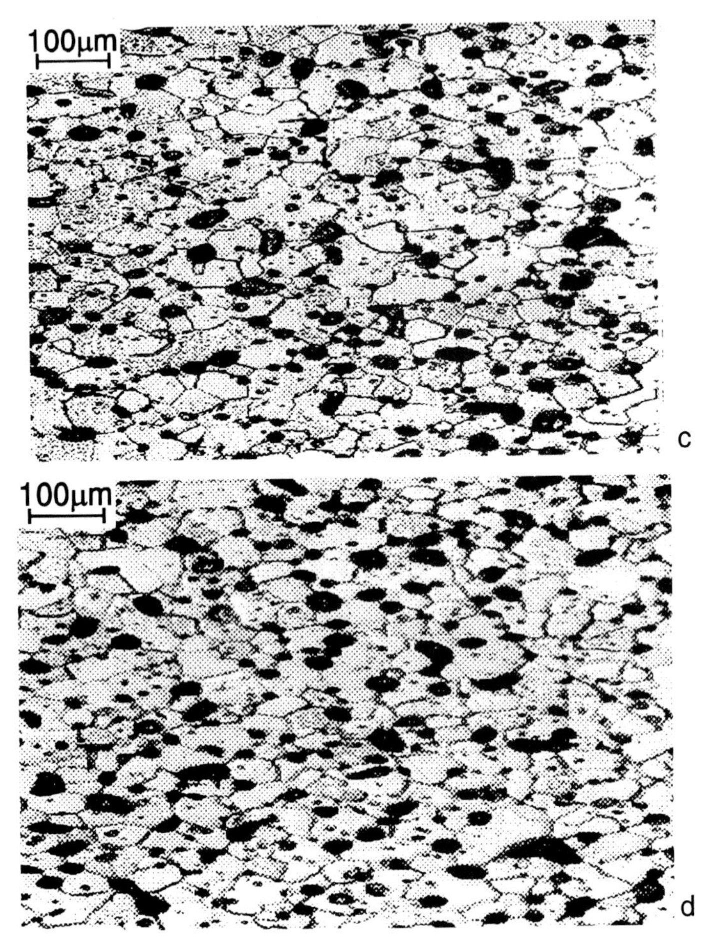

Figure 8.5 (*Continued*)

This equation can be transformed further into the following:

$$S_V = 4\,N_L = 2\,P_L \tag{8.14}$$

where P_L is the density of intersection points of the 3-D random test line array with the interphase boundaries of the particles studied.

Implementation of this formula depends again on the isotropy/anisotropy of particles. If they are isotropic, the lines can be drawn on any section and sets of parallel lines can be used.

If the particles are anisotropic, or the isotropy cannot be taken for granted, the test lines need to be randomly oriented in space and more sophisticated methods have to be employed to draw such a system of lines. Some solutions to this problem are outlined in Chapter 3 under the heading of the vertical sectioning, orientator, and trisector methods.

Example 8.4

Consider the microstructure shown in Figure 8.6. Let us assume that this is a representative micrograph of the particles, which are randomly and isotropically distributed in the volume of the material. In this case the surface area of the particle boundaries can be estimated using a system of parallel lines. The following parameters were obtained for the microstructure shown:

$$N_L = 2.39 \times 10^{-2}\ \mu m^{-1} = 2.39 \times 10^4\ m^{-1},$$
$$P_L = 4.78 \times 10^{-2}\ \mu m^{-1} = 4.78 \times 10^4\ m^{-1}.$$

These two yield the following estimate of S_V:

$$S_V = 9.56 \times 10^{-2}\ \mu m^{-1} = 9.56 \times 10^4\ m^{-1}.$$

The estimation method described in Example 8.4 is based on the assumption that the particles are isotropic. If this assumption is not valid the test lines should be randomly oriented in space. The practical implementation of this approach is described by the concept of vertical sectioning (see also Chapter 3). Vertical sectioning is carried out through the following steps:

1. selection of a vertical axis;
2. selection of the sections for observation;
3. probing with cycloids or circles.

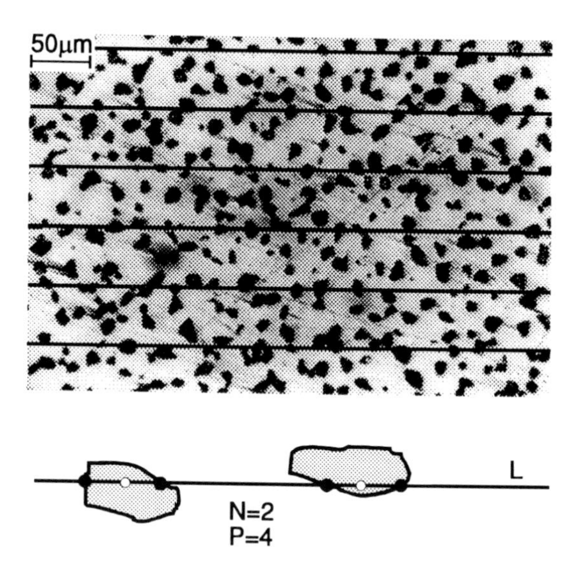

Figure 8.6 A system of particle sections tested with a system of lines.

In the first step, an axis **t** is selected to which all subsequently cut sections are parallel. The orientation of this axis is without restriction. Obviously, choosing a certain orientation may lead to a simplification in the measurements and, if possible, it is desirable to choose the axis to be aligned with some specific direction with respect to the specimen geometry or microstructure. In the case of cylindrical specimens, or where there is a fiber texture, the axis can be taken to be parallel to the axis of the specimen or texture.

The sections for observation, C_i, should be oriented randomly with respect to a reference direction **p** perpendicular to **t**. This can be achieved by a random selection of the orientation of the first section and then systematic orienting of the subsequent ones by rotating them by a constant angle around **t**. It is important to ensure that the vertical axis is known for each section subsequently examined.

Two types of probing elements—(a) a system of cycloids and (b) a system of circles—are used to quantify the traces of the boundaries revealed in sections into an estimate of their surface area in the volume of the material. A system of cycloids typically used in the studies is exemplified by the lines depicted in Figure 8.7. The cycloids are organized into a rectangular pattern with the shorter edge being necessarily oriented parallel to the vertical axis **t**. When such a system of cycloids is placed on subsequent vertical sections, they form in space a system of 3-D curves, which, if broken into linear segments, have the property of a uniform orientation.

Grids of cycloids typically contain also arrays of testing points which are used to estimate the area of the fields studied with the cycloids, and in turn to estimate the volume, V, to which the surface of the interfaces, S, is referred.

The number of intersection points with cycloids, I_j, is determined for each vertical section cut, C_j. Also, the fraction of testing points, P_i, falling into the

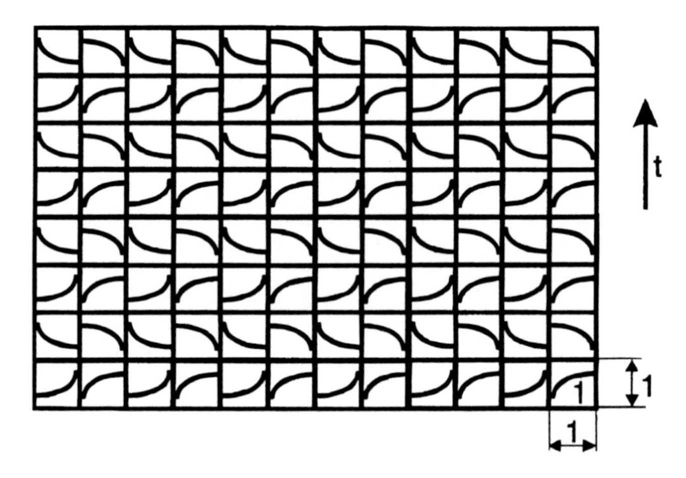

Figure 8.7 A system of cycloids used for estimation of the particle surface area in unit volume.

reference space is counted. An estimator of S_V, defined as the ratio S to the volume V of the reference space is given as:

$$S_V \approx 2 \frac{M}{\sum P_i} \frac{\sum I_j}{L} \tag{8.15}$$

where M is the final magnification of the microstructural images and L is the total length of the cycloids used. The intersection points from a grid of cycloids can be counted with respect to the area of the material examined on a particular cross-section. In this case, the reference space is the volume of material and in most practical cases $P_i = 1$, which means that all the testing points are within the image of the microstructure. The reference space can also be the volume of the particles studied and then P_i describes the volume fraction occupied by the particles.

Example 8.5

Specimens of a cast iron, in the form of rods, were subjected to heat treatment, which resulted in the microstructure depicted in Figure 8.8. Vertical sections of the specimens were cut parallel to the rod axis. These sections were at an angle 60° to each other. The sections were polished and etched to reveal the sections of the particles. A microstructure typical of the vertical sections is shown in Figure 8.8. A system of cycloids was imposed on the images of the microstructure. The following result was obtained:

$$S_V = 12.3 \times 10^4 \text{ m}^{-1}.$$

Instead of cycloids, a system of circles can be used. The circles, of radius R, are arranged into a regular array with their centers forming a square grid

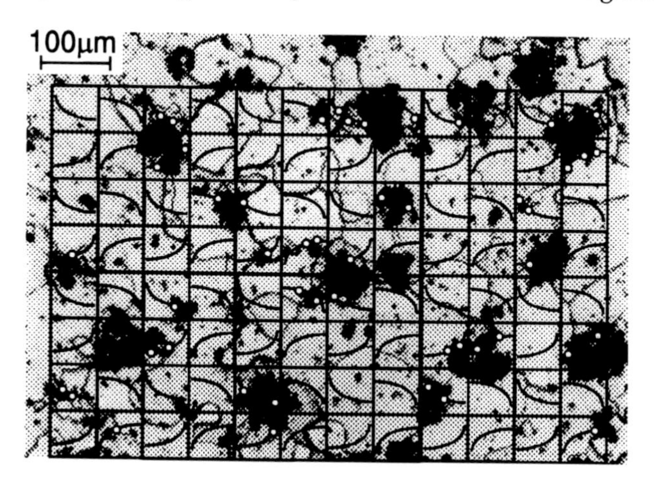

Figure 8.8 Microstructure of a cast iron revealed on a section of parallel to the axis of the rod-like specimen. A system of cycloids is superimposed that was used to measure the surface area of the particles in unit volume.

(Figure 8.9). For each circle, intersection points with the traces of the interfaces are determined, B_i. These points divide the circle into a series of arcs, B_kB_{k+1}. In order to estimate the parameter S_V, the distance of the intersection points, B_i, from the horizontal diameter of the circle, d_i, is measured for each intersection. Also the arcs, which are placed inside the reference space, are projected on to the horizontal diameter and the total length of projections, L_j, is established in a way that takes into account possible overlaps. Finally, the surface-to-volume ratio, S_V, is estimated from the following formula:

$$S_V \approx \frac{2}{R} M \frac{\sum d_i}{\sum L_j} \tag{8.16}$$

where the summation is over all intersections and all vertical sections. This method is certainly more amenable to automatic image analysis than to manual processing.

Gokhale and co-workers[2] have shown recently that in some applications the number of vertical sections can be reduced to three, arranged symmetrically about a selected direction (such sections making angles of 120°). They called this version a trisector and suggested that the following formula can be used:

$$S_V \approx 2.012 \left(P_L\right)_{cycl} \tag{8.17}$$

where $(P_L)_{cycl}$ is the density of the intersection points per unit length of the cycloids.

It should be also pointed out that in some situations it might be more convenient to refer the surface area of the particle boundaries to their volume

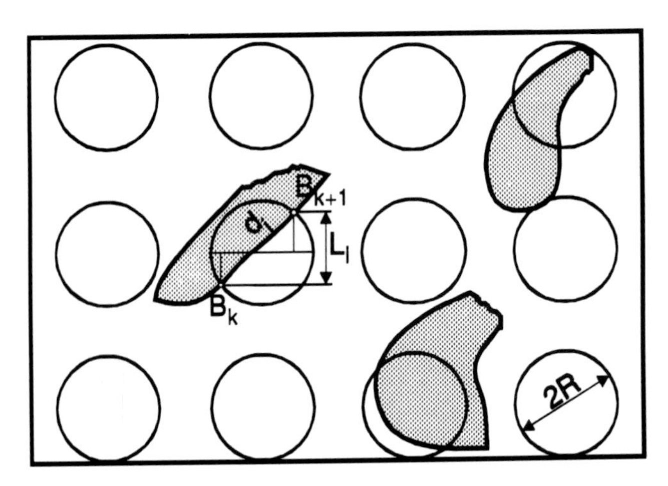

Figure 8.9 A system of circles that can be used to estimate the surface area of the particles in unit volume.

instead of referring them to the volume of the specimen. Such a particle volume ratio, $(S_V)_\alpha$, can be computed for given set of α-particles from the following formula:

$$\left(S_V\right)_\alpha = \frac{S_\alpha}{V_\alpha} \tag{8.18}$$

Example 8.6

The system of particles described in Example 8.5 is characterized by the surface area per unit volume, S_V, given as:

$$S_V = 12.3 \times 10^4 \text{ m}^{-1}.$$

The microstructure was also studied on the sections inclined at different angles to the specimen axes. Particle area measurements yielded the following estimate of the volume fraction of the particles:

$$V_V = 17\%.$$

These two estimates give the following value for the particle boundary per unit volume:

$$(S_V)_\alpha = 68.3 \times 10^4 \text{ m}^{-1}.$$

The surface area of particles can also be estimated from parallel sections using the method described in Reference 3. In this case, the sections of particles are sampled with a set of uniformly distributed points, Q_i. Then from each point inside a particle cross-section, 2-D random intercepts from the point Q_i to the boundary of the particle section are drawn and the boundary is intersected at the point I_i. In the case of a convex particle cross-section, there is one intersection point and one intercept, l_i^o, formed. This intercept is assumed to make an angle β_i with the tangent to the particle boundary at I_i. Using this notation, the mean surface, S, averaged over the particles sampled can be estimated from the following formula:

$$E_V(S) \approx \frac{4\pi}{N} \sum \left[\left(l_i^o\right)^2 \left(1 + \cot \beta_i \left(\frac{\pi}{2} - \beta_i\right)\right)\right] \tag{8.19}$$

where N is the number of particles sampled and the summation is over all the intercepts, which equals the number of particles sampled. The symbol E_V underlines the fact that the particles measured are sampled in proportion to their volume.

This formula can be extended for non-convex particle intersections, if the summation is performed over all intercepts l_i^o, l_i^1 . . . from the points Q_i inside particles to the intersection points I_i^o, I_i^1. . . .

8.2.3 Mean curvature of particles

The concept of curvature of lines has been described in Chapter 2. It can be used to define the curvature of a surface.

The curvature of the surface at a point is determined in terms of the curvature of lines on sections parallel to the normal to the surface. Among these lines there are those that show minimum and maximum curvature, k_{min} and k_{max}. These are the so called main curvatures of the surface at the point.

The mean curvature, H, is the mean of these two extreme curvatures:

$$H = \frac{1}{2}\left(k_{min} + k_{max}\right) \tag{8.20}$$

This value is equal to zero for a plane and has a constant value for a sphere:

flat surface H = 0 for any point;
sphere of radius R H = R for any point.

For more complicated surface geometry, the curvature varies from point to point, and the total and the mean curvatures can be computed according to the following equations:

$$M = \int_S H\, dS \tag{8.21}$$

$$M_S = \frac{M}{S} \tag{8.22}$$

respectively, where S is the area of the surface studied. The ratio of surface curvature to the volume, M_V, is given as:

$$M_V = \frac{M_S}{V} \tag{8.23}$$

where V is the volume and M_S the mean curvature of all the surface in this volume. This parameter, the mean volume curvature, M_V, is related to the mean curvature, k_A, of the lines of the intersections of these surfaces with the plane of observation by:

$$M_V = k_A \tag{8.24}$$

It can be shown (see Chapter 2) that k_A can be estimated via counting tangent points with a sweeping line passing the image of the feature on the section as illustrated in Chapter 2.

In the case of convex surfaces, the number of tangent points per section is always two and the following simplified equation can be derived:

$$M_V = 2\, N_A \tag{8.25}$$

where N_A is the planar density of the intersections. In this case the curvature of a system of features can be estimated by a simple counting procedure of the profiles revealed in the plane of observation.

Example 8.7

Two types of graphite particles in iron are shown on the micrographs in Figure 8.10b and Figure 8.10c—representative of microstructures A and B, respectively. These particles differ significantly in their shape. Particles in microstructure A are of a much more regular shape while particles in microstructure B are elongated. These two types of particles show the same mean volume curvature:

$$(M_V)^A = 0.004\ (\mu m^2)^{-1},$$
$$(M_V)^B = 0.004\ (\mu m^2)^{-1}.$$

However, the particles differ in the values of the mean surface curvature:

$$(M_S)^A = 2 \times 10^{-5}\ (\mu m^2)^{-1},$$
$$(M_S)^B = 2 \times 10^{-6}\ (\mu m^2)^{-1}.$$

It should be noted also that if the surface-to-volume ratio, S_V, is known then the volume curvature density, M_V, can be turned into the surface curvature density according to the equation:

$$M_S = \frac{M_V}{S_V} \tag{8.26}$$

Measurements of particle curvature have proved to be useful in studies of the kinetics of phase transformations.[4]

8.2.4 *Density of particles*

The density of particles according to stereological notation is designated N_V. This defines the number of particles in a given reference volume, V. In the case of materials in which particles are distributed more or less uniformly, this parameter defines the number of particles in a unit volume.

Despite its apparent simplicity, the methods of N_V estimation are not quite obvious and for years were the subject of intensive theoretical studies. This difficulty was related to the lack of proper sampling procedures for particles distributed in space. The most frequently used technique for particle observations is an examination of 2-D sections. From the point of view of stereology this is equivalent to sampling particles with a plane probe. In this case a plane is drawn through the system studied and observations are made of the particles hit by this plane. However, this manner of particle selection introduces a systematic error due to the fact that large particles are hit more frequently than the smaller ones as illustrated schematically in Figure 8.11. In fact, the particles are sampled (hit by the plane) with a probability

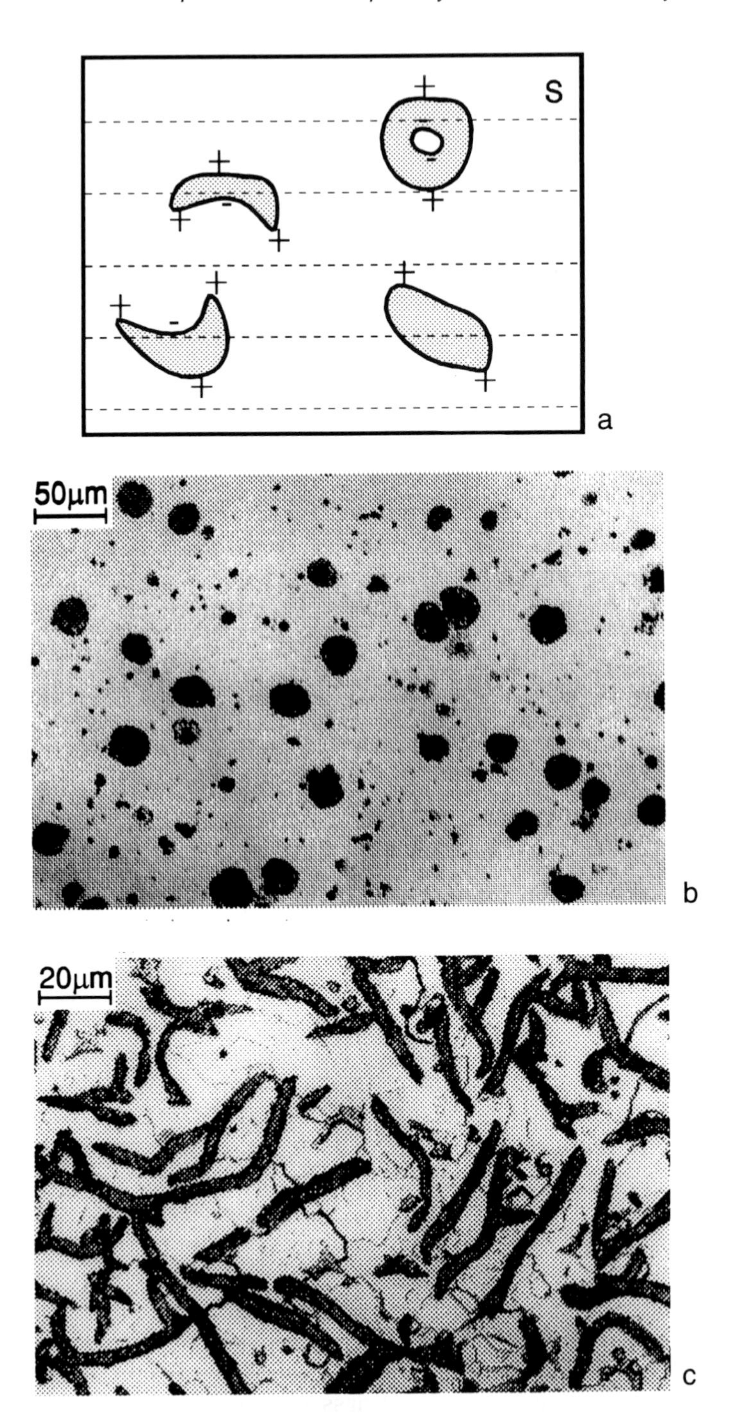

Figure 8.10 Schematic representations of the measurements (a) used to estimate the mean curvature of the graphite particles revealed in cast iron samples A and B seen in Figures 8.10b and c, respectively.

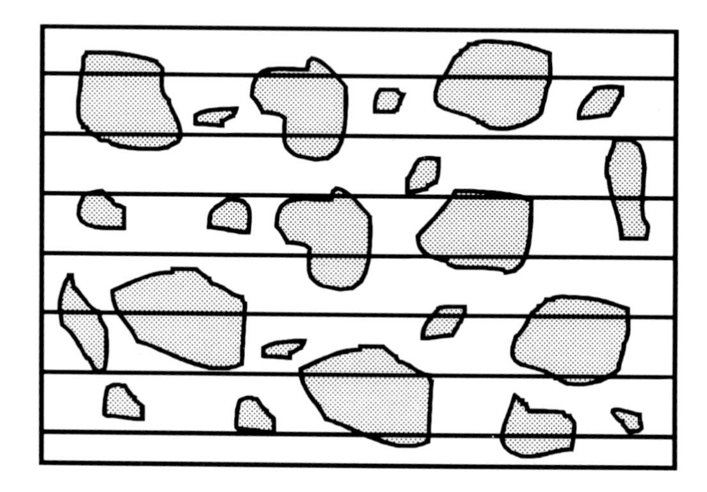

Figure 8.11 Schematic explanation of the effect of the particle size and shape on the probability for a particle to be sampled (hit) in the measurements on the microstructure of a two-phase material.

proportional to their height, h_i. With such a sampling procedure, whatever measurements are made, there is a likelihood of introducing a bias.

This problem of particle sampling was resolved in the past by making assumptions with regard to the particle shape. Actually, in order to obtain sensible results, it was necessary to describe beforehand the shape of particles quite precisely. Simple solutions have been obtained for features which have shapes such as spheres, rods, disks, and ellipsoids. Some of them are quoted in Chapter 2 and more can found in Reference 5.

The classical procedure starts from the assumption that we deal with a system (material) containing particles of a known shape, for example, in the form of cubes. It is also assumed that the particles have sizes, measured, for example, by volume or edge length, in the present case designated as L, varying from 0 to some maximum value L_{max}. This range of sizes is divided into a finite number of classes, L_i, where i = 1,2,3, . . ., N. The relative number of particles in each class is designated N_i.

Knowing the shape of particles, it is possible for each size class to obtain theoretical distributions of particle sections. This can be done either experimentally, by studying 3-D model figures, or with the help of computer simulations. Usually the analysis starts with the largest particles in the class L_N, i.e., the range of particle sections is obtained from 0 to some maximum section size, A_{max}. This range of particle section sizes is divided into M classes, A_j (j = 1,2, . . ., M), with the biggest sections in class A_M. In the next step the relative numbers of sections in each class, n_{jN}, are established for the particles in class L_N. By proceeding to the classes of smaller particle size, a matrix is obtained, n_{jK}, which defines potential contributions of particles of size K to the sections of size j.

The next part of the procedure is to produce a model of the experimental process of particle sampling or, in other words, a decision is made as to what are the chances of revealing on the cross-section particles of a given size. In most cases, the chances of cutting a particle are proportional to its height and in the case of cubes it is proportional also to the length of the edge (this means that particles with an edge length ten times bigger are ten times more frequently cut by the section of observation). In this way the expected relative number of sections in a given class can be expressed in the following way:

$$n_j = N_1 L_1 n_{j1} + N_2 L_2 n_{j2} + \ldots + N_N L_N n_{jN} \tag{8.27}$$

or in matrix notation:

$$\begin{bmatrix} n_1 \\ n_2 \\ \cdots \\ n_m \end{bmatrix} = \begin{bmatrix} A_{11} & A_{12} & \cdots & A_{1N} \\ A_{21} & A_{22} & \cdots & A_{2N} \\ \cdots & \cdots & \cdots & \cdots \\ A_{m1} & A_{m2} & \cdots & A_{mN} \end{bmatrix} \begin{bmatrix} N_1 \\ N_2 \\ \cdots \\ N_N \end{bmatrix} \tag{8.28}$$

where

$$A_{ij} = n_{ij} L_j \tag{8.29}$$

This is a linear equation that can be solved to find one solution if N = m. More about this procedure can be found in Reference 6. Application of the equations to the present case of cubes is given in the following example.

Example 8.8

Particles in the shape of cubes are characteristic elements of the microstructure of some steels. Figure 8.12a shows a microstructure revealed in a low-alloy steel and Figure 8.12b gives the distribution function of the sections of the particles, which have been assumed to be cube-shaped. Figure 8.12c presents a distribution function of the intercepts for a cube of unit volume. By comparing the plots in Figure 8.12b and c, it can be concluded that the material contains cubes of volume $V_c = l_c^3 = 79.5 \times 10^3\ \mu m^3$.

Consider a system of N such cubes in a material of volume $V = A \times H_o$. Cut the volume of the material with parallel sections H apart with $H > l_c$, where l_c is the length of cube edge. In the slab of the material of volume $A \times H$ there are $N_h = N \times H/H_o$ cubes. The probability that a cube in this volume is cut by a section plane is proportional to the ratio l_c/H. Thus, the number of cube sections per section area is given as:

$$N_A = \alpha \frac{l_c}{H} N \frac{H}{H_o} \frac{1}{A} \tag{8.30}$$

(α is a constant), which is equivalent to the following:

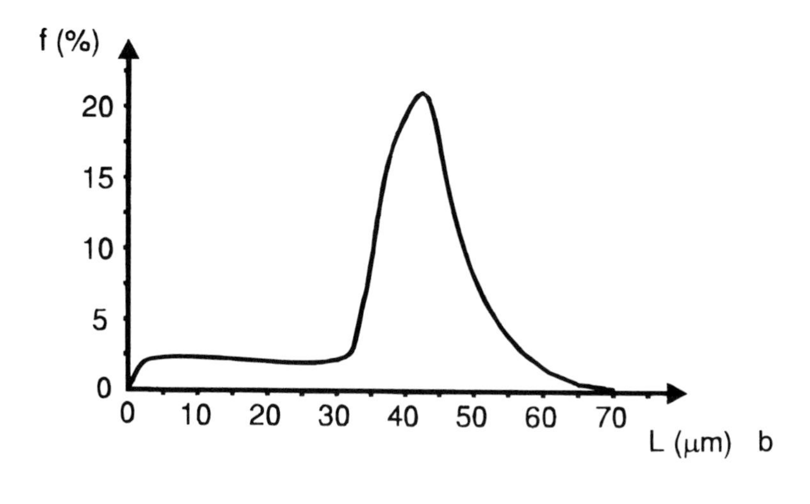

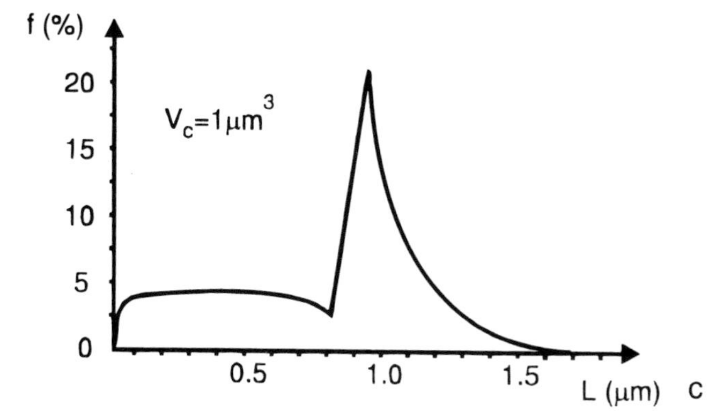

Figure 8.12 Typical image of the microstructure revealed on a section of a steel (a); the data on the distribution function of the particle section intercept length (b); and the intercepts for a unit cube (c).

$$N_A = \alpha N_V l_c \tag{8.31}$$

More detailed studies of cube geometry show that $\alpha = 0.667$.

A number of shapes has been analyzed in detail and a comprehensive summary is given by Underwood.[5] His analysis covers both regular spheres as well as elongated particles. For the case of fixed shape and size particles the general formula can be simplified into the form:

$$N_V = \kappa \frac{N_A^2}{N_L} \tag{8.32}$$

with the values of κ being dependent on the projected area and height of the particles. The value of κ for selected particle shapes are given in Table 8.1.

For the case of a system characterized by a particle size distribution function, Equation 8.32 needs to be modified and additional parameters have to be measured. For a system of spherical particles of different sizes this additional parameter is E(m)—the mean value of the reciprocal of the particles' cross-section diameter, d_i:

$$E(m) = \frac{1}{N} \sum_{i=1}^{i=N} \frac{1}{d_i} \tag{8.33}$$

The differences in the analysis between constant- and variable-size systems of particles are demonstrated using a system of spheres in the following example.

Table 8.1 Values of the Coefficient κ for a System of Constant Size Particles of Different Shapes

Value of the coefficient κ	Shape of the particles
$\frac{\pi}{4} = 0.785$	Spheres
$\frac{2}{\pi} = 0.637$	Disks
0.732	Cylinders (radius=height)
0.739	Hemispheres
0.667	Cubes
$4\frac{a}{c}$	Square rods (a × a × c)
0.744	Truncated octahedra
0.784	Pentagonal dodecahedra

Example 8.9

Consider the micrograph shown in Figure 8.13. Suppose that the particles revealed in the section of observations are spherical and have the same volume V. Based on this assumption one can use the following equation for estimating particle density:

$$N_V = \frac{\pi N_A^2}{4N_L} = 21{,}187\ mm^{-3} \tag{8.34}$$

The distribution function of particle section areas for the particles shown in Figure 8.13 is depicted in Figure 8.14 together with a modelled distribution for monosized spheres. These two distribution functions differ significantly. The assumption of a constant size of particles is not justified. In order to estimate the density of particles one has to use the following equation:

$$N_V = \frac{2E(m)N_A}{\pi} = 20{,}120\ mm^{-3} \tag{8.35}$$

where

$$E(m) = \frac{1}{N}\left(\frac{1}{d_1} + \frac{1}{d_2} + \ldots + \frac{1}{d_N}\right) \tag{8.36}$$

This equation takes into account the fact that the particles of a different size are cut by the intersection plane with different probability. In fact, the probability for a particle of diameter d being cut, and thus the density of intersections of such particles, is proportional to d.

It can be noted that neither Equation 8.34 nor 8.35 can be used if the particles are not spherical.

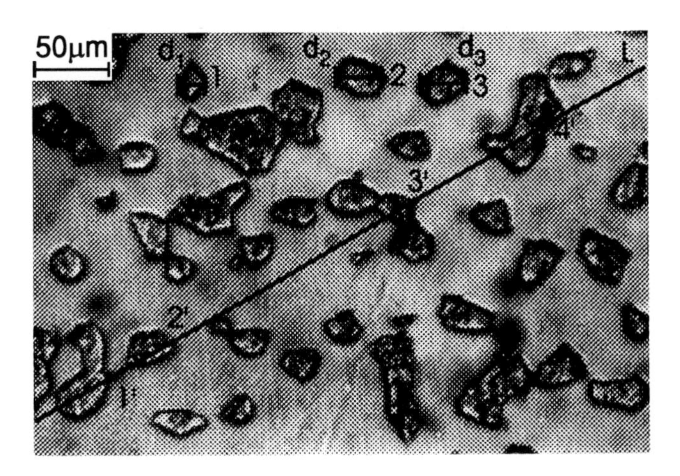

Figure 8.13 Microstructure of a two-phase material. The microstructure was tested by counting the number of sections per unit cross-sectional area and the density of the particles per unit length of a test line (one of the test lines is shown).

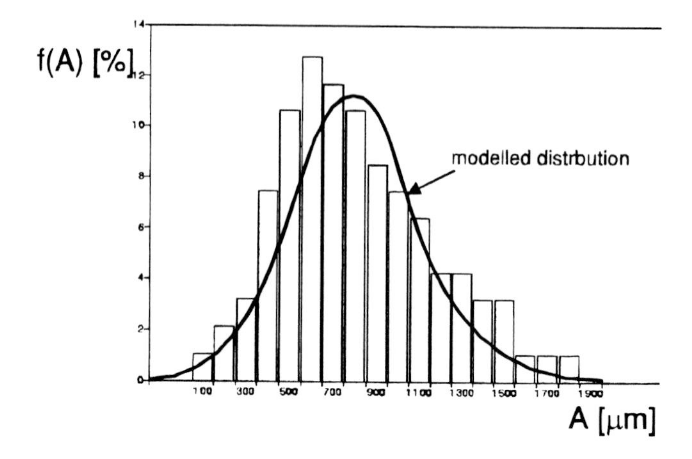

Figure 8.14 The distribution function of the particle section areas for a model microstructure consisting of monosized particles together with this function for the particles in Figure 8.13.

Table 8.2 Formulae for Systems of Mono- and Multisized Spherical Particles

	Mono	Multi	Mean value
V	$\frac{4}{3}\pi r^3$	$\frac{\pi V_V}{2E(m)N_A}$	$E(V)$
S	$4\pi r^2$	$\frac{2\pi N_L}{E(m)N_A}$	$E(S)$
r	$\frac{2N_L}{\pi N_A}$	$\frac{\pi}{4E(m)}$	$E(r)$
N_V	$\frac{\pi N_A^2}{4N_L}$	$\frac{2E(m)N_A}{\pi}$	N_V

It may be noted from the example given above that the radius of particles in a monosized system is directly related to the densities of their intersections with test lines and planes. In a system of particles with a variety of sizes the mean radius is a function of the reciprocals of the diameters of the particle sections.

Similar expressions to those given for systems of spherical particles in Table 8.2 can be derived for particles of more complicated shapes. However, in these cases more parameters are necessary in order to describe the geometry of the particle cross-sections observed on sections of the material.

Particles of disk shape are generally defined by their radius, r, and thickness, t. On cross-sections such disks are seen as elongated convex figures characterized by a length, L, and a width-to-length ratio, G. For a system of disks of a variety of sizes one can derive the following equations:

$$E(r) = \frac{\pi}{4E(L^{-1})}; \quad E(t) = \frac{\pi E(G)}{4E(L^{-1})} \tag{8.37}$$

and

$$E(V) = \frac{\pi V_V}{8E(L^{-1})N_A} \tag{8.38}$$

$$N_V = \frac{8}{\pi^2} E(L^{-1})N_A \tag{8.39}$$

where V_V is the volume fraction of the particles, $E(L^{-1})$ the mean of the reciprocals of L (the disk lengths on the section plane), and E(G) is the mean value of the ratio of the width-to-length dimensions of disk sections.

Example 8.10

A microstructure of a two-phase material studied on randomly oriented sections revealed characteristic micrographs exemplified by Figure 8.15. The needle-like appearance of the particles seen in Figure 8.15 has helped to identify them as sections of particles with disk shape. The following parameters have been measured for the sections of the particles:

E(L)=2.1 μm,
E(G)=27.

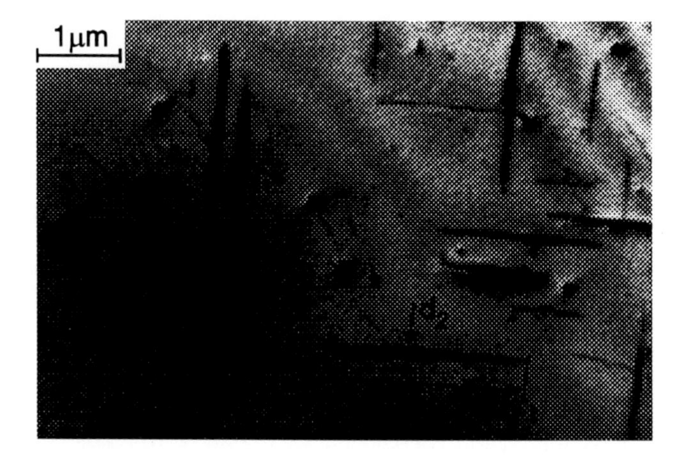

Figure 8.15 An example of a section of a two-phase material.

The analysis presented so far leads to the conclusion that the particle density can be estimated if data are available about the shape and dimensions of the particles. The parameters measured on cross-sections of the material are first used to estimate the size of particles and then their density. The size of particles appears to account for differences in geometrical sampling of the large and small particles. In fact, the formula for N_V can be generally given in the form:

$$N_V \sim N_A \frac{1}{E(D)} \tag{8.40}$$

where E(D) is the mean true diameter of the particles. This parameter can be estimated from a 2-D image of the microstructure only if the shape of particles is known.

Modern stereological procedures, derived in the eighties, make it possible to estimate the density of particles, their mean size and coefficient of variation without any assumption about the their shape.

The density of particles can be estimated as a single parameter, i.e., without the need to compute size distributions, which may provide redundant and not very precise information, using the disector method. In this case, each of the particles is reduced to a point (for convenience it can be assumed that this is the center of gravity of the particle of interest) and these points are counted in some reference volume, V. The density of the particles is then computed directly from the equation:

$$N_V = \frac{N}{V} \tag{8.41}$$

where N is the number of counted points in the volume V.

The concept of the disector has been outlined in Chapter 3. The idea is to count those particles that have their highest points **below** the surface of the initial observations—the so called "look-up" plane—and not on the subsequent sections. The image of the first section, I_0, is recorded and then the first layer removed and the particles are counted in section I_1 which were not seen on I_0. (See Figure 8.16.) This is repeated with the subsequent layers so as to ensure that any particle is counted only once. The disector provides a method for proper sampling of the particles disregarding their shape and size.

Example 8.11

Two schematic micrographs are shown in Figure 8.17 that were obtained by a serial sectioning technique with the sections placed 2 μm apart. Particles revealed in section I_{i+1} and not seen in I_i were counted. The number of such particles between I_i and I_{i+1} was equal to 1. Since the area of observation was $A = 2.4 \times 10^5\ \mu m^2$, the density, $N_V = 2.089 \times 10^3\ mm^3$.

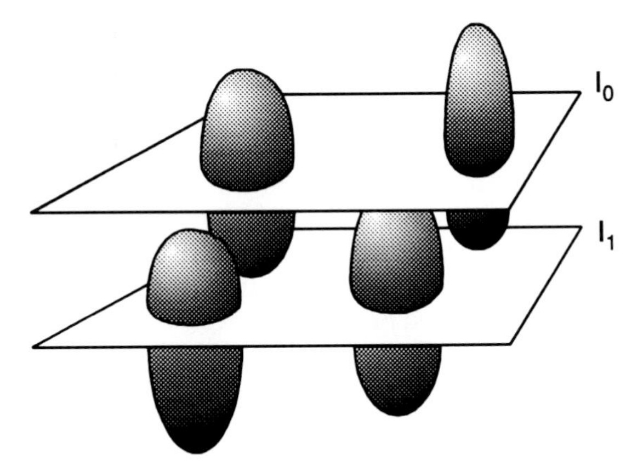

Figure 8.16 A schematic explanation of the concept of the disector method (see Chapter 3).

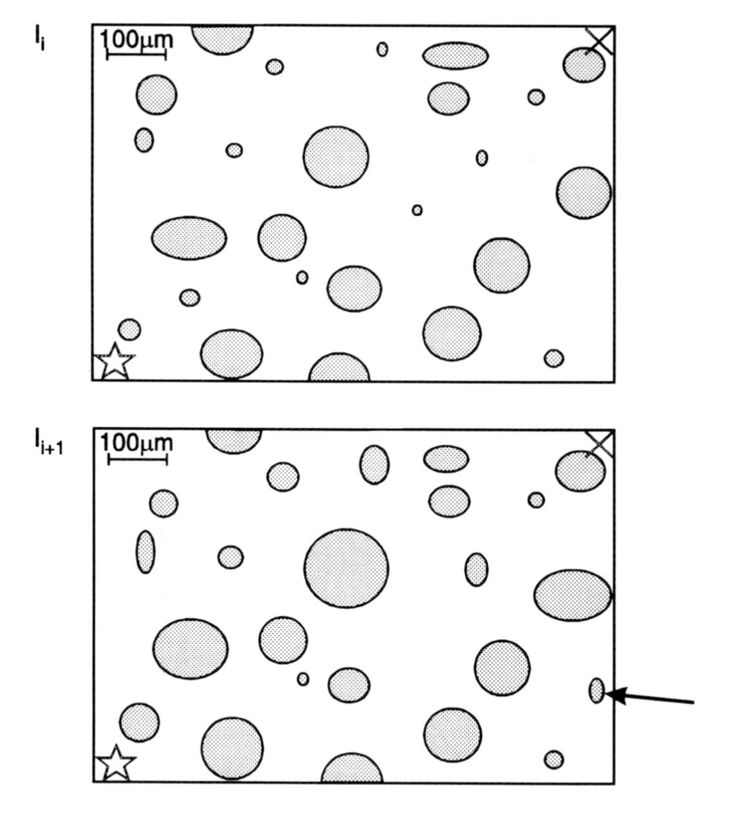

Figure 8.17 Schematic representation of the microstructures revealed on two parallel sections in a multiphase material. The particle indicated by the arrow was hit by the second section and not by the first. Such particles are counted in the estimation of the density of particles.

By its nature the disector approach is a volume probing method since it requires a study of the particles in some sampled volume of the material. It should be stressed that in this procedure the sampled volume needs to be known, which means in practice that we have to measure the total thickness of the layers removed, t, and the area of observations, A ($V = At$). In most cases t can be measured with sufficient precision mechanically and A is obtained from the known magnification. For better alignment of the area of observations on consecutive sections and better control of the layers removed, one may mark the surface of the specimen with microhardness indentations. This method also permits a calibration of the thickness of the material removed from the known geometry of a hardness indent (see Chapter 3).

8.3 Individual properties of particles: mean volume and coefficient of variation

In certain cases the global characteristics of particles in terms of volume fraction, specific interfacial area, etc. proves to be insufficient to explain the properties of materials. For example, the strengthening effect of particles is known to be dependent on the degree of their dispersion, or in other words on their size and, in particular, their average volume.

As in the case of their density, the size of particles in the material can be approached either with some specific assumptions about their shape or without such assumptions at the expense of making more sophisticated measurements. If the first method is chosen, one has to follow the procedures already described, based on the division of the particles of a model shape into size classes. The focus of the following discussion is therefore to be placed on assumption-free techniques. The discussion starts with the description of the method for determining the mean particle size and the coefficient of variation of this, and then moves to an estimation of the size distribution function.

It has been shown recently that the mean particle volume revealed in an observation section can be obtained by a combination of point, line and plane sampling. The idea is to draw a randomly placed and oriented line piercing a particle and measure the length of the intercept, or intercepts if the particle is not convex. In this method the following steps are taken:

Step 1
random sectioning planes, A_i, for observations
Step 2
random selection of points, Q_j, on the sections of the particles
Step 3
random intersecting with lines, L_k, drawn through the points Q_j

As discussed in Chapter 3 where more details can be found about the method, random sectioning with planes can be executed using the concept of the orientator. This requirement for random sections can be dropped if the particles are randomly distributed over the volume of the specimen. In such a case, a system of parallel sections can be used or one large section.

A system of randomly selected points can be superimposed on the image of the microstructure relatively easily. The coordinates of these points can be determined using random numbers. One can also use a randomly placed grid of points. The lines, one through each point, are drawn through all the points inside particle sections. They may be chosen to be parallel to some reference direction if the particles are randomly oriented.

The lines intersect particle profiles in at least two points, A_0^+ and A_0^-, at a distance $l_0^\pm$ from the point Q. If the particles are concave more pairs of intersection points, $A_j^\pm$, are formed. The distances from the sampled point, Q_i, hitting the i-th particle, to the intersection point pairs denoted as $l_j^\pm$, are used to estimate the mean volume of the particles.

With these steps taken, the mean weighted volume, $E_V(V)$, can be computed from the following equation:[7]

$$E_V(V) = E\left[\frac{2\pi}{3}\left[\left(l_o^+\right)^3 + \left(l_o^-\right)^3 + \sum_{j=1}^{k-1}\left|\left(l_j^+\right)^3 - \left(l_j^-\right)^3\right|\right]\right] \tag{8.42}$$

where k is the total number of intercepts created by one line.

It should be noted that the equation given above provides an estimate of the mean weighted volume, $E_V(V)$, simply because of the way the particles are sampled. By sectioning a specimen we reveal the large particles more frequently than the small ones. The fact that this is the weighted mean does not imply that this is a less useful parameter for application in materials science. Evaluation of the usefulness of a given parameter can only be made with respect to some specific model for the role of particles in controlling the properties of the material. Existing theories are not particularly precise from this point of view and the weighted mean volume may prove to be quite satisfactory in most cases. It should also be pointed out that Equation 8.42 is suitable for computer-aided automatic analysis.

Gundersen[8] has further shown that the equation can be reduced to the form:

$$E_V(V) = \frac{4\pi}{3} E\left(l^3\right) \tag{8.43}$$

where l is the distance from the sampled point inside a particle profile to the intersection of the line with the particle profile and this averaging is over all intersection points and all particles sampled.

Example 8.12

The system of particles shown in Figure 8.18a was studied using the concept of point sampled intercepts. The distribution function of the intercept length is given in Figure 8.18b. The mean value E(l) was found to be 17 μm. Figure 8.18c shows a distribution function of $f(l^3)$. The mean value $E(l^3)$ was found to be equal to 4.9×10^3 μm³. In this way an estimate of the mean weighted value of the particle volume, $E_V(V)$, is equal to 2.06×10^5 μm³.

The equations cited can be applied also to:

a) a single particle cut by a number of sections;
b) particles sampled with the use of a disector.

In these two cases one can obtain an estimate of the particle volume, case (a) and the mean number volume, case (b) as the disector samples particles with the same probability, irrespective of their sizes. The following formula may then be used:

$$E(V) = E\left[\frac{2\pi}{3}\left[\left(l_o^+\right)^3 + \left(l_o^-\right)^3 + \sum_{j=1}^{k-1}\left|\left(l_j^+\right)^3 - \left(l_j^-\right)^3\right|\right]\right] \tag{8.44}$$

if the points are sampled inside a profile of a given particle and averaging is taken over all such points:

$$E_N(V) = E\left[\frac{2\pi}{3}\left[\left(l_o^+\right)^3 + \left(l_o^-\right)^3 + \sum_{j=1}^{k-1}\left|\left(l_j^+\right)^3 - \left(l_j^-\right)^3\right|\right]\right] \tag{8.45}$$

if the particles probed with points and lines are those sampled with the help of a disector.

The methods described so far have dealt with a system of particles, V_i, in a volume of material, V. In this situation it is assumed that on a cross-section, C, through the specimen a large number of particle sections, S_j, are observed. These sections are the objects of interest and their sampling with a grid of points is combined with line intercepts. As a result the mean weighted particle volume and the mean surface area are obtained, $E_V(V)$ and $E_V(S)$, for the population of particles, V_i. However, if the particles are sampled by means of a disector the same procedures give the number means, $E_N(V)$ and $E_N(S)$.

Particles in multiphase systems usually differ in their volume and the same mean volume can be ascribed to geometrically different populations of particles. Differences in the volume of particles can be described in a similar way by the procedures described in Chapter 6, for the case of grains in polycrystals. One of the parameters introduced there was based on the minimum and maximum volume, V_{min} and V_{max}. The degree in the diversity (spread) in the particle size is described by a parameter C_1 defined as:

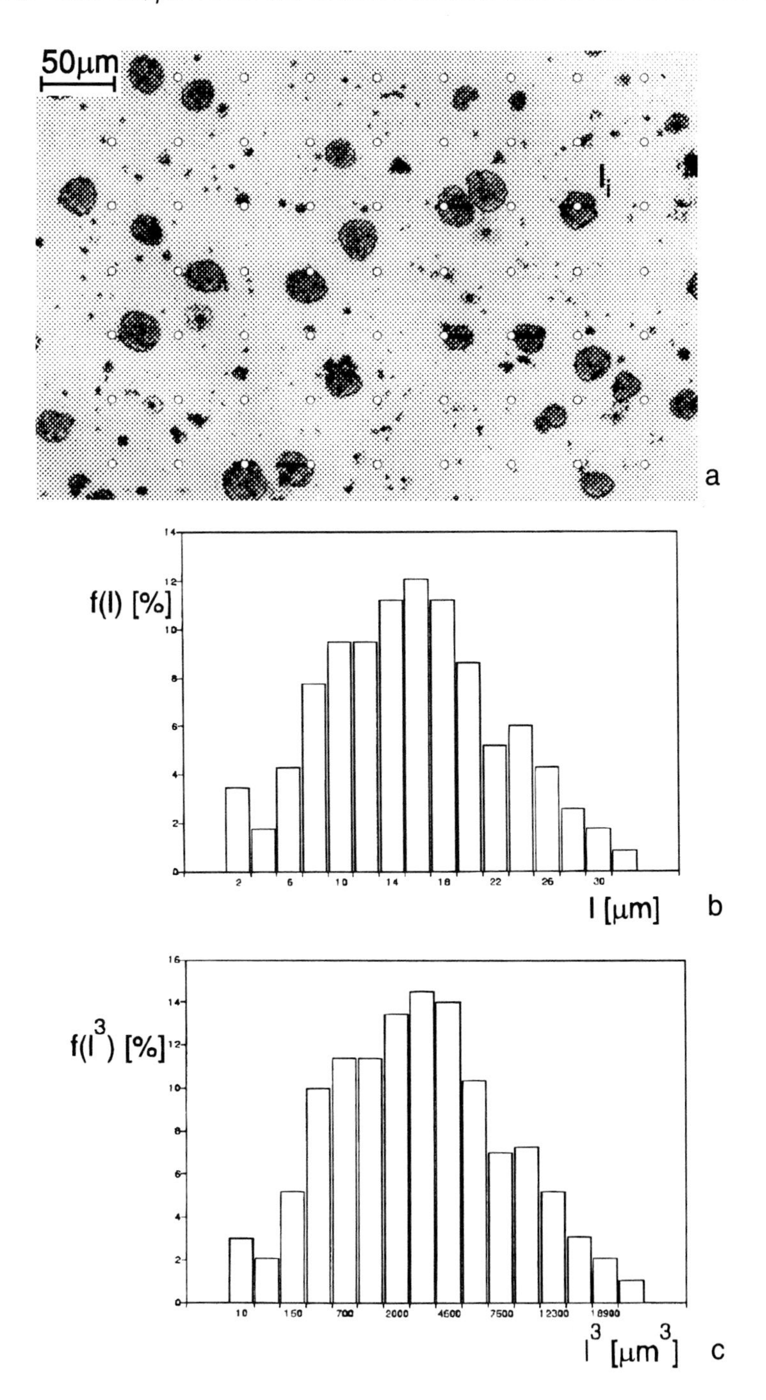

Figure 8.18 A system of particles sectioned with a randomly oriented plane that has been tested with point sampled intercepts (a); the distribution function of the point sampled intercepts, f(l), is shown in (b); the distribution function of $f(l^3)$ is shown in (c).

$$C_1 = V_{max} - V_{min} \tag{8.46}$$

Another parameter C_2 can be defined such that:

$$C_2 = \frac{V_{max} - V_{min}}{E(V)} \tag{8.47}$$

The diversity (spread) in the particle size can also be quantified in terms of the variation, VAR(V), standard deviation, SD(V), or coefficient of variation, CV(V). These second order moments carry information on particle size uniformity.

The procedure proposed by Jensen and Gundersen,[9] used previously to determine the mean weighted volume of grains, can also be used to quantify the degree of particle volume uniformity. This is related to the following equation:

$$VAR_V(V) = E_V(V^2) - (E_V(V))^2 \tag{8.48}$$

and the relationship derived in Reference 10:

$$E_V(V^2) = 4\pi k E(A^3) \tag{8.49}$$

In this relationship A is the area of the particles hit by the P_i points and k is a constant with a value in the range from 0.071 to 0.083.[10] The value of k depends on the particle shape. Its value needs to be determined either experimentally or by appropriate modelling.

From the last two equations one can obtain the following relationships:

$$VAR_V(V) = 4\pi k E(A^3) - \frac{\pi^2}{9}\left[E(l^3)\right]^2 \tag{8.50}$$

and

$$CV_V^2(V) = \frac{36kE(A^3)}{\pi[E(l^3)]^2} - 1 \tag{8.51}$$

Experimental data show that the distribution function, $f_N(V)$, frequently may be well approximated by a log-normal distribution, $LN_N(V)$.[11] This observation can be used to simplify the experimental studies of the volume distribution parameters described so far. Such a distribution considerably simplifies the relationships between the weighted and true variance of particle volume and in turn the true and weighted values of CV(V). For a log-normal distribution one can obtain:

$$CV_N(V) = CV_V(V) \tag{8.52}$$

This equation shows that for a log-normal volume distribution, $LN_N(V)$, the coefficients of variation for true volume and weighted volume distribution functions are equal.

Taking into account the physical aspects of the procedures of microscopical measurements, one can derive the following relationships:

$$CV_N(V) = \sqrt{\frac{36kE(A^3)}{\pi\left[E(l^3)\right]^2} - 1} \tag{8.53}$$

and

$$E_N(V) = \left[\frac{\pi^4}{972k}\frac{\left(E(l^3)\right)^5}{E(A^3)}\right]^{\frac{1}{3}} \tag{8.54}$$

These two equations can be used to estimate the true volume and the coefficient of variation for a log-normal distribution of particle volume from the relatively simple measurement procedures described earlier.

The coefficient CV(V) can also be computed if the two complementary estimates of the mean volume are known: $E_N(V)$ and $E_V(V)$. It should be remembered that the following equations are valid:

$$E_V(V) = A\int V^2 f(V)\,dV = E_N(V^2)\cdot E_N(V)^{-1} \tag{8.55}$$

$$CV_N(V)^2 = \frac{VAR_N(V)}{E_N(V)^2} = \frac{E_N(V^2) - E_N(V)^2}{E_N(V)^2} = \frac{E_V(V)}{E_N(V)} - 1 \tag{8.56}$$

From the last of these equations it can be concluded that an estimation of CV(V) is possible if the mean volume and weighted mean volume are known. The first mean, the number mean, can be estimated using the concept of the disector. On the other hand, the volume-weighted mean volume can be approached by the point sampling method described in Chapter 3 and the present chapter.

Example 8.13

Figure 8.19 shows a system of particles revealed in a section of a two-phase material. Mean values $E_N(V)$ and $E_V(V)$ were measured using the disector method and point sampled intercepts. Using Equation 8.56 the following value of $CV_N(V)$ was estimated:

$$CV_N(V) = 1.51 \quad (E_V(V) = 0.88;\ E_N(V) = 0.27\ [10^5\mu m^3]).$$

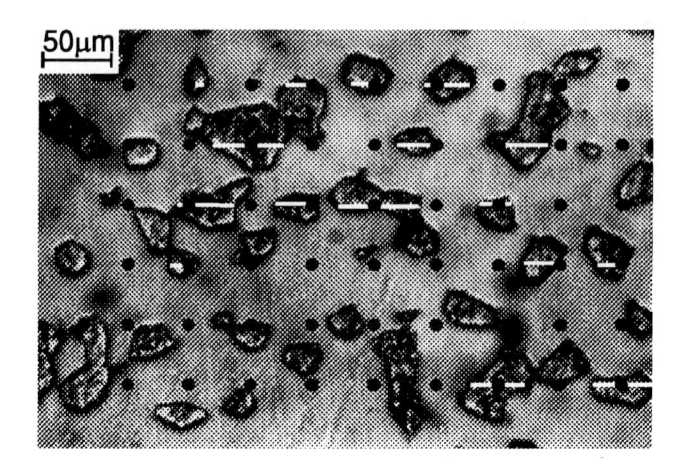

Figure 8.19 A microstructure revealed on a section of a two-phase material with a superimposed grid of test points. Intercepts are drawn through the points hitting sections of the particles.

The mean squared volume, $E(V^2)$, required for estimation of CV(V), can also be determined by a combination of two sampling procedures with test points. The following steps should to be taken:

a) point sampling with a system of points, Q_i, superimposed on the sections of the particles;
b) further point sampling with two more points randomly selected inside the intersections selected in (a).

For the particle intersections revealed on a cross-section of the specimen this procedure can be used to estimate the volume-weighted variance $VAR_V(V)$ from Equation 8.48 using the following formula for $E_V(V^2)$:

$$E_V\left(V^2\right) \approx 4\pi E\left(a_o^2 \Delta\right) \tag{8.57}$$

where a_o is the area of the section sampled in step a) and Δ is the area of the triangle formed by two additional random points selected inside the section.

8.4 Particle arrangements and orientations

Within the framework of a general approach to the arrangement of particles, their spatial distribution can be described in terms of the free separation between the particles and their mean random spacing. The mean free separation of particles, λ, is defined as the mean value of the distance between the particles measured along a randomly placed test line. The value of λ can be estimated from the following equation:

$$\lambda = \frac{1 - V_V}{N_L} \tag{8.58}$$

where V_V is the volume fraction of the particles studied and N_L is their linear density, i.e., the number of particles per unit length of a randomly placed and oriented test line.

Using the concepts of the mean free distance and the mean intercept length, L_3, for an individual particle one can also define the mean random spacing, δ, as the mean uninterrupted center-to-center distance between all possible particles. The following relationship can be derived for δ:

$$\delta = \lambda + L_3 \tag{8.59}$$

This relationship involves the parameters λ and δ that determine the distance between any two particles measured on randomly chosen lines (δ determines center-to-center distance, while λ determines edge-to-edge length of intercepts).

Example 8.14

Figure 8.20 shows a micrograph typical of the random sections of a two-phase material from which the following parameters were derived:

$V_V = 0.23$;
$N_L = 1.167 \times 10^{-2}\ \mu m^{-1}$;
$\lambda = 66\ \mu m$;
$\delta = 306\ \mu m$;
$L_3 = 240\ \mu m$.

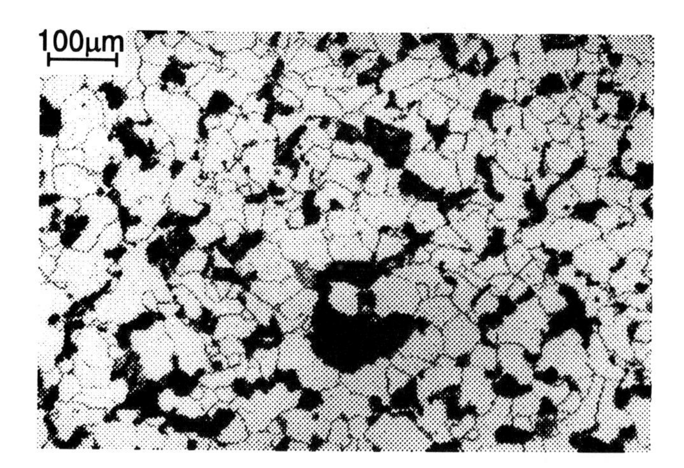

Figure 8.20 A microstructure of a two-phase material used in Example 8.14. The dark areas are sections of the particles distributed in the matrix recorded here as light areas. It can be noted that the matrix is in the form of a polycrystal. This microstructure was tested with a system of points and test lines.

In a large number of applications the relevant parameter characterizing a system of particles is found to be the distance between the nearest neighbors. This distance can be measured directly, as discussed later in the text. The description can also concentrate attention on the mean value of the distance, which is designated as Γ.

The formulae that can be used to determine the mean distance between nearest neighbors have been derived for two specific cases: the distance in a plane Γ_2 and in space Γ_3. In both cases an assumption is made that the particles are small, which effectively means that they are much smaller than the distance between them. With this condition met, the distances depend only on the particle densities and are given as:

$$\Gamma_2 = \frac{1}{2}(P_A)^{-\frac{1}{2}} \tag{8.60}$$

$$\Gamma_3 = 0.554 P_V^{-\frac{1}{3}} \tag{8.61}$$

Example 8.15

Consider the systems of particles revealed in Figure 8.21a, b. The values of the basic parameters measured for these microstructures are given in Table 8.3.

The parameters introduced so far, λ, δ, and Γ, are linear measures of particle dispersion. They represent slightly different characteristics of the average distance between the particles. As such they are insensitive to the position of individual particles as illustrated in the example given above. On the other hand, in a number of materials science applications it is observed that particles are distributed in a nonuniform way and the description of this tendency for particle segregation is a major task for the quantitative characterization of materials.

Particles in a number of systems are known to form clusters of significantly higher density than the density in other regions. Before this effect is analyzed in more detail, a general comment should be made that the density of particles in a volume of a material should refer to particle number, N(B), in some specified volume B. The parameter N(B) is determined as a function of spatial coordinates (x_1,x_2,x_3) or shortly x_i. Two types of coordinate systems can be used:

a) a system related to the geometry of the specimen studied, for instance, related to its surface;
b) a system positioned at the center of a particle.

The first approach means that the possible nonuniformity or clustering are located at some characteristic positions inside the artifact studied. The

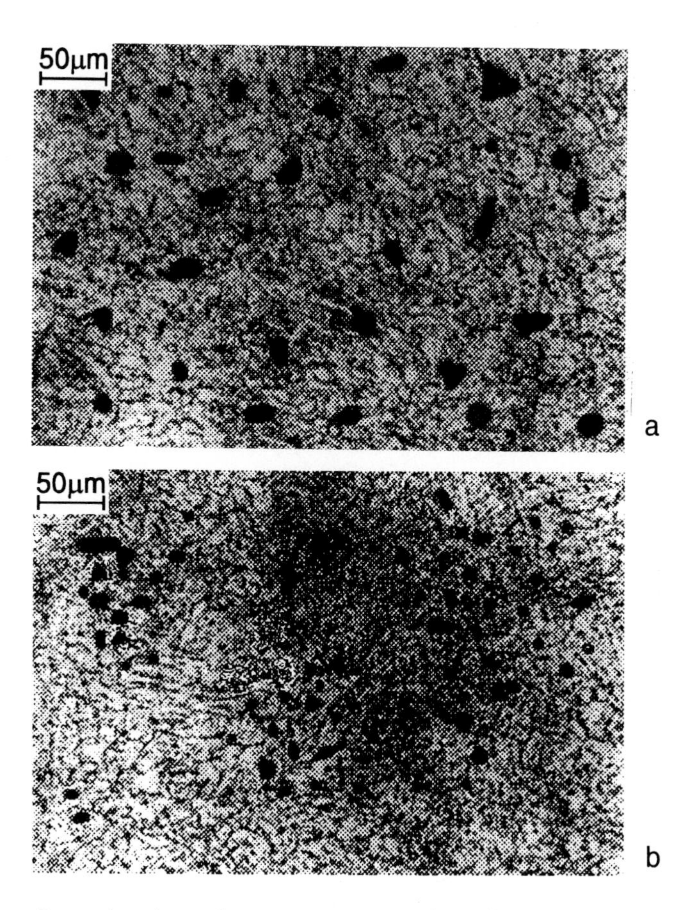

Figure 8.21 Examples of two-phase microstructures that differ in the spatial arrangement of the particles: (a) uniform distribution A and (b) distribution characterized by the presence of clusters B. Example 8.15 shows that these two microstructures yield similar values of the basic stereological parameters used to describe the dispersion of the particles.

Table 8.3 Values of the Basic Parameters Characterizing the Particle Distributions Shown in Figure 8.21a,b

Microstructure	$\lambda(\mu m)$	$\delta(\mu m)$	$\Gamma_2(\mu m)$	$\Gamma_3(\mu m)$
A	120	131	35.4	28.1
B	136	143	28.4	24.3

Note: The values of Γ_3 were computed assuming that the particles are of spherical shape.

second approach leads to the concept of radial distribution functions, which describes properties of the particle neighborhoods.

The density function $N_B(x_1,x_2,x_3)$ defined for some volume B or density function $N_A(x_1,x_2)$ defined for area A changes discrete functions of particle positions into continuous functions of their density. This effect is demon-

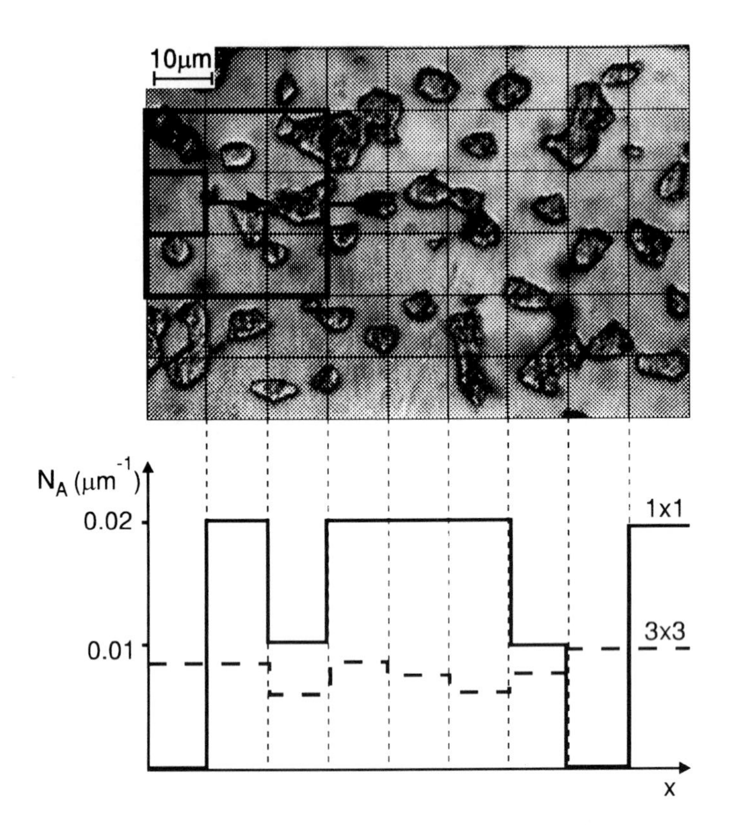

Figure 8.22 Discrete distribution of particle sections and its representation in the form of a continuous particle section density function.

strated for the system of particles shown in Figure 8.22. The particle density functions obtained in this way may be studied further using the special tools developed for analytical functions, an example of which is given below.

Example 8.16

Figure 8.23a shows particles revealed in a transverse section of a surface-treated material. The number of particles was studied in strips of the material at some distance, x, from the surface of the specimen. The function $N_A(x)$ is shown in Figure 8.23b.

An alternative approach to the problem assumes a division of the matrix containing the particles into finite volume elements, B_{ijk}, into which the studied specimen is thought to be divided. If the studies are restricted to an examination of 2-D sections then the area of observations is divided into A_{ij} elements. The density of particles or their volume fraction, $V_V = X$, is measured in each element and a matrix X_{ij} is formed such that:

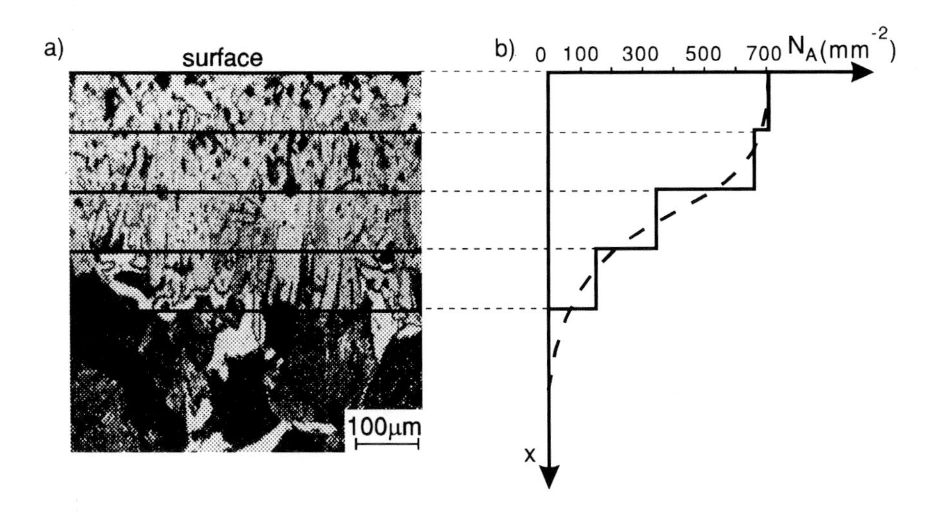

Figure 8.23 Microstructure on the cross-section of a material subjected to a surface treatment (a) and the function describing the volume fraction of the second phase particles in the strips of the material parallel to the surface of treatment (b). There is a clear tendency for the volume fraction of the particles of interest to decrease with distance from the surface.

$$X_{ij} = X\left(A_{ij}\right) \tag{8.62}$$

With this approach the density function is of a discrete nature and represented by a matrix. This matrix can be analyzed subsequently in terms of possible variations in the values of X_{ij} as a function of i and j; that is with respect to the position in the division of the field of observation. It can also be visualized by assigning to the values of X_{ij} colors or gray levels. The resulting image can also be studied using the tools of image analysis and mathematical morphology.

Patterns in the position of the particles can also be studied with the help of the formalism called cluster analysis, which is described in Chapter 4. This approach assumes that a population of particles is defined by the positions, X_i, of some specified points, for example, the centers of gravity, in the reference system (x_1,x_2,x_3). The position of the particles in the population studied is given in the form of a matrix X_{ij}:

$$X_{ij} = \begin{matrix} x_{11} & x_{12} & x_{13} & \\ x_{21} & x_{22} & x_{23} & \\ \cdots & \cdots & \cdots & \cdots \\ x_{k1} & x_{k2} & x_{k3} & \end{matrix} \tag{8.63}$$

In this matrix the rows x_i (i = 1,2,3) define the coordinates measured of a given particle and the columns x_m (m = 1,2, . . ., l) define the values of the particle coordinates with respect to a given axis.

The distance between the particle, X_i and X_j, is marked with the symbol $d(X_i,X_j)$ or an abbreviated form—d_{ij}.

Example 8.17

Fig. 8.24a shows particles revealed in a section of a two-phase material. The positions of the particles were established in the coordinate system indicated. These positions were subjected to cluster analysis. The results are given in Figure 8.24b.

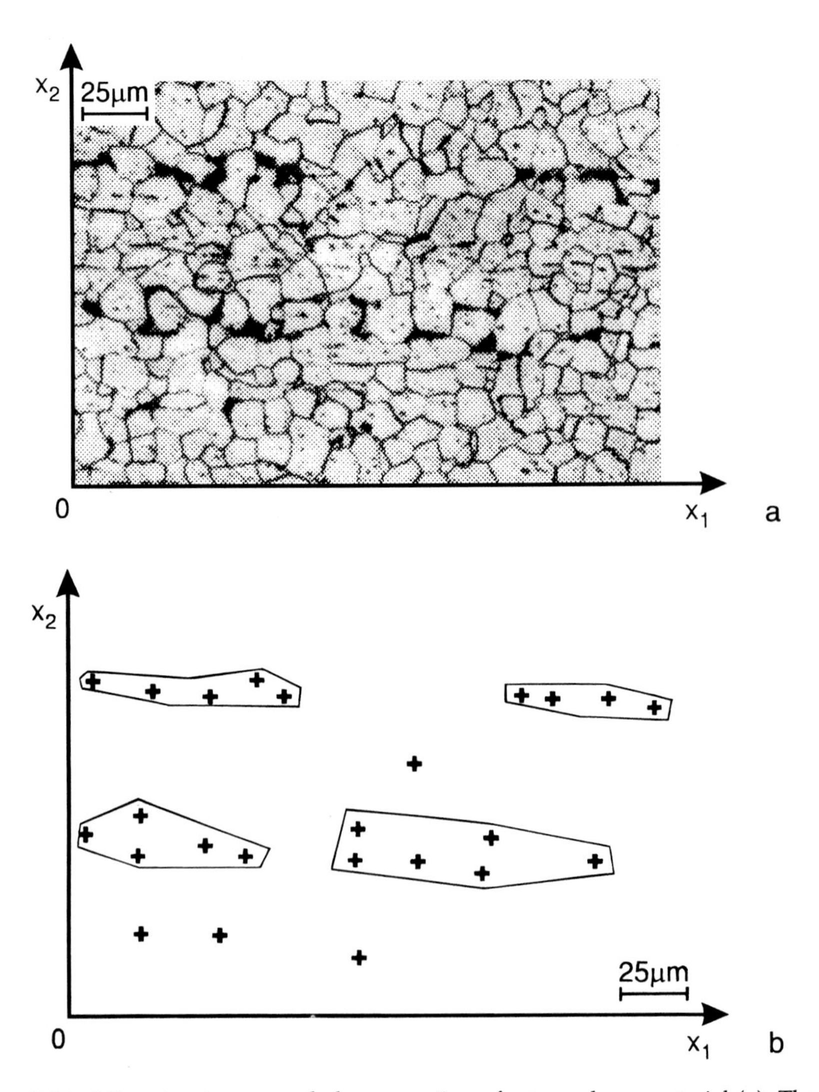

Figure 8.24 Microstructure revealed on a section of a two-phase material (a). The positions of the particle sections are displayed in (b). Clusters of the particle sections are indicated.

Examples of systems showing different types of clustering are given also in Figure 8.25.

As has been shown in the case of cluster analysis, the spatial positions of the particles in a microstructure may be characterized in terms of the distances between them. If there are N particles in a certain volume, V, of the material there would be a symmetric matrix, $N \times N$, of the distances, d_{ij}, between any two particles of the system. However, there is also another way of representing the character of particle spatial distribution, which is based on the concept of the radial distribution function. This is the function that defines the average number of particles in a spherical shell of inner radius r and thickness dr with its center positioned at the center of any particle. For the 3-D situation, the radial distribution function, $f_V(r)$, is defined in the following way:

$$f_V(r) = \frac{1}{4\pi r^2} \frac{dN_V(r)}{dr} \tag{8.64}$$

where $N_V(r)$ is the average density of particles in a spherical shell, of inner and outer radii r and r + dr, respectively. For the 2-D situation the radial distribution function $f_A(r)$ is defined in the following way:

$$f_A(r) = \frac{1}{2\pi r} \frac{dN_A(r)}{dr} \tag{8.65}$$

where $N_A(r)$ is the average density of particles in a circular annulus of inner and outer radii r and r + dr, respectively. These two distribution functions are related one to another if the function $f_A(r)$ describes the case for particle sections. Some details of such a relationship can be found in References 12 to 14.

However, the details of these relationships depend on the model of the particles and will not be considered here in more detail.

Example 8.18

In studies of a microstructure a question has been raised about a tendency for the pores to form some specific pattern in their spatial distribution. Figure 8.26 shows a schematic example of this tendency. The results of a cluster analysis are given below:

N_A (number of particles per unit of area) = 1.68×10^{-4} $(\mu m^2)^{-1}$,
N_A^c (number of particles per unit of cluster area) = 6.4×10^{-4} $(\mu m^2)^{-1}$.

The concept of the radial function can be extended to analyze the situation of multi-phase systems. Consider a microstructure containing two types of particles, α and β. In such a system four distribution functions can be defined which are for simplicity given in 2-D form:

$$f_A^{\alpha}(r) = \frac{1}{2\pi r} \frac{dN_A^{\alpha}(r)}{dr} \tag{8.66}$$

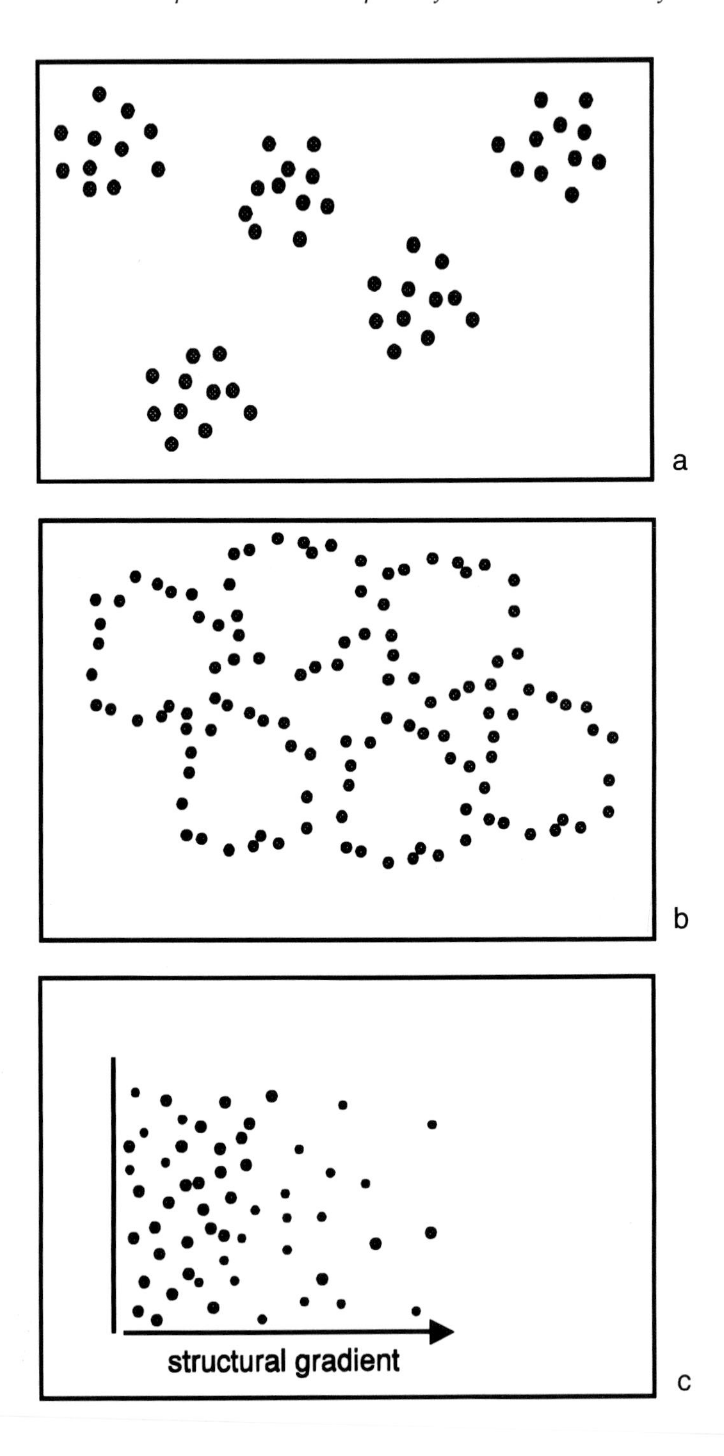

Figure 8.25 Schematic representation of the possible types of clusters in two-phase materials: (a) localized clusters; (b) cellular segregation; (c) gradient in the distribution.

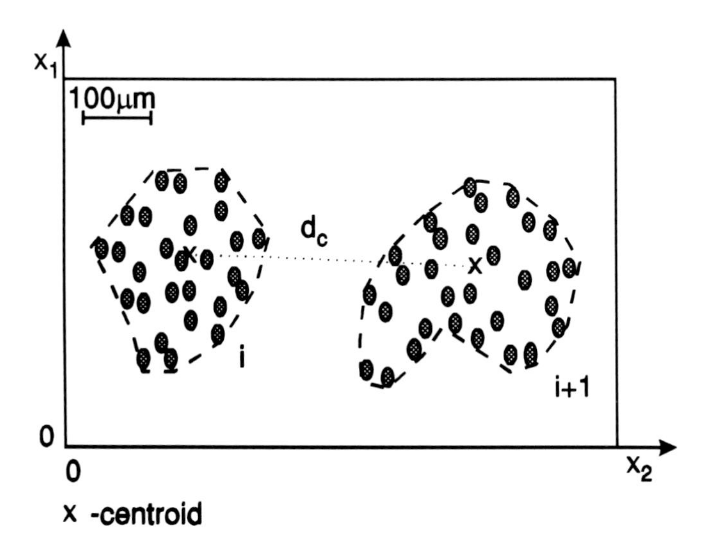

Figure 8.26 Schematic representation of the microstructure of a material containing clustered pores. Each cluster of pores is characterized individually by the position of its center and local density of pores.

$$f_A^\beta(r) = \frac{1}{2\pi r}\frac{dN_A^\beta(r)}{dr} \tag{8.67}$$

$$f_A^\alpha(r) = \frac{1}{2\pi r}\frac{dN_A^\alpha(r)}{dr} \tag{8.68}$$

$$f_A^\beta(r) = \frac{1}{2\pi r}\frac{dN_A^\alpha(r)}{dr} \tag{8.69}$$

Among these functions, the first two describe the spatial distribution function of particles α and β, respectively. This means that each of these functions is defined for a model structure that contains particles of only one type. These two functions can be used to detect a tendency for clustering of particles among particles α and β separately.

The other two radial distribution functions describe the tendency for particles of different types to be located at a small distance from one another. If this segregation is prominent a particle, say α, tends to be surrounded by a higher density of β particles. If there is no segregation these mixed distribution functions, with the exception of their values at small distance r, are constant. It should be noted that the distance r can be expressed as a multiplication of r^*, which is the mean distance to the nearest neighbor.

The concept of a radial distribution function can be modified to describe the tendency for particles to be located at grain boundaries (segregation of particles to grain boundaries). Here the separation distance of particles from the nearest boundaries is recorded, from which the distribution function may be derived.

Wiencek and Stoyan[15–16] have studied the spatial distribution of particles in terms of a covariance function, C(r). This function defines the probability that two random points at a distance r lie within the particle. This function possesses the following properties:

$$\begin{cases} C(0) = V_V \\ C(\infty) = (V_V)^2 \\ \dfrac{dC}{dr} \quad for \quad r = 0 \quad is \quad equal \quad \dfrac{S_V}{4} \end{cases} \tag{8.70}$$

This correlation function reflects the tendency of particles to be separated by a specific distance. For some systems this can be approximated by polynomial functions.

Example 8.19

Figure 8.27a shows a microstructure revealed on a section of a ceramic composite material. The matrix contains strengthening particles revealed as bright areas and pores, revealed as black regions. Spatial distributions of the particles and pores were studied using the methods based on radial distribution function and covariance. The results are shown in Figure 8.27b–e. Each figure shows the experimental data as well as the reference curves, which were obtained for random distributions of particle and pore centers. It can be noted that the pores show a much higher tendency for clustering than the particles. This is reflected by the higher values of covariance for the distance approaching 0.

The methods discussed take into account positions of the particles disregarding, however, possible patterns in their size, shape, and in the case of elongated particles, orientation. In a more precise approach, particles observed in multiphase materials frequently can be characterized in terms of:

1. position—x_i;
2. size—m_i;
3. specific orientation of their axis—ω.

The position of a given particle is described with respect to a reference coordinate system and for simplicity of further analysis it can be assumed that x_i defines the position of the particle center. The size of the particle, on the other hand, can be expressed by its volume, area or using some linear measure (for instance the length of the maximum chord). Particle orientation is given as the angle between a reference direction and the elongation axis of the particles. This parameter cannot be defined uniquely for spherical and near spherical particles and it can be assumed that such particles have an infinite number of specific orientations.

In short, the relevant characteristics of particles can be given as a point in a multidimensional space: $Q_i(x_i, m_i, \omega_i)$.

Each of the parameters included in this description can be analyzed separately in terms of the appropriate distribution function. However, the problem of particle size distribution functions has been discussed earlier and here attention is focused on the position and orientation of particles.

Consider a system of elongated particles distributed uniformly in the volume and observed on a representative cross-section of the material. If

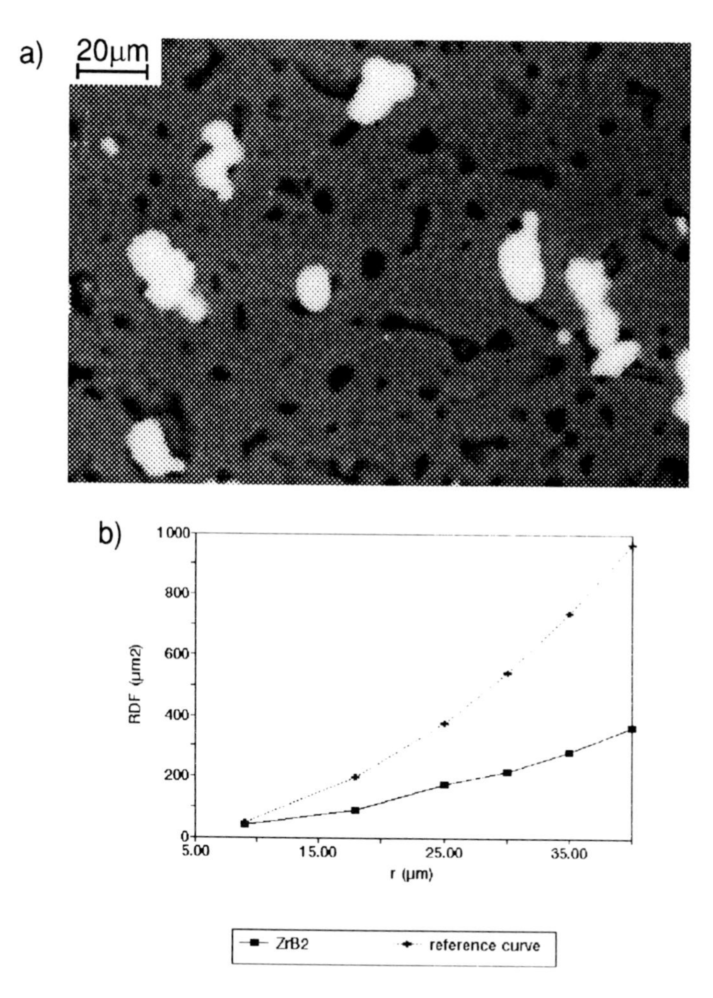

Figure 8.27 Microstructure of a ceramic composite (a). The matrix contains bright strengthening particles and black pores. This microstructure has been studied in terms of the radial distribution function of the particles (b) and pores (c). The covariance functions are given in (d) and (e). (Figures 8.27c–e can be found on the following page.)

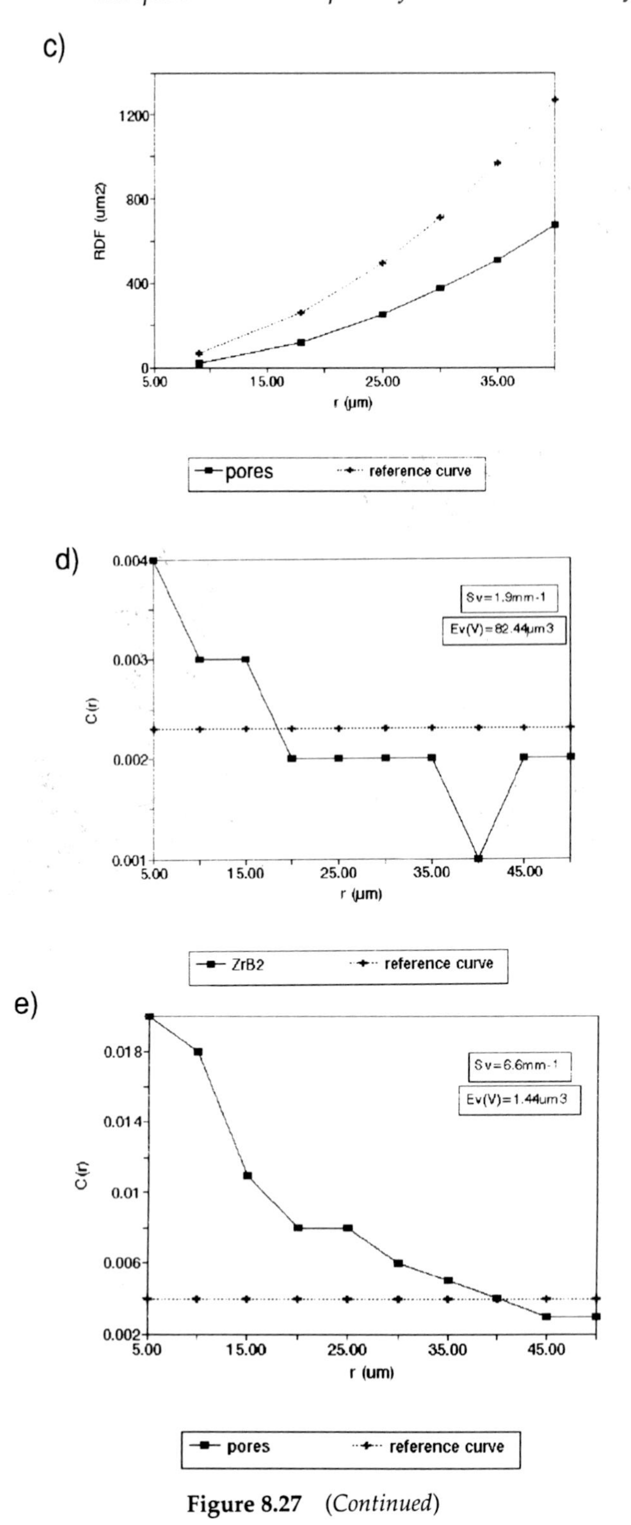

Figure 8.27 *(Continued)*

attention is focused on the orientation of the particles they can be represented by two parameters $[m_i, \omega_i]$ with the first defining the size, for instance, maximum dimension, and the second the orientation of the i-th particle. In this case the particles are represented by segments as schematically shown in Figure 8.28. The orientation of a system of particles can be described in terms of a distribution orientation function, $R(\beta)$, which defines the total length of segments with an orientation angle, ω, smaller than β. Assuming that the analysis covers n particles, the function $R(\beta)$ can be estimated in the following way:

$$R(\beta) = \frac{1}{M}\sum_{i=1}^{i=n} m_i s(\omega_i, \beta) \tag{8.71}$$

where

$$M = \sum_{i=1}^{i=n} m_i \tag{8.72}$$

is the total length of the segments representing the particles and $s(\omega_i,\beta)$ is defined as:

$$s(\omega_i, \beta) = \begin{cases} 1 & for \quad \omega_i < \beta \\ 0 & otherwise \end{cases} \tag{8.73}$$

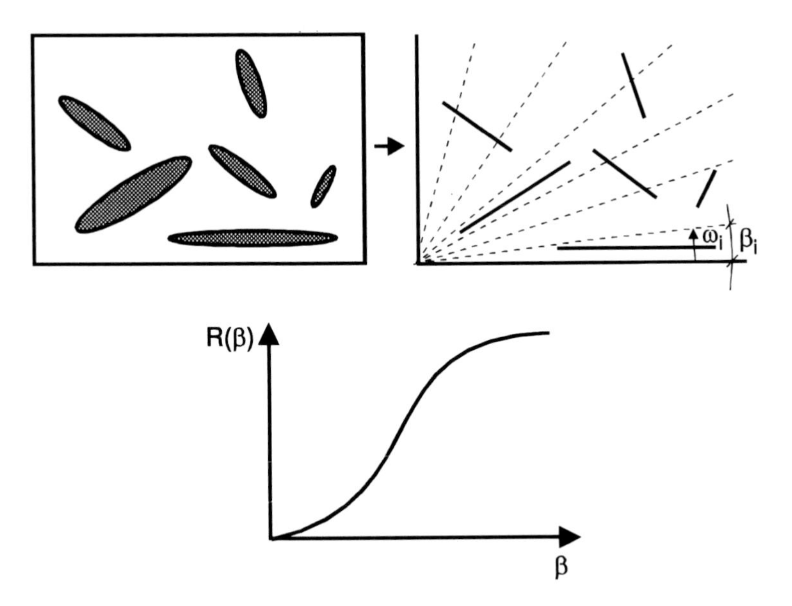

Figure 8.28 Schematic representation of particle sections and their representation used in the studies of order in the position and orientation of particles.

The function R(β) describes the cumulative weighted orientation distribution function for the particles represented by segments. The corresponding probability distribution function, r(β), can be estimated using the concept of a kernel estimator. In particular, one can use a method known in the literature as the Epanechnikov kernel, which leads to the following expression:

$$r(\beta) = \frac{1}{M} \sum_{i=1}^{i=n} u(\beta - \omega_i)\, m_i \tag{8.74}$$

with a u-kernel that for $-h < (\beta - \omega_i) < h$ is given as:

$$u(\omega_i - \beta) = \frac{3}{4h}\left(1 - \frac{(\omega_i - \beta)^2}{h^2}\right) \tag{8.75}$$

and is equal to zero otherwise. The parameter h in this equation is called the band width. The higher the value of h, the smoother is the estimating function r(β).

An application of the R and r functions is shown in the following example.

Example 8.20

A two-phase material with the initial microstructure shown in Figure 8.29a was strained in uniaxial tension. As a result of straining the particles started to show a tendency for elongation, with the elongation axis preferentially aligned to the direction of straining Figure 8.29b. Also shown are the relations R(β) and r(β) for these two types of structure.

The procedure described above can be modified by changing the way particles are selected and represented by two parameter sets. Instead of the measurements being conducted on all the particles in a selected area, a grid of points can be overlaid on the image of the particles and the longest chords, l_i, passing through the points established. Later the same analysis is carried out with l_i as the size parameter m_i.

Another possible anisotropy of the particle arrangements concerns the position of their centers. In this type of study, the particles can be represented by points positioned at their centers (centroids) together with a quantity indicating their size. Using the notation previously introduced, the particles in a system are represented by the pairs (m_i, x_i) or (m_i, α_i), $i = 1, 2, \ldots, n$. Then the question is raised as to the existing transitional and directional invariance. In other words, to what extent do the parameters describing the geometry of particles depend on their position and their orientation.

One possible approach to detecting anisotropy of the orientation centers is via studying the following set of functions:

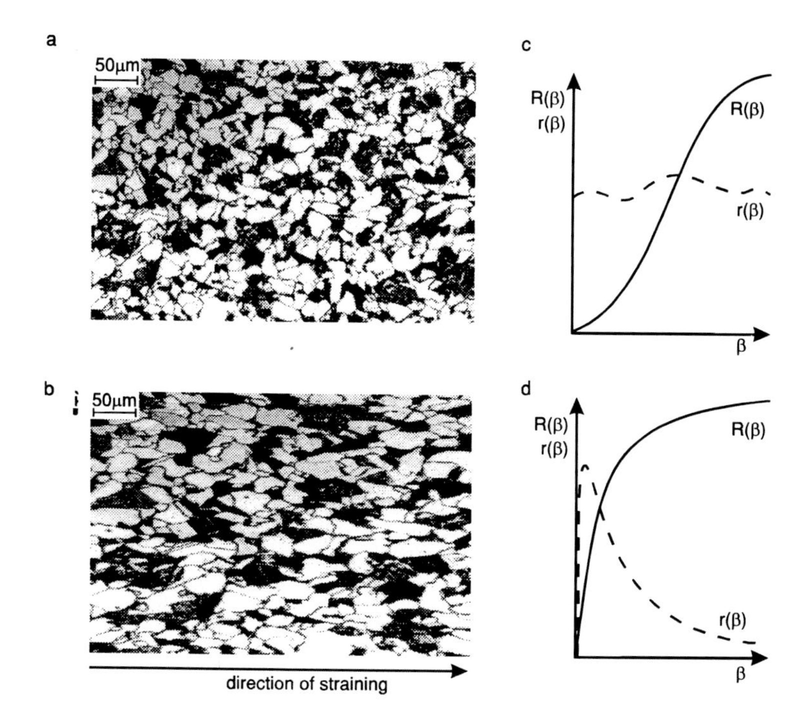

Figure 8.29 Microstructures used in Example 8.20: (a) initial isotropic distribution of particles; (b) anisotropic distribution of the particles; (c) and (d) the respective functions R(β) and r(β).

1. point-pair rose density, $o_{r1,r2}(\alpha)$;
2. orientation correlation function, $k_d(r)$;
3. nearest neighbors mean angle, c_d.

The point-pair rose density function $o_{r1,r2}(\alpha)$ is based on measurements of the orientation of lines passing through a pair of points representing particle centers and chosen in such a way that the distance between them, $d_{i,j}$, meets the condition:

$$r_1 < d_{i,j} < r_2.$$

An estimator of the $o_{r1,r2}(\alpha)$ is given by the following:

$$o_{r_1,r_2}(\alpha) = \frac{1}{A\left(W_{x_i} \cap W_{x_j}\right)} \sum_{i=1}^{i=m} \sum_{j=1}^{j=m} 1_{r_1,r_2}\left(\left\|x_i - x_j\right\|\right) m_i m_j u\left(\alpha - \alpha_{ij}\right) \qquad (8.76)$$

In this equation $1_{r1,r2}$ is an indicator that is equal to 1 for d_{ij} meeting the above-given condition and 0 otherwise. The function u(x) is the same kernel function as before and α_{ij} is the angle that the line makes passing through the chosen pair of points with the reference axis, defined as the smaller of the two angles the two lines define. The presumming expression in the equation cited takes into account the possible edge effects related to the pairs of particles placed close to the frame of observations. For an observation area in the form of a rectangle $a \times b$ and $x_i = (x_i, y_i)$ this term is given as:

$$A\left(W_{x_i} \cap W_{x_j}\right) = \left(a - \left|x_i - y_i\right|\right)\left(b - \left|x_j - y_j\right|\right) \tag{8.77}$$

An orientation correlation function, $k_d(r)$, can be used to describe the tendency in the system for alignment of particles showing some degree of elongation. This function is determined on the basis of the measurements of the difference in the angles $\alpha_i - \alpha_j$ for pairs of particles placed a distance r apart. The difference in angles $\alpha_i - \alpha_j$ is measured in a way that makes it always less than a right angle and for a given r the function defines the mean angle, α_d.

The anisotropy characterization of nearest neighbors is constructed generally along the same lines as described so far, with the difference that the analysis is confined to the pairs of particles that are nearest neighbors.

Consider a particle defined by the parameters (x_i, m_i, α_i) and its nearest neighbor $(x_i^n, m_i^n, \alpha_i^n)$ which is characterized by the smallest particle center-to-center distance. The parameter c_d is the mean value of the difference $\alpha_i - \alpha_i^n$ measured in the same way as in the previous example. This parameter can be estimated from the following:

$$c_d = \frac{1}{n}\sum_{i=1}^{i=n}\left(\alpha_i - \alpha_i^n\right) \tag{8.78}$$

The existence of order in the arrangement of the particle centers can be visualized in a graphical form similar to that used in the case of crystalline structures. The data for such a presentation is collected in the analysis of neighbors of all particles. In this approach, the neighbors are ranked based on their distance from a given particle and then a specified number of them, say six or eight, are taken into consideration. This is an extension of the previous concept in the sense that the angles are measured with respect to the line joining the particle analyzed to its nearest neighbor. The data are presented in the form of a map depicting the probability of finding another particle in different locations around the particle analyzed.

8.5 Description of particle shapes

A description of particle shape, in general, can be carried out along the same lines as the description of grain shape discussed in Chapter 6. However, it should be noted, that particles are not subject to the requirement of space filling, as is the case with grains, and as a result the restrictions on their shape are less obvious.

Particle shape is a much more complicated property than size. This is especially evident when attempts are made to quantify the shape by means of some number of parameters. As a result, in many cases the shape of objects is described by a sketch or illustrated in a series of images taken from different angles of observation without any attempt to transform them into a formal mathematical description. Thus, particles are merely classified into some broad categories represented by a conventional geometrical object. The following categories of particle are frequently distinguished:

a) spheres;
b) cubes;
c) disks;
d) needles;
e) plates.

The first two from this list, i.e., spheres and cubes, have their shape uniquely defined and do not need further specification. This is not true for the next three. The shape of disks can be more precisely described by the ratio of their thickness, t, to the radius of the maximum section, R:

$$\text{disks—}t/R.$$

The shape of needles on the other hand depends on the ratio of the needle length, l, to the maximum radius in the direction normal to the elongation axis, r:

$$\text{needles—}l/r.$$

The shape of plates can be differentiated in terms of the ratio of the true surface area, S, to the projected surface area in the direction approximately normal to the plate, S_p:

$$\text{plates—}S/S_p.$$

This list is far from being complete and more geometrical objects can be studied and described in terms of some simple dimensionless parameters. The set of such objects used in studies of a given microstructure forms **a reference shape set**.

Particle shape can also be described in terms of geometrical properties such as number of faces, F, number of edges, E, etc. In this case a cube exemplifies a shape that is coded as F = 6 and E = 12. On the other hand, a spherical particle is described by F = 1 and E = 0.

From the point of view of materials science, important geometrical properties are those which describe:

a) the degree of elongation;
b) the degree of surface complexity.

The importance of information on the possible elongation of particles is related to the potential anisotropy of the material properties. The degree of surface complexity is important in analyses of the phenomena taking place at interphase boundaries, such as segregation and phase transformations. These properties, elongation and surface complexity, can be quantified by means of the following ratio:

$$R_S/R_V,$$

where R_S and R_V are equivalent surface and equivalent volume radii respectively, i.e.:

$$R_S = \sqrt{\frac{S}{4\pi}} \qquad (8.79)$$

and

$$R_V = \sqrt[3]{\frac{3V}{4\pi}} \qquad (8.80)$$

One can also use the ratio of R_S^c/R_S, where R_S^c is the surface radius of the minimum volume sphere that contains the volume of a given particle. The values of the ratios defined in this way are given in Table 8.4 for some simple geometrical forms. More such parameters can be found in Reference 5.

Another approach to particle shape concentrates attention on the properties of the system instead of looking at the geometry of individual elements. The degree of particle convexity can be described using the concept of the so called star volume, V^*. This star volume is the volume of the portion of space inside a particle that is directly seen from a point inside the particle, for example, its center of gravity. For convex particles the star volume V^* is the same as the true volume:

$$V^* = V \qquad (8.81)$$

On the other hand, for concave particles:

$$V^* < V \qquad (8.82)$$

Table 8.4 The Values of Ratio R_s/R_v for Simple Geometrical Forms

Geometrical forms	S	V	R_s/R_v
Sphere	$4\pi r^2$	$(4/3)\pi r^3$	1
Cube	$6l^2$	l^3	1.114
Disk	$2\pi r^2 + 2\pi rd$	$\pi r^2 d$	$0.778(r/d+1)^{1/2}\cdot(d/r)^{1/6}$
Needle	$2\pi r^2 + 2\pi rl$	$\pi r^2 l$	$0.778(r/l+1)^{1/2}\cdot(l/r)^{1/6}$

The ratio of the mean values of the two parameters, V*/V, is therefore a measure of particle convexity.

The method for estimation of the mean weighted particle volume based on point sampled intercepts has been described in the preceding sections of this chapter. The mean volume-weighted particle star volume, $E_V(V^*)$, can be estimated using the same procedure, if only those intercepts are taken which end at the sampled point. Gundersen and Jensen[17] defined the costar coefficient:

$$costar = \frac{E_V(V) - E_V(V^*)}{E_V(V)} \qquad (8.83)$$

This coefficient is equal to 0 for convex particles and it increases with an increasing degree of irregularity in the shape of particles. The upper limit of the costar is equal to 1.

8.6 *Shape of particle sections*

Particle shape in a material may be examined in full via systematic sectioning or as a result of extracting the particle from the matrix. On the other hand, some information about the shape can be retrieved from randomly taken cross-sections of the specimen and an analysis of the shape of particle sections. This can be done using qualitative or quantitative approaches.[18]

In the qualitative approach an attempt is made to reconstruct the particle shape via analysis of types of section shapes. The section shapes are classified into groups represented by simple 2-D objects such as:

a) circles;
b) ellipses;
c) strips;
d) polygons (triangles, squares).

These shapes are linked to the shape of particles in the way summarized in Table 8.5.

Table 8.5 Relationships Between the 3-D Shape of Particles and the Shape of their 2-D Sections

3-D	2-D
Spheres	Circles
Cubes	Triangles, squares, hexagons
Disks	Ellipsis, circles
Needles	Circles, ellipses
Flakes	Strips, irregular circles

From such a table some deductions can be made with respect to the shape of particles. It should be noted that any conclusions are justified only if the analysis has a quantitative character.

In the quantitative analytical method a model of the particle population studied is constructed. Depending on its complexity, this model may assume that:

a) particles are of the same shape and size and are randomly oriented in space;
b) particles have the same shape and are randomly oriented but differ in size;
c) particles have the same shape and size but are preferentially oriented with respect to some reference directions;
d) particles differ not only in size but also in shape and are possibly preferentially oriented.

The model system of particles is subsequently intersected with a cross-section plane and the intersected shapes and sizes are collected. Later in the process of particle shape analysis, these shapes of particle sections are compared with the distribution of the sections recorded on the images of the microstructure studied. In most practical cases these quantitative procedures become quite complicated and their implementation requires powerful computers and precise input data in the form of a large number of recorded particle sections.

The situation is very much simpler if the studies are aimed at detecting changes in the shape of particles without identifying their shape. In this case, the analysis concerns exclusively the shapes of the particle sections. These sections form a set of 2-D figures, C_i, which generate distribution functions, $f(x_s)$, where x_s is one of the shape factors such as:

p/d_2 perimeter to equivalent diameter ratio;
d_{max}/d_2 maximum chord to equivalent diameter.

Other dimensionless shape factors can also be used and some of them are given in Chapter 6.

This method of analysis requires that the distribution functions f(x) are recorded for the different microstructures of the material studied. For each microstructure, appropriate distribution functions of shape parameters are produced and these are used to identify differences in the shape of particles in the microstructures. Such differences can be judged on the basis of variation in the mean values, E(x), or higher order moments of the distribution functions.[18]

Particles observed in multiphase materials usually differ in the details of their shape and size. In an attempt to characterize the shape of particles forming a population in which individual elements show to some extent different properties, frequently it is necessary to define average properties of the system. This can be done directly by averaging over the properties of all particles measured, or taking into account the size of the particles.

Suppose that the particle shape can be quantified by a parameter x (it may be the ratio of the maximum chord to equivalent diameter) and their size by a parameter y (it may be, for example, the area of the particle section). The population of particles is described by the distribution function f(x,y). The average shape factor for the system of particles can be obtained from the following equation:

$$E(x) = \frac{1}{N}\left(x_1 + x_2 + \ldots + x_N\right) \tag{8.84}$$

The size-weighted shape factor is given as:

$$E_y(x) = \frac{1}{N}\frac{\left(y_1x_1 + y_2x_2 + \ldots + y_Nx_N\right)}{y_1 + y_2 + \ldots + y_N} \tag{8.85}$$

These two average shape factors can be used alternatively or together. The first one, E(x), has more fundamental meaning. The second, on the other hand, gives a description which is usually found to correlate better with the human perception of the average particle shape. This is related to the fact that apparently the biggest particles seen in the image of microstructures attract more attention.

Example 8.21

Figure 8.30 shows microstructures of ferrite in a series of low-alloy steels. The shape of ferrite particle sections has been quantified by the ratio d_{max}/d_2, where d_{max} is the maximum chord and d_2 is the equivalent diameter. The results of the particle shape measurements are given in Table 8.6.

8.7 Fourier analysis of section shapes

The shape of particle sections can also be analyzed by means of Fourier transforms.[19] According to the Fourier theorem, any continuous and periodic function, F(x), with period, T, can be represented in the form of a series of sine and cosine functions. These series are defined as the Fourier series of the function F(x) and are given by the following formulae:

$$F(x) = A_0 + \sum_{n=1}^{\infty} A_n \cos\frac{2\pi nx}{T} + \sum_{n=1}^{\infty} B_n \sin\frac{2\pi nx}{T} \tag{8.86}$$

$$F(x) = A_0 + \sum_{n=1}^{\infty} C_n \cos\left(\frac{2\pi nx}{T} - \varphi_n\right) \tag{8.87}$$

where

$$C_n = \sqrt{A_n^2 + B_n^2} \tag{8.88}$$

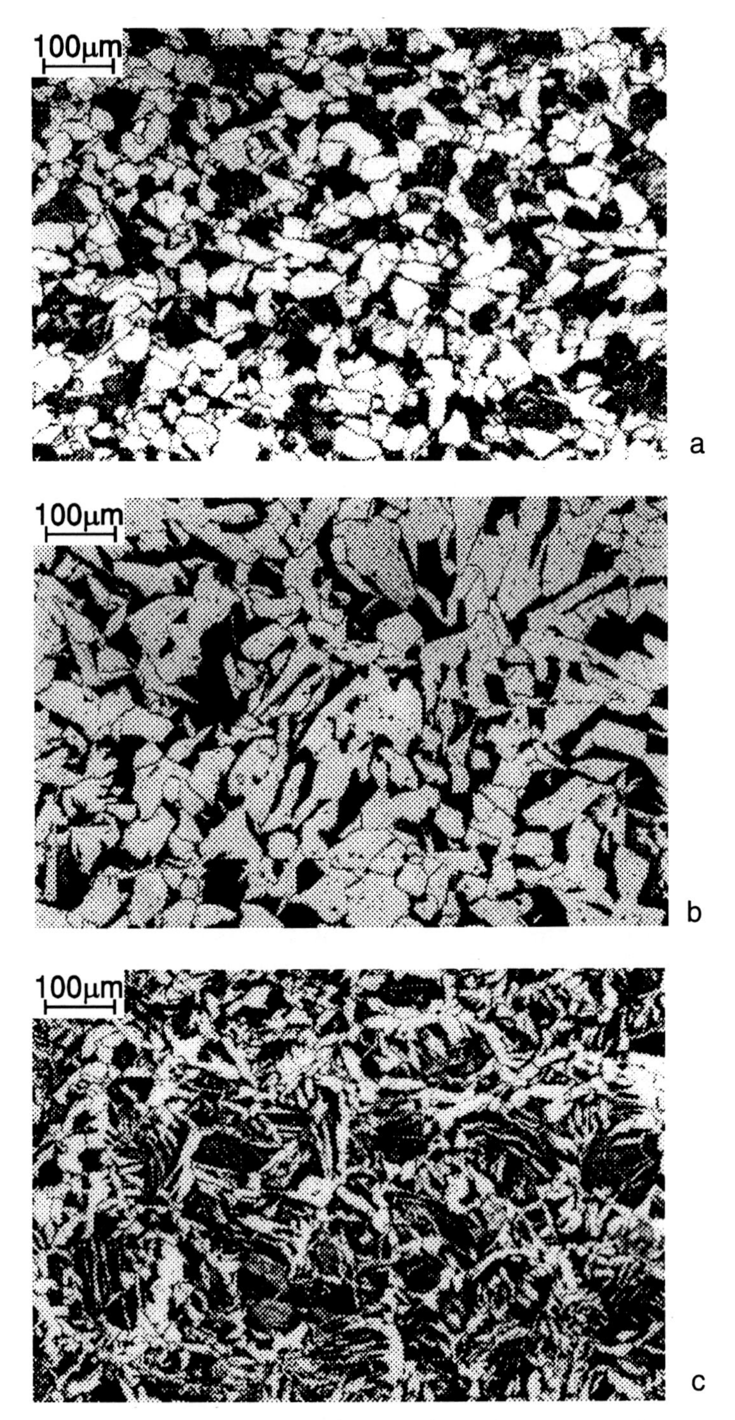

Figure 8.30 Various microstructures of ferrite in low-alloy steels used in Example 8.21. Shape data for these microstructures are given in Table 8.6.

Table 8.6 The Shape Factors of the Ferrite Particles Shown in Figure 8.30

	Figure 8.30a	Figure 8.30b	Figure 8.30c
d_{max}/d_2	1.37	1.76	2.31

$$\varphi = \arctan\frac{B_n}{A_n} \tag{8.89}$$

$$A_0 = \frac{1}{T}\int_0^T F(x)dx \tag{8.90}$$

$$A_n = \frac{2}{T}\int_0^T F(x)\cos\frac{2\pi nx}{T}dx \tag{8.91}$$

$$B_n = \frac{2}{T}\int_0^T F(x)\sin\frac{2\pi nx}{T}dx \tag{8.92}$$

The Fourier representation of a given function F(x) can be limited to the first n elements of the Fourier series. This yields a function that approximates F(x). The higher the value of n the better is the precision of the approximation.

In the quantitative metallography approach particle shape is usually studied via examination of 2-D sections. In this case the particles are represented as single or multiple loops. In the case of convex contours each loop can be described by the vector function R(θ). This is the function defining the position of a particle contour in a cylindrical system of coordinates (r,θ) attached to the center of gravity of a given particle (Figure 8.31).

The function R(θ) is periodic, with a period 2π, and can be represented in the form:

$$R(\theta) = R_0 + \sum_{n=1}^{\infty} R_n\cos(n\theta - \varphi_n) \tag{8.93}$$

where R_0 is the radius of the circle of the same area as the particle, R_n—n-th harmonic figure of n bulges, φ_n—orientation of the harmonic figure with respect to the references axis.

In the case of nonconvex particles the contours can be defined in terms of the angle Θ between the tangent to the particle section contour and some reference direction (Figure 8.32). This angle changes with the position of the tangent point and can be analyzed as a function of a distance along the line, l, from a starting point, P_0. In this way the shape is described by a θ(l) function. This function is, however, not periodic. To make it independent of L, the contour length, it is transformed in the following way:

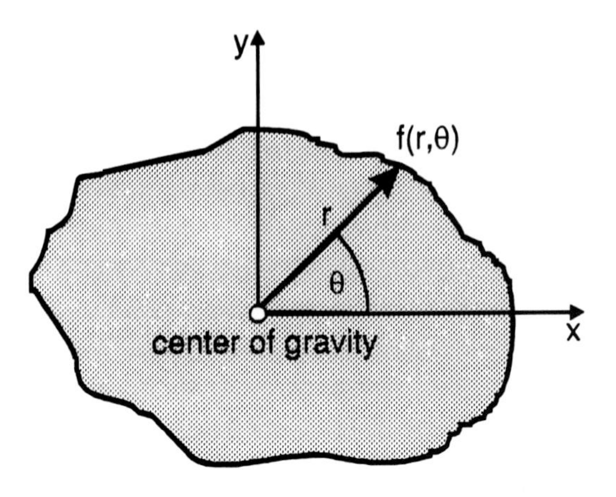

Figure 8.31 Particle contour in a cylindrical system of coordinates.

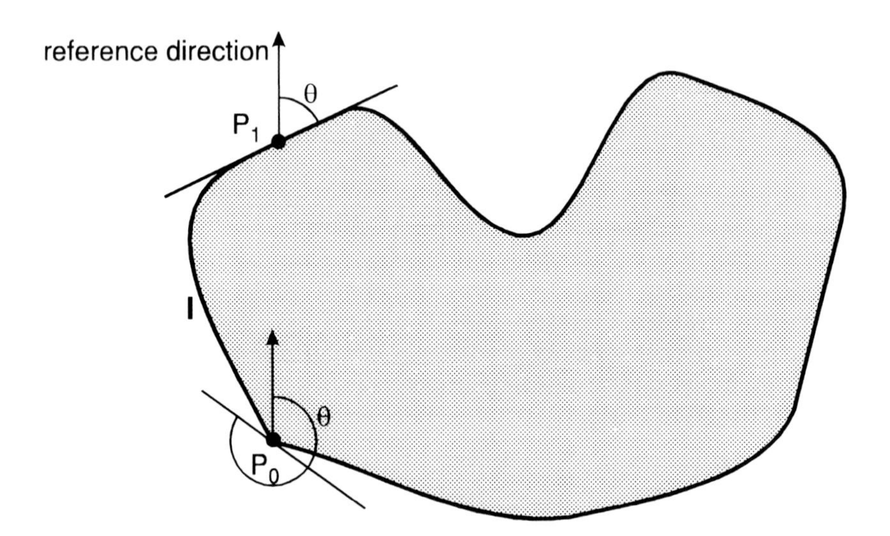

Figure 8.32 Definition of the angle θ as the angle between the local tangent to a contour and a reference direction.

$$\theta^*(t) = \theta\left(\frac{Lt}{2\pi}\right) + t \tag{8.94}$$

where

$$t = \frac{2\pi L}{T} \tag{8.95}$$

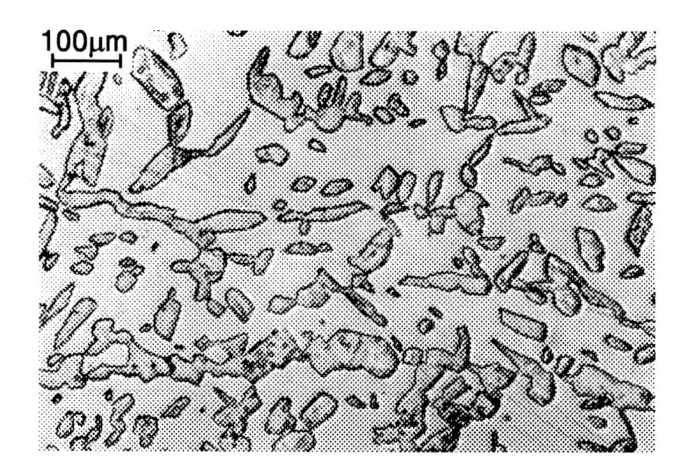

Figure 8.33 Particle sections revealed in a duplex austenitic ferritic stainless steel in the as-annealed state.

The two methods described may be implemented by software in a microcomputer. More details can be found in the following example.

Example 8.22

The Fourier analysis method has been utilized in studies of particles in a two-phase material (a duplex stainless steel). The material was plastically deformed by cold working and subsequently annealed at 1050°C for 1 h. Cylindrical specimens of the material have been strained to fracture in tensile tests carried out at room temperature. Microstructural observations were conducted on longitudinal sections. The fields of observations were located:

a) within shoulders of the specimens;
b) within the gauge section at some distance from the neck;
c) inside the neck.

The heat treatment applied produced a microstructure consisting of austenitic particles in a ferritic matrix, exemplified by the micrograph shown in Figure 8.33. It may be observed that the particles of austenite are agglomerates of austenite grains and frequently assume nonconvex shapes. Consequently, the second method described above has been used to quantify their shape. Figure 8.34 shows the particle shape in the deformed sections of the specimens. The particles are significantly more elongated than in the as-annealed material. Some of the results of the particle shape study are depicted in Figure 8.35. The amplitudes of the harmonic functions for individual particles studied have been averaged over the number of particles

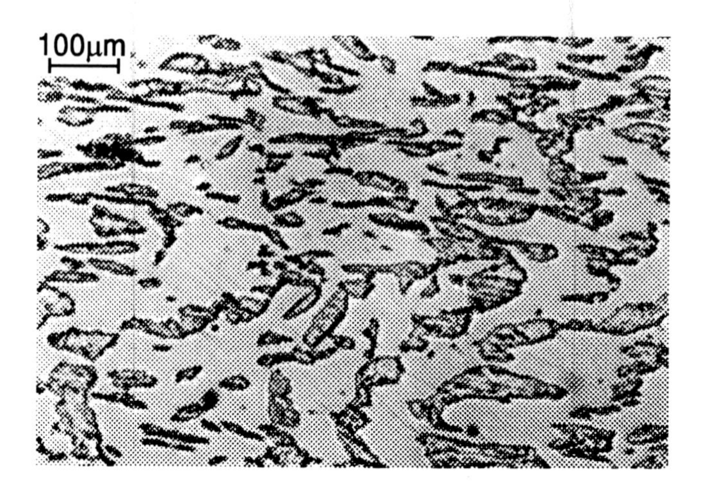

Figure 8.34 Particle sections revealed in a duplex austenitic ferritic stainless steel after plastic deformation by uniaxial straining.

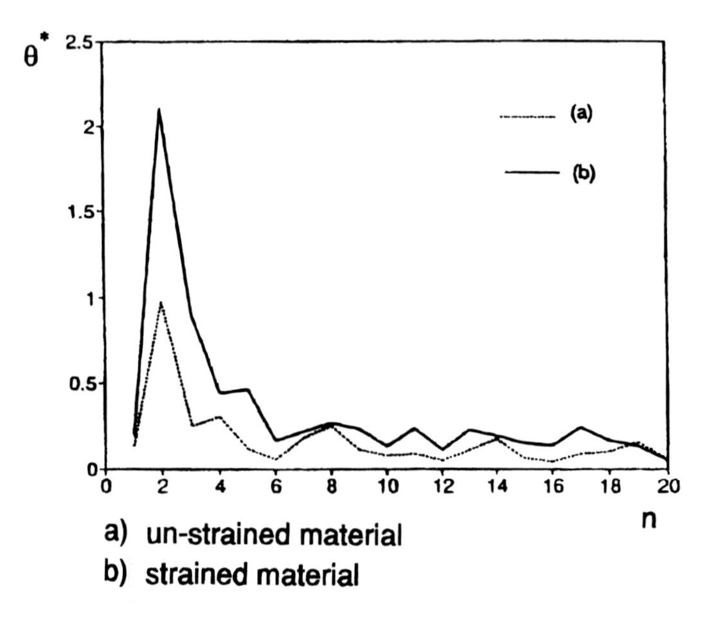

Figure 8.35 Fourier series analysis of the particle section contours for the microstructures shown in Figures 8.33 and 8.34.

studied in order to obtain typical shapes of particles as a function of the absorbed plastic strain. On this basis the contribution of austenite to the deformation of the material has been found to be the factor controlling the plasticity of the duplex steel.

References

1. Rhines, F. N. and DeHoff, R. T., *Quantitative Microscopy*, McGraw-Hill, New York, 1968.
2. Gokhale, A., Drury, W. J., and Whited, B., Quantitative microstructural analysis of anisotropic materials, *Materials Characterization*, 31, 11, 1993.
3. Jensen, E. B. and Gundersen, H. J., Stereological estimation of surface area of arbitrary particles, *Acta Stereologica*, 6, 25, 1987.
4. Wiencek, K. and Ryś, J., Coagulation of cementite particles in carbon-steels, *Neue Huette*, 4, 137, 1981.
5. Underwood, E. E., *Quantitative Stereology*, Addison Wesley, Massachusetts, 1970.
6. Coleman, R., The statistics of stereological unfolding, *Acta Stereologica*, 6, 255, 1986.
7. Gundersen, H. J. G., quoted by Cruz-Orive, L. M., Stereology recent solutions to old problems and glimpse into the future, *Acta Stereologica*, 6, 3, 1987.
8. Gundersen, H. J. G., Unbiased estimators of number volume and surface area of arbitrary particles, a review of very recent developments, *Acta Stereologica*, 6, 3, 1987.
9. Jensen, E. B. and Gundersen, H. J. G., Fundamental stereological formulae based on isotropically oriented probes through fixed points with application to particle analysis, *Journal of Microscopy*, 153, 249, 1989.
10. Jensen, E. B. and Sorensen, F. B., A note on stereological estimation of the volume-weighted 2nd moment of particle-volume, *Journal of Microscopy*, 164, 21, 1991.
11. Bucki, J. J. and Kurzydłowski, K. J., Measurements of grain volume distribution parameters in polycrystals characterized by a log-normal distribution function, *Scripta Metallurgica et Materialia*, 28, 689, 1993.
12. Tanemura, M., On the stereology of spatial structure of interacting particle systems, *Acta Stereologica*, 6, 129, 1987.
13. Robine, H., Jernot, J. P., and Chermant, J. L., Stereological determination of the radial distribution of spheres, *Acta Stereologica*, 6, 159, 1987.
14. Hanish, K. H. and Stoyan, D., Stereological estimation of the radial distribution function of centers of spheres, *Journal of Microscopy*, 122, 131, 1981.
15. Wiencek, K. and Stoyan, D., Spatial correlation in metal structures and their analysis, the covariance I, *Materials Characterization*, 26, 167, 1991.
16. Wiencek, K. and Stoyan, D., Spatial correlation in metal structures and their analysis, the covariance II, *Materials Characterization*, 31, 47, 1993.
17. Gundersen, H. J. G. and Jensen, E. B., Stereological estimation of the volume-weighted mean volume of arbitrary particles observed on random sections, *Journal of Microscopy*, 138, 127, 1985.
18. Slater, J., *Quantitative Investigation of the Transformation Kinetics of Low-Alloy Steels*, Ph.D. thesis, Cambridge University, 1976.
19. Reti, T. and Czinege, I., Shape characterization of particles via generalized Fourier analysis, *Journal of Microscopy*, 156, 15, 1989.

Subject Index

H

I

K

L

M

V

W

Y